Master Essential

ALGEBRA

Skills Practice Workbook
with Answers

$$6x^2 - x - 12 = 0$$

$$(3x + 4)(2x - 3) = 0$$

$$3x + 4 = 0 \quad \text{or} \quad 2x - 3 = 0$$

$$x = -\frac{4}{3} \quad \text{or} \quad x = \frac{3}{2}$$

Chris McMullen, Ph.D.

Master Essential Algebra Skills Practice Workbook with Answers
Chris McMullen, Ph.D.

Copyright © 2020 Chris McMullen, Ph.D.

www.improveyourmathfluency.com
www.monkeyphysicsblog.wordpress.com
www.chrismcmullen.com

Zishka Publishing
ISBN: 978-1-941691-34-2

Mathematics > Algebra
Study Guides > Workbooks > Math

1.2 Essential Vocabulary

In order to learn algebra, you will first need to understand some important words. It wouldn't be helpful to tell you, "The next step is to divide by the coefficient of the variable," if you have no idea what the words "coefficient" and "variable" mean. If you want to understand what is going on when we discuss algebra, you need to study the following words and definitions. The sooner you can remember these definitions, the better. If you come across a mathematical word in this book that you don't understand, you can look it up in the handy glossary at the back of the book.

An **unknown** refers to a letter, like x or y, that you are trying to solve for in a problem.

A **variable** refers to a letter, like x or y. We call it a "variable" because it doesn't have the same value for different problems. For example, you might find that x equals 3 for one problem, but that x equals 12 in another problem. The value of x "varies" from one problem to another.

The terms "unknown" and "variable" both refer to letters like x, y, t, etc. that we don't know the values for (until we solve for them).

A **constant** has a fixed value. All real numbers, like 5, $-\frac{3}{2}$, 418.27, and even $2\sqrt{3}$ are constants. Where it can get confusing is when we use letters to represent constants as well as variables. For example, in the formula $h = \frac{1}{2}gt^2$, we consider h and t to be variables, but consider g to be a constant. Why? Because near the surface of the earth, g has a constant value (with a magnitude of 9.8 m/s^2).

A **coefficient** is a number that multiplies a variable. For example, in $6x$ the coefficient is the number 6, while in $9y^4$ the coefficient is 9.

The terms "coefficient" and "constant" aren't interchangeable. Although a coefficient may be constant, a coefficient has a very specific role: it must multiply a variable. For example, in $4x^2 = 36$, the coefficient is 4, while 4 and 36 are both constants (but we would call 36 the constant "term"; see below for the definition of "term").

An **equation** is easy to spot because it has an equal ($=$) sign. For example, $7x + 2 = 30$ is an equation. If it doesn't have an equal sign, like $3x - 8$, it isn't an equation.

An **expression** doesn't have an equal sign ($=$) or inequality (like $<$ or $>$). For example, $3x - 8$ is an expression.

You can **solve** an equation. For example, $x = 4$ solves the equation $7x + 2 = 30$ since $7(4) + 2 = 28 + 2 = 30$. (If you didn't follow the math in this paragraph, don't worry. We will learn this in Sec. 1.13.)

You can **simplify** an expression (but you can't *solve* it). For example, $5x - 4 + 3x - 2$ simplifies to $8x - 6$. (We'll understand why in Chapter 2. For now, you should be able to see that $8x - 6$ is indeed simpler than $5x - 4 + 3x - 2$.) To simplify an expression means to find an equivalent expression that has a simpler form. (When we learn how to simplify expressions, you'll see concrete examples of what this means.)

The **terms** of an equation, expression, or inequality are separated by $+$ signs, $-$ signs, $=$ signs, or inequal signs (like $<$ or $>$). For example, $2x^2 + 9x + 4$ has 3 terms (which are $2x^2$, $9x$, and 4) and $7x + 5 = 3x + 25$ has 4 terms (which are $7x$, 5, $3x$, and 25).

Example 1. Is $(x - 2)^2 + 5$ an expression or an equation?
It is an expression because it doesn't contain an equal ($=$) sign.

Example 2. Is $\frac{24}{x} = 6$ an expression or an equation?
It is an equation because it contains an equal ($=$) sign.

Example 3. What are the terms of $x^3 - 2 = 6$?
There are three terms: x^3, 2, and 8. Terms are separated by $+$, $-$, and $=$ signs.

CONTENTS

INTRODUCTION

The goal of this workbook is to help students master essential algebra skills through practice.

- The first chapter is an essential preparatory chapter. It explains what algebra is, defines key vocabulary terms, discusses the language of algebra, shows how multiplication and division are expressed in algebra, and shows how to plug numbers into an equation. It also reviews the order of operations, fractions, and negative numbers.

- The remaining chapters cover essential algebra skills, such as combining like terms, distributing, factoring, the FOIL method, variables in the denominator, cross multiplying, ratios, rates, the quadratic formula, powers, roots, substitution, simultaneous equations, rationalizing the denominator, inequalities, and word problems.

- Each section concisely introduces the main ideas, explains essential concepts, and provides representative examples to help serve as a guide. A full solution is given for every example.

- Answer key. Practice makes permanent, but not necessarily perfect. Check the answers at the back of the book and strive to learn from any mistakes. This will help to ensure that practice makes perfect.

1 GETTING READY

1.1 What Is Algebra?

Algebra is a highly practical branch of mathematics for the following reason:

Algebra uses letters (like x, y, or t) to represent unknown quantities, and provides a system of rules for determining the unknowns.

This system of rules makes algebra very useful for solving a wide variety of problems. Following are a few examples where it is useful to represent numbers with letters.

- If a car travels with constant speed, the distance traveled equals the speed of the car times the elapsed time. By using the letter d to represent the distance traveled, the letter r to represent the speed (which is a rate), and the letter t to represent the elapsed time, we can express this relationship with the formula $d = rt$. If we know any two of these letters, the rules of algebra allow us to solve for the unknown quantity. (Although this example is simple enough that you could solve such problems without algebra, there are many formulas that would be very difficult to solve without using algebra, such as $\sqrt{\frac{y^2}{b^2} - \frac{x^2}{a^2}} = c$.)

- Word problems can be solved systematically and efficiently by using letters to represent unknowns and applying the rules of algebra. For example, if we know that five times a number minus forty equals eighty, algebra lets us write the equation $5x - 40 = 80$ and provides a prescription for determining that x is equal to 24, which we will learn in Chapter 2. (Again, this simple example can be solved without algebra, but for more challenging problems applying algebra makes the solution much more straightforward and efficient.)

- Using letters to represent numbers allows us to express mathematical rules in a general form. For example, note that $5 \times (6 + 4) = 5 \times 10 = 50$ has the same answer as $5 \times 6 + 5 \times 4 = 30 + 20 = 50$. Similarly, $7 \times (5 + 3) = 7 \times 8 = 56$ has the same answer as $7 \times 5 + 7 \times 3 = 35 + 21 = 56$. Using letters, we can express this rule in the general form $a(b + c) = ab + ac$ (where ab means a times b). This is known as the distributive property.

1.3 Multiplying and Dividing in Algebra

We almost never use the standard times symbol ($\times$) in algebra. Why not? It's because x is the most commonly used variable in algebra. If you wrote down an equation using both the variable x and the times symbol $\times$, these could easily be confused (especially when writing by hand). It is a good habit to avoid using the times symbol ($\times$). When you read algebra, you need to be aware of the different ways that multiplication may be represented.

A common way to multiply numbers is to use parentheses like one of these examples:
- $(3)(4)$ means 3 times 4.
- $3(4)$ also means 3 times 4.
- $(3)4$ is less common, but still means 3 times 4.
- $(3)(4)(5)$ means 3 times 4 times 5.
- $3(4)(5)$ also means 3 times 4 times 5.
- $(3)4(5)$ and $(3)(4)5$ are less common, but still mean 3 times 4 times 5.

An alternative is to use a middle dot ($\cdot$):
- $3 \cdot 4$ means 3 times 4.
- $3 \cdot 4 \cdot 5$ means 3 times 4 times 5.
- $(3) \cdot (4)$ unnecessarily uses both parentheses and a middle dot, but it still means 3 times 4. We recommend avoiding this, but beware that you may encounter it.

When a variable multiplies another quantity, no multiplication symbol is used:
- $5x$ means 5 times x.
- xyz means x times y times z.
- $4x^2y$ means 4 times x^2 times y.
- $2(x - 3)$ means 2 times the quantity $x - 3$.

You should avoid the following (but beware that you may encounter them):
- $3 \cdot x$ should instead be written as $3x$.
- $2(x)$ should instead be written as $2x$. Compare these unnecessary parentheses to the example above where parentheses are needed (with $x - 3$).

Example 4. For $5x^2 - 4 = 3y$, what are the variables and what are the coefficients? The variables are x and y. The coefficients are 5 and 3. Coefficients multiply variables.

Exercise Set 1.2

Directions: Apply the definitions from this section to answer the following questions.

1) Is $(x - 1)^2 = 16$ an expression or an equation?

2) Is $\frac{3}{x^2} - \frac{4}{x} + \frac{1}{8}$ an expression or an equation?

3) What are the terms of $x^3 + 8x^2 - 3x + 6$?

4) What are the terms of $9 - x = 4$?

5) What are the terms of $5xy^2 - 7y^3 + 3$?

6) For $3x - 8 = 7$, what are the variables and what are the coefficients?

7) For $5x^2 - 4 + 2y^2$, what are the variables and what are the coefficients?

Repeated multiplication is best expressed using an exponent:

- xx is best written as x^2 (just like $5 \cdot 5$ is the same as 5^2).
- xxx should be written as x^3 (just like $5 \cdot 5 \cdot 5$ is the same as 5^3).

A coefficient or exponent of one is unnecessary. You almost never see a coefficient of one or an exponent of one written.

- x is the same as $1x^1$. This is almost always written as just x.
- $x^2 y^3$ is the same as $1x^2 y^3$. The preferred form is $x^2 y^3$.
- $5x^4 y$ is the same as $5x^4 y^1$. The preferred form is $5x^4 y$.

Similarly, with division, we almost never use the standard division symbol ($\div$) when doing algebra.

In algebra, division is most commonly expressed as a fraction:

- $\frac{4}{3}$ is equivalent to dividing 4 by 3.
- $\frac{18}{6}$ means 18 divided by 6. (The answer is 3.)
- $\frac{x}{2}$ means x divided by 2.
- $\frac{3}{x}$ means 3 divided by x.
- $\frac{x^2 - 1}{5}$ means the quantity $x^2 - 1$ is divided by 5.

The main alternative to writing a fraction is to use the slash (/) symbol. The slash isn't as common, but the slash can help to avoid writing a fraction within a fraction.

- 4/3 is equivalent to dividing 4 by 3. We would usually write this as $\frac{4}{3}$.
- 18/6 means 18 divided by 6. (The answer is 3.)
- $x/2$ means x divided by 2. We would usually write this as $\frac{x}{2}$.
- $\frac{x/3}{x/4}$ means $\frac{x}{3}$ divided by $\frac{x}{4}$. Here the slashes are helpful since the alternative is $\frac{\frac{x}{3}}{\frac{x}{4}}$.

A denominator of 1 is unnecessary and should be avoided.

- $\frac{4}{1}$ is the same as 4. It is simpler to just write 4.
- $\frac{x}{1}$ is the same as x. It is simpler to just write x.

Example 1. $(4)(6) = 24$ **Example 2.** $3 \cdot 7 = 21$

Example 3. $5(4)(3) = (20)(3) = 60$ **Example 4.** $42/6 = 7$

Example 5. $\frac{32}{4} = 8$ **Example 6.** $xxx = x^3$

Example 7. $3xy$ means 3 times x times y.   **Example 8.** $\frac{x}{4}$ means x divided by 4.

Exercise Set 1.3

Directions: Carry out arithmetic to determine the answer.

1) $9 \cdot 7 =$ 2) $\frac{48}{8} =$

3) $27/3 =$ 4) $7(8) =$

5) $(4)(2)(3) =$ 6) $\frac{12}{1} =$

7) $\frac{32}{4} =$ 8) $5(6)(3) =$

Directions: Simplify each expression.

9) $1xx =$ 10) $\frac{x^4}{1} =$

Directions: Write out what each expression means using words.

11) $5x^2y$

12) $\frac{4x^3}{7}$

13) $\frac{x/2}{y/5}$

1.4 Algebra in English

In algebra, we use a letter such as x or y to represent a variable, which as an unknown quantity that we would like to solve for. When algebra is applied to solve a practical problem, the variable represents something specific. For example, a variable might be used to represent the amount of money that a person owes, a person's age in years, or the time it takes to drive from one city to another.

A problem that is expressed in words can be expressed using algebra by recognizing how the language relates to the math. For example, if x represents a person's age in years today, then the expression $x + 5$ represents the person's age in five years. There are many ways to describe arithmetic operations in English. The keywords tabulated below illustrate some common ways to describe arithmetic operations with words.

- addition: sum, total of, combined, together, in all, increased by, gained, greater than, more than, raised to
- subtraction: difference, minus, left over, taken away, fewer, decreased by, lost, less than, smaller than
- multiplication: multiplied by, times, product, twice, double, triple, increased by a factor of, decreased by a factor of (or even the single word "of")
- division: divided by, per, out of, half, third, fourth, split, average, equal pieces, fraction, ratio, quotient, percent
- powers and roots: squared, cubed, square root, cube root, raised to the power
- equal sign: equals, is, was, makes, will be

For example, consider the following problem: "The square of a number is twelve more than the number." We can represent this problem with the equation $x^2 = x + 12$. (In Chapter 6, we'll learn how to solve such an equation to see that one solution is $x = 4$. You can check that this works since $4^2 = 16$ and $4 + 12 = 16$.) In this section, we will focus on the relationship between algebra and language.

Example 1. Let x represent the number of boxes. There are nine more lids than there are boxes. What represents the number of lids?

Add 9 to the number of boxes. The number of lids is $x + 9$.

Example 2. Let y represent the height of a fence. A ladder is twice as tall as the fence. What represents the height of the ladder?

Multiply the height of the fence by 2. The height of the ladder is $2y$.

Example 3. Eight more than a number is three times the number. Represent this with an equation.

Add 8 to the number and set this equal to 3 times the number: $x + 8 = 3x$.

Exercise Set 1.4

Directions: Write the indicated algebraic expression or equation.

1) Let x represent the length of a log. The log is then cut in half. What represents the length of each piece of the log?

2) Let t represent the time in seconds that a red ball has been rolling. A blue ball started rolling four seconds after the red ball started rolling. What represents the time that the blue ball has been rolling?

3) Let y represent Bill's paycheck in dollars. Pat's paycheck is fifty dollars less than twice Bill's paycheck. What represents Pat's paycheck?

4) Let x represent a number. What represents the square root of the number?

5) Two consecutive even numbers have a product of eighty. Represent this with an equation.

1.5 Order of Operations

The acronym PEMDAS helps to remember the order of arithmetic operations:
- P stands for Parentheses. Simplify expressions in parentheses first.
- E stands for Exponents. Deal with exponents and roots after parentheses.
- MD stands for Multiplication and Division. After dealing with parentheses and exponents, multiply and divide from left to right.
- AS stands for Addition and Subtraction. After multiplying and dividing from left to right, add and subtract from left to right.

To understand why the order of operations matters, consider the expression $3 + 2 \cdot 4$. According to PEMDAS, we should multiply before we add. This gives the correct answer of $3 + 8 = 11$. If instead you add before you multiply, you would get $5 \cdot 4 = 20$, which is different (and incorrect).

Example 1. $2 + 3(7 - 3)^2 = 2 + 3(4)^2 = 2 + 3(16) = 2 + 48 = 50$
We treated the parentheses first, then found the exponent, then multiplied, and added last.

Example 2. $7 + 3 \cdot 2^2 - 6/2 = 7 + 3 \cdot 4 - 6/2 = 7 + 12 - 3 = 19 - 3 = 16$
We found the exponent first, then multiplied and divided left to right, and added and subtracted left to right last.

Exercise Set 1.5

Directions: Evaluate each expression by following the order of operations.

1) $2 + 8^2/4 - 3 \cdot 2^2 =$

2) $3(12 - 2 \cdot 4) - 2(3^2 - 2^2) =$

3) $6 + 4(20 - 5 \times 3) =$

4) $(3^2 - 5)^3(8 - 2 \cdot 3)^2 =$

5) $6\sqrt{8 \cdot 3 + 5^2} - 3\sqrt{6 \cdot 4 - 5 \cdot 3 + 4^2} =$

6) $4(5)(2)^3 - 7(3)(2)^2 =$

7) $(9 \cdot 8 - 7 \cdot 6)/3 + 6 =$

8) $\frac{64}{4} + \frac{36}{6} =$

9) $\frac{8 \cdot 6 \cdot 4}{4 \cdot 3 \cdot 2} =$

10) $\frac{12 + 6 \cdot 3}{2 \cdot 8 - 6} =$

11) $\sqrt{(5 + 2 \cdot 4)^2 - (9 - 2 \cdot 2)^2} =$

12) $\left(\frac{4 \cdot 7 + 8}{3^2}\right)^3 =$

13) $(3^3 - 7 \cdot 3)^{2 \cdot 5 - 2^3} =$

14) $\sqrt{\frac{6 \cdot 7 - 5 \cdot 2}{2 \cdot 7 - 3 \cdot 2}} =$

1.6 Algebra Operations

Addition and subtraction are represented with the standard plus (+) and minus (−) signs. For example, $x^2 - 2x + 8$ means to first square x, subtract two times x, and add eight. Addition is **commutative**, which means that it doesn't matter in which order the numbers are added. The commutative property of addition can be expressed in the form $x + y = y + x$. For example, if $x = 5$ and $y = 3$, this becomes $5 + 3 = 3 + 5 = 8$. Note that subtraction isn't commutative. For example, $5 - 3 = 2$ is positive whereas $3 - 5 = -2$ is negative.

As discussed in Sec. 1.3, the standard symbols for multiplication (×) and division (÷) are avoided in algebra. Multiplication between numbers is expressed with parentheses like $7(8) = 56$ or with a middle dot like $7{\cdot}8 = 56$, and no symbol is used to multiply variables like $5xy$ (which means 5 times x times y). Division is usually expressed as a fraction like $\frac{x}{2}$, though the slash symbol (/) is occasionally used like $x/2$. Like addition, multiplication is also commutative, since it doesn't matter in which order numbers are multiplied. The commutative property of multiplication can be expressed as $xy = yx$. For example, $4(6) = 6(4) = 24$. Like subtraction, division isn't commutative. Order matters for subtraction and division. For example, $\frac{12}{3} = 4$ whereas $\frac{3}{12} = \frac{1}{4}$.

Another property that relates to the order in which numbers are added or multiplied is the **associative** property. When adding or multiplying three or more numbers, the associative property states that it doesn't matter how the numbers are grouped. The associative property of addition may be expressed as $(x + y) + z = x + (y + z)$ and the associative property of multiplication may be expressed as $(xy)z = x(yz)$. To add three numbers, it doesn't matter which pair of numbers is added first. For example, $(2 + 3) + 4 = 5 + 4 = 9$ is the same as $2 + (3 + 4) = 2 + 7 = 9$. Similarly, to multiply three numbers, $(3{\cdot}4)5 = (12)5 = 60$ is the same as $3(4{\cdot}5) = 3(20) = 60$.

The **distributive** property involves both multiplication and addition. The distributive property may be expressed as $x(y + z) = xy + xz$. For example, compare $4(3 + 6) =$

$4(9) = 36$ with $4(3) + 4(6) = 12 + 24 = 36$. The distributive property is particularly useful in algebra, as we will explore in Chapter 4.

The **identity** property of addition states that adding zero has no effect. The identity property of multiplication states that multiplying by one has no effect. These properties may be expressed as $x + 0 = x$ and $1x = x$. For example, $7 + 0 = 7$ and $1(4) = 4$.

The **inverse** property of addition states that any number plus its negative counterpart equals zero. The inverse property of addition may be expressed as $x + (-x) = 0$. The inverse property of addition basically states that subtraction is the opposite of addition since $x - x = 0$. (Adding a negative number has the same effect as subtraction, as we will review in Sec. 1.10.) As an example, $6 + (-6) = 0$. The quantity $-x$ is called the **additive inverse** of x.

The inverse property of multiplication states that when any number is multiplied by its **reciprocal**, the result is one. The reciprocal of a number is found by dividing one by the number. For example, the reciprocal of 2 is equal to $\frac{1}{2}$. The inverse property of multiplication may be expressed as $x\frac{1}{x} = 1$ (which means that x times $\frac{1}{x}$ equals one). The inverse property of multiplication basically states that division is the opposite of multiplication since $\frac{x}{x} = 1$. As an example, $3\left(\frac{1}{3}\right) = \frac{3}{3} = 1$. The reciprocal $\frac{1}{x}$ is also called the **multiplicative inverse** of x. The reciprocal may also be written as x^{-1}, as we will review in Sec. 1.11.

The **reflexive** property states that any quantity is equal to itself. The reflexive property may be expressed as $x = x$. For example, $5 = 5$. The reflexive property may seem to be quite trivial, yet it has important uses. For example, the reflexive property allows us to swap the two sides of an equation. Consider the simple equation $x = y$. Since x equals y, it must also be true that y equals x. That is, if $x = y$ is true, it must also be true that $y = x$. This allows us to swap the two sides of *any* equation. For example, the equation $3x - 4 = 12$ may rewritten as $12 = 3x - 4$. If the equation $3x - 4 = 12$ is true, it must also be true that $12 = 3x - 4$. If we let $w = 3x - 4$ and let $y = 12$, we can write these equations as $w = y$ and $y = w$. For an example of this using numbers, consider that $3(4) - 5 = 7$ may also be expressed as $7 = 3(4) - 5$, and both equations are true since $3(4) - 5 = 12 - 5 = 7$. Both equations state that $7 = 7$.

The **transitive** property states that if $x = y$ and if $x = z$, then it follows that $y = z$. The transitive property has an important use in algebra because it serves as the basis for substitution. For example, if $y = 5x - 8$ and if $y = 3x + 2$, the transitive property lets us write $5x - 8 = 3x + 2$.

Exercise Set 1.6

Directions: Apply the properties of this section to answer the following questions.

1) Which property is expressed by $xy + xz = x(y + z)$? What is different about this equation compared to that given in the text? Explain why we may write the equation in this form.

2) Which property is expressed by $0 + x = x$? What is different about this equation compared to that given in the text? Explain why we may write the equation in this form.

3) Which property is expressed by $(x + y) + z = x + (y + z) = y + (x + z)$? What is different about this equation compared to that given in the text? Explain why we may write the equation in this form.

4) Which properties are expressed by $x - y = -y + x$? Explain why we may write the equation in this form.

1.7 Fractions

A fraction represents division. For example, $\frac{3}{4}$ means 3 divided by 4. Similarly, $\frac{24}{8} = 3$ because 24 divided by 8 equals 3. A fraction may also include variables. For example, $\frac{x}{2}$ means x divided by 2.

In algebra, if a fraction is greater than one, write it as an improper fraction. Don't write it as a mixed number. For example, to write two and one-half in algebra, write $\frac{5}{2}$. Don't write it as a mixed number with 2 beside $\frac{1}{2}$ because, in algebra, that would represent multiplication. You could write it as $2 + \frac{1}{2}$, using a plus sign. When a number appears beside a fraction in algebra, it means to multiply. For example, $3\frac{x}{5}$ means to multiply 3 by $\frac{x}{5}$. (It is not a mixed number.)

Fractions are common in algebra. For example, consider the simple equation $3x = 4$. We can solve this simple equation by dividing each side of the equation by 3. When we do this, the 3's cancel on the left and we get $x = \frac{4}{3}$. We can check that this solves the equation by plugging it into the original equation: $3\left(\frac{4}{3}\right) = \frac{3(4)}{3} = \frac{12}{3} = 4$. Since fractions are common in algebra, it will be useful to review some rules regarding fractions.

If the numerator and denominator share a common factor, the fraction can be **reduced**. To reduce a fraction, divide the numerator and denominator by their greatest common factor. For example, $\frac{9}{12}$ can be reduced by dividing 9 and 12 each by 3, as shown below.

$$\frac{9}{12} = \frac{9/3}{12/3} = \frac{3}{4}$$

In algebra, when an answer is a fraction, if the fraction is reducible, it is conventional to reduce the answer. For example, if you solve an equation to get $x = \frac{16}{20}$, you should reduce this answer as follows: $x = \frac{16}{20} = \frac{16/4}{20/4} = \frac{4}{5}$.

The way to add or subtract fractions is to find a **common denominator**, like the example below where $\frac{3}{4}$ was multiplied by $\frac{3}{3}$ to make $\frac{9}{12}$ and $\frac{1}{6}$ was multiplied by $\frac{2}{2}$ to make $\frac{2}{12}$.

$$\frac{3}{4} - \frac{1}{6} = \frac{3\cdot 3}{4\cdot 3} - \frac{1\cdot 2}{6\cdot 2} = \frac{9}{12} - \frac{2}{12} = \frac{7}{12}$$

The same idea applies if a variable is present, as shown below.

$$\frac{x}{2} + \frac{x}{3} = \frac{3x}{2\cdot 3} + \frac{2x}{3\cdot 2} = \frac{3x}{6} + \frac{2x}{6} = \frac{5x}{6}$$

Multiplying fractions is easier than adding or subtracting fractions since you don't need to find a common denominator. Multiply their numerator and denominators together as shown below. Note that two fractions beside one another indicate multiplication.

$$\frac{3}{7}\frac{4}{5} = \frac{3\cdot 4}{7\cdot 5} = \frac{12}{35}$$

The **reciprocal** of a fraction is found by swapping the numerator and denominator. For example, the reciprocal of $\frac{5}{8}$ is equal to $\frac{8}{5}$. The reciprocal of an integer is one divided by the integer. For example, the reciprocal of 4 is equal to $\frac{1}{4}$.

To divide by a fraction, multiply by its reciprocal, like the following example. Although we don't normally use the $\div$ symbol in algebra, we made an exception here in order to help make this clear. Note that $\frac{2}{3}\frac{2}{1}$ means two-thirds times $\frac{2}{1}$. Two fractions beside one another are multiplying.

$$\frac{2/3}{1/2} = \frac{2}{3} \div \frac{1}{2} = \frac{2}{3}\frac{2}{1} = \frac{2\cdot 2}{3\cdot 1} = \frac{4}{3}$$

Example 1. Reduce $\frac{15}{25}$.

$$\frac{15}{25} = \frac{15/5}{25/5} = \frac{3}{5}$$

Example 2. Subtract $\frac{1}{3}$ from $\frac{1}{2}$.

$$\frac{1}{2} - \frac{1}{3} = \frac{1\cdot 3}{2\cdot 3} - \frac{1\cdot 2}{3\cdot 2} = \frac{3}{6} - \frac{2}{6} = \frac{1}{6}$$

Example 3. Multiply $\frac{2}{3}$ by $\frac{9}{4}$.

$$\frac{2}{3}\frac{9}{4} = \frac{2\cdot 9}{3\cdot 4} = \frac{18}{12} = \frac{18/6}{12/6} = \frac{3}{2}$$

We reduced $\frac{18}{12}$ to $\frac{3}{2}$ by dividing 18 and 12 each by 6. In the previous example, if you note that $\frac{9}{3} = 3$ and $\frac{2}{4} = \frac{1}{2}$, you can arrive at the answer of $\frac{3}{2}$ more simply.

Example 4. What is the reciprocal of $\frac{2}{7}$?

Swap the numerator and denominator to get $\frac{7}{2}$.

Example 5. Divide $\frac{5}{8}$ by $\frac{3}{7}$.

$$\frac{5/8}{3/7} = \frac{5}{8} \div \frac{3}{7} = \frac{5 \, 7}{8 \, 3} = \frac{5 \cdot 7}{8 \cdot 3} = \frac{35}{24}$$

Exercise Set 1.7

Directions: Apply properties of fractions discussed in this section.

1) Reduce $\frac{8}{20}$.

2) Reduce $\frac{16}{14}$.

3) Reduce $\frac{28}{42}$.

4) Reduce $\frac{27}{36}$.

5) Reduce $\frac{64}{48}$.

6) Reduce $\frac{56}{40}$.

7) $\frac{4}{5} + \frac{7}{3} =$

8) $\frac{5}{6} - \frac{2}{9} =$

9) $\frac{8}{15} + \frac{9}{20} =$

10) $\frac{7}{4} - \frac{5}{6} =$

11) Find the reciprocal of $\frac{8}{9}$.

12) Find the reciprocal of $\frac{1}{3}$.

13) Find the reciprocal of 6.

14) Find the reciprocal of $\frac{7}{11}$.

15) $\frac{9}{5}\frac{7}{2} =$

16) $\frac{8}{9}\frac{5}{4} =$

17) $\frac{2}{9}\frac{3}{8} =$

18) $\frac{2}{3}6 =$

19) $\frac{7/5}{9/2} =$

20) $\frac{4/3}{2/5} =$

21) $\frac{12/5}{4} =$

22) $\frac{9}{2/3} =$

1.8 Decimals

Decimals are common in certain applications of algebra, such as finance (since dollars and cents are expressed in decimal values like \$12.75) and science (where measured values tend to include decimals like 6.24 kg). When solving an equation that involves decimals, you can remove all of the decimals from the equation by multiplying both sides of the equation by a sufficiently large power of ten. For example, for the equation $0.225x + 0.075 = 3.5$, if we multiply both sides of the equation by 1000, the equation becomes $225x + 75 = 3500$.

However, if working with an expression like $5.62x - 8.3 + 3.815x - 0.78$ which does not have an equal sign, the previous trick won't work.

To add or subtract two decimals, align the decimals at their decimal points. If the top decimal has fewer decimal places, add trailing zeros to it until it has the same number of decimal places as the bottom decimal. For example, the problem $1.4 - 0.68$ may be rewritten as $1.40 - 0.68$.

$$\begin{array}{r} 1.40 \\ -0.68 \\ \hline 0.72 \end{array}$$

To multiply two decimals, first multiply the numbers as if they were integers and then add a decimal point so that the result has the same number of decimal places as the two numbers combined together. For example, 0.4×1.32 has three decimal places all together. Since $4 \times 132 = 528$, it follows that $0.4 \times 1.32 = 0.528$. Note that 0.528 has three decimal places.

To divide two decimals, multiply the numerator and denominator by the same power of ten needed to remove the decimals, like the example below. If the numerator and denominator share a common factor, reduce the fraction by dividing the numerator and denominator by the greatest common factor.

$$\frac{2.43}{0.009} = \frac{2.43(1000)}{0.009(1000)} = \frac{2430}{9} = \frac{2430/9}{9/9} = \frac{270}{1} = 270$$

Example 1. Rewrite $0.75 - 1.5x = 0.24$ without any decimals.

None of the decimals extends further than the hundredths place. Multiply every term of the equation by 100 to get $75 - 150x = 24$.

Example 2. Rewrite $0.7x^2 - 1 = 3$ without any decimals.

There is only one decimal and it extends into the tenths place. Multiply every term of the equation by 10 to get $7x^2 - 10 = 30$. Even though only one number was a decimal value, it was still necessary to multiply all three terms by 10.

Example 3. Add 0.12 to 1.834.

$$
\begin{array}{r}
0.120 \\
+1.834 \\
\hline
1.954
\end{array}
$$

Example 4. Subtract 0.36 from 1.9.

$$
\begin{array}{r}
1.90 \\
-0.36 \\
\hline
1.54
\end{array}
$$

Example 5. Multiply 0.7 by 0.06.

Since $7(6) = 42$ and since $0.7(0.06)$ has three combined decimal places, the answer is $0.7(0.06) = 0.042$.

Example 6. Multiply 1.2 by 1.2.

Since $12(12) = 144$ and since $1.2(1.2)$ has two combined decimal places, the answer is $1.2(1.2) = 1.44$.

Example 7. Divide 0.18 by 3. Express the answer as a reduced fraction.

$$\frac{0.18}{3} = \frac{0.18(100)}{3(100)} = \frac{18}{300} = \frac{18/6}{300/6} = \frac{3}{50}$$

Example 8. Divide 0.025 by 0.75. Express the answer as a reduced fraction.

$$\frac{0.025}{0.75} = \frac{0.025(1000)}{0.75(1000)} = \frac{25}{750} = \frac{25/25}{750/25} = \frac{1}{30}$$

Exercise Set 1.8

Directions: Rewrite each equation below without any decimals.

1) $0.3x - 0.2 = 1.6$

2) $5.2 - 0.024x = 2.39$

3) $0.8x^2 + 0.23x = 4$

Directions: Perform the indicated arithmetic operation.

3) $0.65 + 0.57 =$

4) $2.8 + 0.73 =$

5) $7.2 - 3.45 =$

6) $8 - 1.76 =$

7) $(0.8)(0.9) =$

8) $7(0.125) =$

9) $\dfrac{2.4}{9.6} =$

10) $\dfrac{0.45}{2.5} =$

1.9 Percents

If a percent is needed in an equation or an expression, it may be converted to a decimal or fraction. For example, suppose that an unknown quantity, which we may call x, was invested in a bank that pays simple interest of 3% annually, and that after one year the balance is \$247.20. By converting 3% to the decimal value 0.03, we can determine x by solving the equation $1.03x = 247.2$. (As mentioned in Sec. 1.8, we can then multiply both sides of the equation by 100 to get $103x = 24{,}720$. We'll wait until Chapter 2 to learn how to solve such equations. For now, we'll focus on dealing with percents.)

To convert a percent to a decimal, divide by 100%. For example, $3\% = \frac{3\%}{100\%} = 0.03$.

To convert a percent to a fraction, write the given value over a denominator of 100. If the numerator is a decimal, multiply the numerator and denominator both by the power of ten needed to remove the decimal point, like the example below. If the numerator and denominator share a common factor, reduce the fraction by dividing the numerator and denominator by the greatest common factor.

$$37.5\% = \frac{37.5}{100} = \frac{37.5(10)}{100(10)} = \frac{375}{1000} = \frac{375/125}{1000/125} = \frac{3}{8}$$

Example 1. Convert 64% to a decimal and also to a fraction.

$$64\% = \frac{64\%}{100\%} = 0.64$$

$$64\% = \frac{64}{100} = \frac{64/4}{100/4} = \frac{16}{25}$$

Example 2. Convert 8.32% to a decimal and also to a fraction.

$$8.32\% = \frac{8.32\%}{100\%} = 0.0832$$

$$8.32\% = \frac{8.32}{100} = \frac{8.32(100)}{100(100)} = \frac{832}{10{,}000} = \frac{832/16}{10{,}000/16} = \frac{52}{625}$$

Exercise Set 1.9

Directions: Convert each percent to a decimal and also to a fraction.

1) 45% =

2) 120% =

3) 2.5% =

4) 350% =

5) 87.5% =

6) 76% =

7) 16.8% =

8) 6.25% =

1.10 Negative Numbers

A negative number is the opposite of a positive number. For example, if walking 40 m north is considered positive, then walking 40 m south would be negative. Adding a negative number to a positive number equates to subtraction. For example, $8 + (-3)$ $= 8 - 3 = 5$. Since addition is commutative (Sec. 1.6), the order in which the numbers are added doesn't matter: $8 + (-3) = -3 + 8 = 8 - 3 = 5$. We can express these rules algebraically as

$$x + (-y) = -y + x = x - y$$

The **absolute value** of a number is nonnegative. It represents how far a number is from zero on the number line regardless of direction. When two vertical lines are placed around a number or expression, this indicates the absolute value. For example, $|-4|$ means the absolute value of negative four: $|-4| = 4$.

When a negative number is added to a positive number, the result may be positive or negative. The answer is positive if the positive number has the greater absolute value, and negative if the negative number has the greater absolute value. See the examples below.

$$9 + (-7) = -7 + 9 = 9 - 7 = 2 \text{ because } |9| > |-7|$$
$$7 + (-9) = -9 + 7 = 7 - 9 = -2 \text{ because } |7| < |-9|$$

Adding a negative number to another negative number makes a number that is even more negative. In this case, the absolute values add together and the answer is negative. For example, $-6 + (-5) = -(6 + 5) = -11$. Note that adding a negative number to a negative number is equivalent to subtracting a positive number from a negative number. For example, $-6 - 5 = -(6 + 5) = -11$. These rules can be expressed algebraically as:

$$-x + (-y) = -x - y = -(x + y)$$

Two minus signs in front of a number cancel out. For example, $-(-8) = 8$. This is equivalent to multiplying or dividing a negative number by negative one.

$$-(-x) = \frac{-x}{-1} = x$$

Subtracting a negative number is equivalent to adding a positive number. Two minus signs effectively combine together to make a plus sign. For example, $9 - (-3) = 9 + 3$

= 12. If a negative number is subtracted from another negative number, this equates to adding a positive number to a negative number. For example, $-9 - (-3) = -9 + 3 = -6$. These rules can be expressed algebraically as:

$$x - (-y) = x + y$$
$$-x - (-y) = -x + y$$

When multiplying or dividing numbers, count the minus signs. When an odd number of minus signs are multiplying, the answer is negative. When an even number of minus signs are multiplying, the answer is positive. See the examples below.

$$4(-7) = -4(7) = -28$$
$$-4(-7) = 4(7) = 28$$
$$\frac{-18}{6} = \frac{18}{-6} = -\frac{18}{6} = -3$$
$$\frac{-18}{-6} = -\frac{-18}{6} = -\frac{18}{-6} = \frac{18}{6} = 3$$

These rules can be expressed algebraically as:

$$x(-y) = -xy$$
$$-x(-y) = xy$$
$$\frac{-x}{y} = \frac{x}{-y} = -\frac{x}{y}$$
$$\frac{-x}{-y} = -\frac{-x}{y} = -\frac{x}{-y} = \frac{x}{y}$$

Example 1. $19 + (-7) = 12$

Example 2. $-19 + 7 = -12$

Example 3. $7 + (-19) = -12$

Example 4. $-7 + 19 = 12$

Example 5. $19 - (-7) = 19 + 7 = 26$

Example 6. $-19 - (-7) = -19 + 7 = -12$

Example 7. $-19 - 7 = -26$

Example 8. $5(-4) = -20$

Example 9. $-5(-4) = 20$

Example 10. $\frac{-24}{6} = -4$

Example 11. $\frac{24}{-6} = -4$

Example 12. $\frac{-24}{-6} = 4$

Exercise Set 1.10

Directions: Carry out arithmetic to determine the answer.

1) $18 + (-6) =$

2) $24 - (-8) =$

3) $-32 + 18 =$

4) $-17 - 15 =$

5) $-(-11) =$

6) $9 + (-16) =$

7) $-25 + 40 =$

8) $0 + (-5) =$

9) $-12 - (-4) =$

10) $15 - (-25) =$

11) $48 - (-24) =$

12) $6 + (-6) =$

13) $-9 - (-9) =$

14) $-8 + (-8) =$

15) $-16 - 12 =$

16) $0 - 14 =$

17) $8(-8) =$

18) $-6(-9) =$

19) $-7(-5) =$

20) $9(-7) =$

21) $\frac{56}{-7} =$

22) $\frac{-42}{6} =$

23) $\frac{-81}{-9} =$

24) $-\frac{49}{-7} =$

25) $-2(-3)(-4) =$

26) $-4(5)(-6) =$

27) $\frac{9(-8)}{-6} =$

28) $\frac{-8(-7)}{-4} =$

29) $\frac{(-8)(-6)}{(-2)(-3)} =$

30) $\frac{(6)(-6)}{(-2)(-9)} =$

1.11 Powers and Roots

An **exponent** (which is also called a **power**) is shorthand for repeated multiplication. When the exponent is a whole number, it indicates how many times to multiply the base by itself. For example, $5^3 = 5(5)(5)$ has three fives multiplied together. The 3 in 5^3 is called the exponent (or power) and the 5 in 5^3 is called the **base**. Two common exponents have special names:

- An exponent of 2 is called a **square**. For example, 4^2 is read as "4 squared" and means $4^2 = 4(4) = 16$.
- An exponent of 3 is called a **cube**. For example, 6^3 is read as "6 cubed" and means $6^3 = 6(6)(6) = 36(6) = 216$.

Exponents of zero and one are special:

- If the exponent equals 1, the answer equals the base. For example, $7^1 = 7$.
- If the exponent equals 0, the answer equals one. For example, $5^0 = 1$. When any nonzero value is raised to the power of 0, the answer is 1. We'll explore why in Chapter 3.

When a negative number is raised to an exponent that is a whole number, the answer is negative if the exponent is odd and positive if the exponent is even. For example, $(-2)^3 = (-2)(-2)(-2) = 4(-2) = -8$ and $(-2)^4 = (-2)(-2)(-2)(-2) = 4(4) = 16$.

A **root** is basically the opposite of a power. When a number is placed inside the radical ($\sqrt[n]{}$), it means to find the nth root.

- A **square root**, $\sqrt{}$, asks, "Which number squared equals the value under the radical?" For example, $\sqrt{25} = 5$ because $5^2 = 5(5) = 25$. Technically, -5 is a solution, too, since $(-5)^2 = (-5)(-5) = 25$. However, in algebra, it is common to include only the positive solution for a radical unless a minus sign ($-$) or a plus/minus sign ($\pm$) appears in front of the radical sign.
- A **cube root**, $\sqrt[3]{}$, asks, "Which number cubed equals the value under the radical?" For example, $\sqrt[3]{8}$ because $2^3 = 2(2)(2) = 4(2) = 8$. The cube root of a negative number is negative. For example, $\sqrt[3]{-8} = -2$.

- A general root, $\sqrt[n]{}$, asks, "Which number raised to the nth power equals the value under the radical?" For example, $\sqrt[4]{81} = 3$ because $3^4 = 3(3)(3)(3) = 9(9) = 81$. Although $(-3)^4 = 81$ also, as with square roots, we will only give the positive solution for an even root unless a minus sign or a plus/minus sign appears in front of the radical sign. Odd roots have the same sign as the number inside of the radical. For example, $\sqrt[5]{100,000} = 10$ and $\sqrt[5]{-100,000} = -10$.

An equivalent way to represent a root is with an exponent of $1/n$. For example, the square root of 9 can be written as $9^{1/2}$ or as $\sqrt{9}$. Similarly, the cube root of 27 can be written as $27^{1/3}$ or as $\sqrt[3]{27}$. We'll explore why in Chapter 3.

When the exponent is a fraction, the exponent's numerator indicates a power while the exponent's denominator indicates a root. For example, $4^{3/2}$ combines together 4^3 and $4^{1/2}$. There are two different ways to figure out what $4^{3/2}$ equals. You could first cube 4 to get $4^3 = 64$ and then take the square root to get $64^{1/2} = 8$, or you could first square root 4 to get $4^{1/2} = 2$ and then cube this to get $2^3 = 8$.

$$4^{3/2} = (4^3)^{1/2} = 64^{1/2} = 8$$

$$4^{3/2} = \left(4^{1/2}\right)^3 = 2^3 = 8$$

An exponent of -1 indicates a **reciprocal**. The reciprocal of a fraction has its numerator and denominator swapped. For example, $\left(\frac{5}{7}\right)^{-1} = \frac{7}{5}$. The reciprocal of an integer is one divided by the integer. For example, $6^{-1} = \frac{1}{6}$. We can express these rules algebraically as $\left(\frac{x}{y}\right)^{-1} = \frac{y}{x}$ and $x^{-1} = \frac{1}{x}$. We'll explore exponents of -1 algebraically in Chapter 3.

Any negative exponent can be turned positive by finding the reciprocal of the base. For example, $\left(\frac{8}{27}\right)^{-2/3} = \left(\frac{27}{8}\right)^{2/3} = \left(\frac{27^{1/3}}{8^{1/3}}\right)^2 = \left(\frac{3}{2}\right)^2 = \frac{3^2}{2^2} = \frac{9}{4}$. This rule can be expressed algebraically as:

$$\left(\frac{x}{y}\right)^{-m/n} = \left(\frac{y}{x}\right)^{m/n}$$

We'll explore this algebraically in Chapter 3. For now, we will focus on the arithmetic.

It is common for a variable to have an exponent. For example, x^2 means xx and x^3 means xxx. As mentioned in Sec. 1.3, an exponent of 1 is generally not written. For example, x^1 is equivalent to x. A simple equation like $x^2 = 4$ can be solved by taking the square root of both sides. We get $\sqrt{x^2} = \pm\sqrt{4}$ which simplifies to $x = \pm 2$. The left-hand side states that squaring a number and then square rooting a number effectively leaves the number unchanged. (For example, if you square 5 to get 25 and then take the square root, the answer is 5, which is the original number.) The right-hand side includes a $\pm$ sign because the answer could be positive or negative. In this context, it is customary to consider both the positive and negative solutions. Note that $x = 2$ and $x = -2$ are both solutions to $x^2 = 4$ since $2^2 = 2(2) = 4$ and $(-2)^2 = -2(-2) = 4$. We will consider algebraic equations and expressions with exponents in Chapter 3.

Example 1. $7^2 = 7(7) = 49$ **Example 2.** $3^3 = 3(3)(3) = 9(3) = 27$

Example 3. $9^1 = 9$ **Example 4.** $6^0 = 1$

Example 5. $(-10)^3 = -10(-10)(-10) = 100(-10) = -1000$

Example 6. $(-5)^4 = -5(-5)(-5)(-5) = 25(25) = 625$

Example 7. $\sqrt{16} = 4$ since $4^2 = 16$ **Example 8.** $\sqrt[3]{27} = 3$ since $3^3 = 27$

Example 9. $\sqrt[4]{256} = 4$ since $4^4 = 256$ **Example 10.** $\sqrt[5]{-32} = -2$ since $(-2)^5 = -32$

Example 11. $4^{1/2} = 2$ since $\sqrt{4} = 2$ **Example 12.** $64^{1/3} = 4$ since $4^3 = 64$

Example 13. $8^{2/3} = (8^2)^{1/3} = 64^{1/3} = 4$ **Example 14.** $\left(\frac{3}{5}\right)^{-1} = \frac{5}{3}$

Example 15. $5^{-1} = \frac{1}{5}$ **Example 16.** $4^{-2} = \left(\frac{1}{4}\right)^2 = \frac{1}{4^2} = \frac{1}{16}$

Example 17. $\left(\frac{1}{9}\right)^{-1/2} = 9^{1/2} = 3$ **Example 18.** $\left(\frac{2}{3}\right)^{-3} = \left(\frac{3}{2}\right)^3 = \frac{3^3}{2^3} = \frac{27}{8}$

Exercise Set 1.11

Directions: Carry out arithmetic to determine the answer.

1) $8^2 =$

2) $4^3 =$

3) $5^4 =$

4) $2^5 =$

5) $(-4)^4 =$

6) $(-3)^5 =$

7) $9^0 =$

8) $8^1 =$

9) $\sqrt{49} =$

10) $\sqrt{100} =$

11) $\sqrt[3]{125} =$

12) $\sqrt[3]{512} =$

13) $\sqrt[3]{-64} =$

14) $\sqrt[4]{16} =$

15) $64^{1/2} =$

16) $1000^{1/3} =$

17) $\left(\frac{4}{7}\right)^{-1} =$

18) $7^{-1} =$

19) $27^{2/3} =$

20) $36^{3/2} =$

21) $16^{3/4} =$

22) $5^{-2} =$

23) $4^{-3} =$

24) $\left(\frac{1}{2}\right)^{-4} =$

25) $25^{-1/2} =$

26) $\left(\frac{1}{36}\right)^{-1/2} =$

27) $\left(\frac{4}{9}\right)^{-3/2} =$

28) $\left(\frac{27}{64}\right)^{-2/3} =$

1.12 Number Classification

Whole numbers include 0, 1, 2, 3, 4, 5, 6, 7, 8, 9, 10, 11, 12, etc. Some instructors make a distinction between whole numbers and natural numbers, where whole numbers include the 0 but where natural numbers do not include the zero. (Not everyone agrees with this distinction.) **Integers** include the negative whole values $-1, -2, -3, -4$, etc. along with the whole numbers 0, 1, 2, 3, 4, etc. (Some instructors distinguish between integers and whole numbers, where integers include negative values but where whole numbers do not include negative values.)

Rational numbers include fractions (like $\frac{3}{4}$ or $\frac{29}{5}$) and decimals (like 0.0032 or 14.18) in addition to integers. Like integers, rational numbers may be positive, negative, or zero. Rational numbers include repeating decimals like $\frac{4}{11} = 0.\overline{36} = 0.3636363636...$ (with the 36 repeating forever) because a repeating decimal can be expressed as a fraction. However, rational numbers don't include decimals that go on forever without repeating. For example, $\sqrt{2}$ is **irrational** because it goes on forever without repeating (starting with 1.41421356...). Similarly, the constant pi, $\pi = 3.14159265...$, which is the ratio of the circumference of any circle to its diameter, is irrational because it goes on forever without repeating.

Real numbers include rational and irrational numbers. Real numbers don't include square roots of negative values like $\sqrt{-4}$ because it isn't possible for a squared number to be negative. For example, if you square negative two, you get $(-2)^2 = -2(-2) = 4$, which is positive. The square root of a negative number is called an **imaginary number**. We use a lowercase i to represent the imaginary number $\sqrt{-1}$. A **complex number** has both real and imaginary parts, like $3 + 2i$ or like $\sqrt{5} + \sqrt{-5}$.

Note the distinction between positive, negative, nonnegative, nonpositive, and nonzero. For example, the distinction between positive and nonnegative is that a nonnegative number could be zero whereas a positive number can't be zero (of course, neither can be negative). A nonzero number could be positive or negative.

Example 1. Indicate if each number is real, rational, irrational, and integer.

- 42 is a real rational integer.
- 5.2 is real and rational.
- $\frac{9}{4}$ is real and rational.
- $\sqrt{3}$ is real and irrational.
- $\sqrt{4}$ is a real rational integer because $\sqrt{4} = 2$.
- $\sqrt{-5}$ is imaginary and irrational.
- -3 is a real rational integer.
- 0 is a real rational integer.

Example 2. Indicate if each number is positive, negative, nonnegative, and nonzero.

- 7 is positive, nonnegative, and nonzero.
- -4 is negative and nonzero.
- 0 is nonnegative.

Exercise Set 1.12

Directions: Indicate if each number is real, rational, irrational, and integer.

1) 0.5

2) 16,300

3) -20

4) $\frac{5}{2}$

5) $\sqrt{25}$

6) $\sqrt{19}$

Directions: Indicate if each number is positive, negative, nonnegative, and nonzero.

7) 7.2

8) 0.01

9) -1.8

10) zero

1.13 Evaluating Formulas

Many calculations in science, engineering, and other fields that apply algebra involve plugging numerical values into a formula. For example, consider the formula for kinetic energy, $K = \frac{1}{2}mv^2$. Given that the mass of an object is $m = 6$ kg and that the speed of the object is $v = 10$ m/s, we can use this formula to determine that the kinetic energy of the object is $K = \frac{1}{2}(6)(10)^2 = \frac{1}{2}(6)(100) = 3(100) = 300$ Joules. We plugged 6 in for m and 10 in for v in the formula $K = \frac{1}{2}mv^2$ to get $K = \frac{1}{2}(6)(10)^2$.

Plugging numbers into formulas will help you get used to the idea that symbols may represent numbers and become acquainted with some basic equations. In this section, the formulas are already solved for the unknown. In Chapter 2, we will start exploring how to manipulate an equation to solve for an unknown.

Example 1. Plug $L = 9$ and $W = 5$ into $P = 2L + 2W$.
$$P = 2L + 2W = 2(9) + 2(5) = 18 + 10 = 28$$

Example 2. Plug $b = 12$ and $h = 6$ into $A = \frac{1}{2}bh$.
$$A = \frac{1}{2}bh = \frac{1}{2}(12)(6) = 6(6) = 36$$

Example 3. Plug $k = 75$ and $m = 3$ into $\omega = \sqrt{\frac{k}{m}}$.
$$\omega = \sqrt{\frac{k}{m}} = \sqrt{\frac{75}{3}} = \sqrt{25} = 5$$

Example 4. Plug $g = 9.8$ and $t = 4$ into $h = \frac{1}{2}gt^2$.
$$h = \frac{1}{2}gt^2 = \frac{1}{2}(9.8)(4)^2 = \frac{1}{2}(9.8)(16) = 4.9(16) = 78.4$$

Exercise Set 1.13

Directions: Plug the given values into each formula.

1) $d = 80, t = 16, v = \dfrac{d}{t}$

2) $C = 20, F = \dfrac{9}{5}C + 32$

3) $v = 6, R = 4, a = \dfrac{v^2}{R}$

4) $a = 4, b = 12, h = 5, A = \dfrac{(a+b)h}{2}$

5) $V = 12, R = 18, P = \dfrac{V^2}{R}$

6) $k = 8, x = 32, c = 12, U = \dfrac{1}{2}k(x - c)^2$

7) $A = 5, B = 5.5, E = \frac{|A-B|}{A} 100\%$

8) $a = 12, b = 5, c = \sqrt{a^2 + b^2}$

9) $c = 6, L = 90, g = 10, T = c\sqrt{\frac{L}{g}}$

10) $v = 40, t = 5, a = 8, x = vt + \frac{1}{2}at^2$

11) $k = 18, q = 4, R = 6, F = k\frac{q^2}{R^2}$

12) $x = 4, y = 12, z = \left(\frac{1}{x} + \frac{1}{y}\right)^{-1}$

2 SOLVING LINEAR EQUATIONS

2.1 One-Step Equations

The equations in this section can be solved in a single step. Identify which arithmetic operation (addition, subtraction, multiplication, or division) is being performed on the variable. Then do the opposite to both sides of the equation, where subtraction and addition are opposites and where multiplication and division are opposites:

- If a constant is **added** to the variable, **subtract** the constant from both sides of the equation. For example, subtract 4 from both sides of $x + 4 = 11$. This gives $x + 4 - 4 = 11 - 4$. The 4 cancels on the left so that we get $x = 7$. Check that $x = 7$ works by plugging it into the original equation: $7 + 4 = 11$.

- If a constant is **subtracted** from the variable, **add** the constant to both sides of the equation. For example, add 3 to both sides of $x - 3 = 9$. This gives $x - 3 + 3 = 9 + 3$. The 3 cancels on the left so that we get $x = 12$. Check that $x = 12$ works by plugging it into the original equation: $12 - 3 = 9$.

- If a constant is **multiplying** the variable, **divide** by the constant on both sides of the equation. For example, divide by 5 on both sides of $5x = 20$. This gives $\frac{5x}{5} = \frac{20}{5}$. The 5 cancels on the left so that we get $x = 4$. Check that $x = 4$ works by plugging it into the original equation: $5(4) = 20$.

- If the variable is **divided** by a constant, **multiply** by the constant on both sides of the equation. For example, multiply by 4 on both sides of $\frac{x}{4} = 3$. This gives $\frac{4x}{4} = 3(4)$. The 4 cancels on the left so that we get $x = 12$. Check that $x = 12$ works by plugging it into the original equation: $\frac{12}{4} = 3$.

Recall that $5x$ means 5 times x and that $\frac{x}{4}$ means 4 divided by x (Sec. 1.3).

Example 1. $x + 6 = 10$

Long solution: Since 6 is added to x, subtract 6 from both sides to get $x + 6 - 6 = 10 - 6$. The reason for this is that 6 cancels on the left. Since $10 - 6$ equals 4, the answer is $x = 4$. Check the answer by plugging it into the original equation: $4 + 6 = 10$.

Short solution: $x = 10 - 6 = 4$ Check the answer: $x + 6 = 4 + 6 = 10$

Example 2. $x - 2 = 5$

Long solution: Since 2 is subtracted from x, add 2 to both sides to get $x - 2 + 2 = 5 + 2$. The reason for this is that 2 cancels on the left. Since $5 + 2$ equals 7, the answer is $x = 7$. Check the answer by plugging it into the original equation: $7 - 2 = 5$.

Short solution: $x = 5 + 2 = 7$ Check the answer: $x - 2 = 7 - 2 = 5$

Example 3. $2x = 12$

Long solution: Since x is multiplied by 2, divide both sides by 2 to get $\frac{2x}{2} = \frac{12}{2}$. The reason for this is that 2 cancels on the left. Since $\frac{12}{2}$ equals 6, the answer is $x = 6$. Check the answer by plugging it into the original equation: $2(6) = 12$.

Short solution: $x = \frac{12}{2} = 6$ Check the answer: $2x = 2(6) = 12$

Example 4. $\frac{x}{3} = 5$

Long solution: Since x is divided by 3, multiply both sides by 3 to get $\frac{3x}{3} = 5(3)$. The reason for this is that 3 cancels on the left. Since $5(3)$ equals 15, the answer is $x = 15$. Check the answer by plugging it into the original equation: $\frac{15}{3} = 5$.

Short solution: $x = 5(3) = 15$ Check the answer: $\frac{x}{3} = \frac{15}{3} = 5$

Exercise Set 2.1

Directions: Solve for the variable in each equation. Check your answer by plugging it into the original equation.

1) $x + 5 = 12$

2) $x - 6 = 8$

3) $4x = 36$

4) $\frac{x}{7} = 9$

5) $x - 6 = 5$

6) $7x = 42$

7) $\frac{x}{9} = 5$

8) $8 + x = 16$

9) $8x = 72$

10) $9 = x - 7$

11) $13 = x + 4$

12) $\frac{x}{5} = 4$

13) $x - 7 = 7$

14) $x + 1 = 1$

15) $6 = \frac{x}{8}$

16) $30 = 5x$

17) $32 = x + 18$

18) $x - 25 = 40$

19) $3x = 0$

20) $\frac{x}{2} = 9$

2.2 Combine Like Terms

The **terms** of an expression are separated by plus (+) and minus (−) signs. (Similarly, the terms of an *equation* are separated by plus, minus, and equal signs.) For example, in the expression $2x^2 + 7x + 4$, there are three terms: $2x^2$, $7x$, and 4.

Like terms have the same exponent of the same variable, like these examples:
- $8x^3$ and $5x^3$ are like terms because they both have x^3.
- $2x^2$ and $\frac{x^2}{3}$ are like terms because they both have x^2.
- $7x$ and $4x$ are like terms because they both have x^1. Recall that $x^1 = x$.
- 12 and 9 are like terms because they are both constant terms. These terms do not have any variables.

In contrast, **unlike terms** either don't have the same variable or the variable doesn't have the same exponent. For example, $3x^2$ and $6x$ are unlike terms because one term has the variable squared while the other does not.

Like terms may be combined together by adding their coefficients. Recall from Sec. 1.2 that a **coefficient** is a number that multiplies a variable. For example, the term $3x^2$ has a coefficient of 3. The examples below show like terms being combined together.
- $8x + 5x = 13x$
- $7x^2 - 4x^2 = 3x^2$
- $12 + 8 - 5 = 15$

If you don't see a coefficient before the variable, the coefficient equals one. For example, x is the same as $1x$ (as mentioned in Sec. 1.3).

When an expression has both like terms and unlike terms, each set of like terms may be combined together by adding their coefficients, like the examples below.
- $2x^2 + 5x + 9x^2 + x = 11x^2 + 6x$
- $4x + 2 + 6x + 3 = 10x + 5$
- $x^2 + 3x + 7 + 2x^2 + 6x + 8 = 3x^2 + 9x + 15$

Unlike terms may **<u>not</u>** be combined together to form a single term. For example, you can't add $3x^2$ and $4x$ together to form a single term. (The best you can do in this case is factor out an x. We'll discuss factoring in Chapter 5.)

To understand why you may combine like terms, but not unlike terms, imagine that you have 12 dollar bills and 7 business cards in your pocket. It wouldn't make sense to say that you have $12 + 7 = 19$ dollars, would it? Dollar bills and business cards are unlike terms. In contrast, if a friend gave you 13 more dollar bills, it would make sense to say that you have $12 + 13 = 25$ dollars. The original dollar bills plus the new dollar bills are like terms.

Example 1. $4x + x = 5x$

Example 2. $5x^2 + 3x^2 - x^2 = 7x^2$

Example 3. $6x + 5 + 4x + 3 = 10x + 8$

Example 4. $2x^2 + 7 + x^2 - 6 = 3x^2 + 1$

Example 5. $3x^2 - 6x - 9 + 2x^2 + 4x - 3 = 5x^2 - 2x - 12$

Exercise Set 2.2

Directions: Simplify each expression by combining like terms together.

1) $9x^2 + 8x^2 =$

2) $8x + 4x + x =$

3) $3x + 6 + 2x + 4 + x =$

4) $5x^2 + 9 - 2x^2 - 3 =$

5) $6x - 4 - 5x - 2 =$

6) $7x^2 + 8x - 5x^2 + 7x =$

7) $8x + 6 - 4x + 4 + 3x + 1 =$

8) $5x^2 - 8 + 4x^2 + 7 - 3x^2 + 5 =$

9) $x^3 + 2x^2 + x^3 - x^2 =$

10) $x^2 + x - x^2 + x =$

11) $6x^2 + 5x + 4 + 3x^2 + 2x + 1 =$

12) $4x^3 + 3x^2 - 2x + 2x^3 - 2x^2 - x^3 - x + x^2 - x =$

13) $8x^4 + 9x^2 + 6 + 7x^4 - 2x^2 - 4 + 5x^4 - 2x^2 - 4 =$

14) $9x^2 - x + 8x^2 - 2x + 7x^2 - 3x + 6x^2 - 4x =$

15) $x + 2 + 3x + 4 + 5x + 6 + 7x + 8 + 9x + 10 =$

2.3 Operating on Equations

The two sides of any equation must always be equal. For example, consider the simple equation $2x + 5 = 17$. Since the right-hand side of the equation equals 17, the left-hand side must also equal 17. When the variable x is multiplied by 2 and this product is added to 5, the result must be 17. Otherwise, the equation would not be true.

We can use the concept that both sides of any equation are always equal in order to manipulate an equation into a different form. Specifically, we can perform operations on an equation provided that we apply the same operation in the same manner to both sides of the equation. Following are examples of operations that might be performed on an equation:

- Add 8 to both sides of an equation.
- Subtract $3x$ from both sides of an equation.
- Multiply both sides of an equation by 4.
- Divide both sides of an equation by 3.

If we perform the same operation in the same manner to both sides of the equation, the equation will still be true.

Consider the trivial equation $8 = 8$. If we add 4 to both sides of this equation, we get $8 + 4 = 8 + 4$ which simplifies to $12 = 12$. If we multiply both sides of this equation by 3, now we get $12(3) = 12(3)$ which simplifies to $36 = 36$. Observe that the equations $8 = 8$, $12 = 12$, and $36 = 36$ are all true.

An equation like $7x - 6 = 15$ works the same way. Presently, each side of the equation equals 15. Otherwise, the equation would not be true. In order for each side of this equation to equal 15, the variable must be $x = 3$. If you plug $x = 3$ into the expression $7x - 6$, you get $7(3) - 6 = 21 - 6 = 15$, which equals 15. If we add 6 to both sides of the equation, we get $7x - 6 + 6 = 15 + 6$ which simplifies to $7x = 21$ because the 6 cancels on the left-hand side. Observe that $x = 3$ solves this new equation since $7(3) = 21$. If we now divide both sides of the equation by 7, we get $\frac{7x}{7} = \frac{21}{7}$ which simplifies to $x = 3$ because the 7 cancels on the left-hand side. We began with $7x - 6 = 15$, added 6 to both sides to get $7x = 21$, and divided both sides by 7 to get $x = 3$. All three of these equations are true. The last equation is the solution to the first equation. This example demonstrates how we can apply operations to both sides of an equation in such a way that the final equation presents the solution to the first equation. We will discuss this powerful technique in Sec. 2.4. For now, we will get some practice operating on both sides of an equation.

It's important that we apply the same operation in the same manner to both sides of the equation. If we only operate on one side of an equation, or if we operate differently on one side of the equation than the other, then the new equation may no longer be true. For example, consider the equation $5 = 5$. If we multiply the right-hand side of the equation by 4, but leave the left-hand side of the equation unchanged, we would get 5 on the left and 20 on the right. Since 5 doesn't equal 20, we can't multiply by 4 only on one side of an equation. Similarly, if we add 6 to the left-hand side and add 9 to the right-hand side, we would get 11 on the left and 14 on the right. Since 11 doesn't equal 14, we can't add different amounts on each side. **<u>Whatever we do to one side of an equation, we must be careful to do the same thing to the other side of the equation if we want the new equation to agree with the previous equation.</u>** This is an important rule of algebra that we will apply in Sec. 2.4 and many other sections of this book.

Example 1. Add 10 to both sides of $3x - 5 = 7$. Does $x = 4$ solve either equation?
- When we add 10 to both sides, we get $3x - 5 + 10 = 7 + 10$.
- Combine like terms to simplify the equation: $3x + 5 = 17$.
- Plug $x = 4$ into the original equation: $3x - 5 = 3(4) - 5 = 12 - 5 = 7$. Since this equals 7, it agrees with $3x - 5 = 7$.
- Plug $x = 4$ into the new equation: $3x + 5 = 3(4) + 5 = 12 + 5 = 17$. Since this equals 17, it agrees with $3x + 5 = 17$.
- According to the two previous bullet points, $x = 4$ solves both equations.

Example 2. Divide by 4 on both sides of $8x = 24$. Does $x = 3$ solve either equation?

- When we divide both sides by 4, we get $\frac{8x}{4} = \frac{24}{4}$.
- Simplify the equation: $2x = 6$.
- Plug $x = 3$ into the original equation: $8x = 8(3) = 24$. Since this equals 24, it agrees with $8x = 24$.
- Plug $x = 3$ into the new equation: $2x = 2(3) = 6$. Since this equals 6, it agrees with $2x = 6$.
- According to the two previous bullet points, $x = 3$ solves both equations.

Example 3. Subtract $3x$ from both sides of $3x + 20 = 7x$. Does $x = 5$ solve either equation?

- When we subtract $3x$ from both sides, we get $3x + 20 - 3x = 7x - 3x$.
- Combine like terms to simplify the equation: $20 = 4x$. Note that $3x$ cancels on the left-hand side.
- Plug $x = 5$ into the original equation. On the left-hand side, we get $3x + 20 = 3(5) + 20 = 15 + 20 = 35$. On the right-hand side, we get $7x = 7(5) = 35$. The two sides are equal when $x = 5$.
- Plug $x = 5$ into the new equation: $4x = 4(5) = 20$. Since this equals 20, it agrees with $20 = 4x$.
- According to the two previous bullet points, $x = 5$ solves both equations.

Exercise Set 2.3

Directions: Operate on both sides of the equation as directed and answer the question.

1) Add 7 to both sides of $4x + 2 = 14$. Does $x = 3$ solve either equation?

2) Divide by 2 on both sides of $6x = 30$. Does $x = 5$ solve either equation?

3) Subtract 8 from both sides of $x + 20 = 30$. Does $x = 10$ solve either equation?

4) Multiply by 6 on both sides of $\frac{x}{2} = 4$. Does $x = 8$ solve either equation?

5) Add $4x$ to both sides of $63 - 4x = 5x$. Does $x = 7$ solve either equation?

6) Subtract $6x$ from both sides of $27 + 6x = 9x$. Does $x = 9$ solve either equation?

7) Divide by 3 on both sides of $30 = 6x$. Does $x = 5$ solve either equation?

8) Subtract 9 from both sides of $9 + x = 11$. Does $x = 2$ solve either equation?

9) Multiply by x on both sides of $\frac{28}{x} = 4$. Does $x = 7$ solve either equation?

2.4 Isolate the Unknown

One of the most common techniques for solving an equation is to <u>isolate the unknown</u>. This means to apply operations to both sides of the equation until the variable appears only by itself on one side of the equation.

- Bring all of the constant terms to one side of the equation and all of the variable terms to the other side of the equation.
- If a term is being added to one side of the equation, you may effectively move it to the other side of the equation by subtracting the term from both sides of the equation. If a term is being subtracted from one side of the equation, you may effectively move it to the other side of the equation by adding the term to both sides of the equation.
- Combine like terms together (Sec. 2.2). For example, $12x - 5x = 7x$.
- Once you have a single variable term on one side of the equation and a single constant term on the other side of the equation, divide both sides of the equation by the coefficient of the variable (like we did in Sec. 2.1). For example, divide both sides of $3x = 12$ by 3 to get $\frac{3x}{3} = \frac{12}{3}$ which simplifies to $x = 4$.

Check your answer by plugging the value for the variable into the original equation. If both sides of the equation equal the same value, the answer is correct.

Note that the second and fourth steps above involve doing the opposite to what is being done. Addition and subtraction are opposites. Multiplication and division are opposites.

Example 1. $4x - 6 = 18$

- We'll leave the variable term on the left. Bring the constant term to the right.
- Since 6 is subtracted, we add 6 to both sides: $4x - 6 + 6 = 18 + 6$.
- 6 cancels on the left side. Simplify the right side: $4x = 24$.
- Divide by the coefficient (4) on both sides: $\frac{4x}{4} = \frac{24}{4}$.
- 4 cancels on the left. Simplify the right side: $x = 6$.

Check the answer. Plug 6 in for x on the left side of the original equation: $4x - 6 = 4(6) - 6 = 24 - 6 = 18$. Since both sides equal 18, the answer is correct.

Example 2. $7x = 8 + 5x$

- We'll leave the constant term on the right. Bring the variable term to the left.
- Since $5x$ is added, we subtract $5x$ from both sides: $7x - 5x = 8 + 5x - 5x$.
- $5x$ cancels on the right side. Combine like terms on the left side: $2x = 8$.
- Divide by the coefficient (2) on both sides: $\frac{2x}{2} = \frac{8}{2}$.
- 2 cancels on the left. Simplify the right side: $x = 4$.

Check the answer. Plug 4 in for x on each side of the original equation:

- $7x = 7(4) = 28$
- $8 + 5x = 8 + 5(4) = 8 + 20 = 28$

Since both sides equal 28, the answer is correct.

Example 3. $13 = 3x - 11$

- We'll leave the variable term on the right. Bring the constant term to the left.
- Since 11 is subtracted, we add 11 to both sides: $13 + 11 = 3x - 11 + 11$.
- 11 cancels on the right side. Simplify the left side: $24 = 3x$.
- Divide by the coefficient (3) on both sides of the original equation: $\frac{24}{3} = \frac{3x}{3}$.
- 3 cancels on the right. Simplify the left side: $8 = x$.
- Note that $8 = x$ is equivalent to $x = 8$ according to the reflexive property (Sec. 1.6).

Check the answer. Plug 8 in for x on the right: $3x - 11 = 3(8) - 11 = 24 - 11 = 13$.
Since both sides equal 13, the answer is correct.

Exercise Set 2.4

Directions: Isolate the unknown, describe the process in words, and check your answer.

1) Isolate the unknown in $5x + 10 = 45$. Describe your solution in words, too.

2) Isolate the unknown in $18 = 6x - 12$. Describe your solution in words, too.

3) Isolate the unknown in $6x = 14 + 5x$. Describe your solution in words, too.

4) Isolate the unknown in $2x + 28 = 9x$. Describe your solution in words, too.

5) Isolate the unknown in $18 + 6x = 90 - 3x$. Describe your solution in words, too.

2.5 Two-Step Equations

The exercises in this section can be solved by isolating the unknown with just two steps of algebra. These simpler problems will help you get the hang of this method before tackling more involved problems.

Example 1. $5x + 7 = 22$

- Step 1: Subtract 7 from both sides: $5x + 7 - 7 = 22 - 7$. This cancels the 7 on the left side. Simplify the right side to get: $5x = 15$.
- Step 2: Divide by 5 on both sides: $\frac{5x}{5} = \frac{15}{5}$. This cancels the 5 on the left side. Simplify the right side to get: $x = 3$.

Check the answer. Plug 3 in for x on the left: $5x + 7 = 5(3) + 7 = 15 + 7 = 22$. Since both sides equal 22, the answer is correct.

Example 2. $9x = 3x + 24$

- Step 1: Subtract $3x$ from both sides: $9x - 3x = 3x + 24 - 3x$. This cancels $3x$ on the right side. Combine like terms on the left side: $6x = 24$.
- Step 2: Divide by 6 on both sides: $\frac{6x}{6} = \frac{24}{6}$. This cancels the 6 on the left side. Simplify the right side to get: $x = 4$.

Check the answer. Plug 4 in for x on each side:

- $9x = 9(4) = 36$
- $3x + 24 = 3(4) + 24 = 12 + 24 = 36$

Since both sides equal 36, the answer is correct.

Example 3. $7 = 4x - 5$

- Step 1: Add 5 to both sides: $7 + 5 = 4x - 5 + 5$. This cancels the 5 on the right side. Simplify the left side to get: $12 = 4x$.
- Step 2: Divide by 4 on both sides: $\frac{12}{4} = \frac{4x}{4}$. This cancels the 4 on the right side. Simplify the left side to get: $3 = x$. Note that $3 = x$ is equivalent to $x = 3$.

Check the answer. Plug 3 in for x on the right: $4x - 5 = 4(3) - 5 = 12 - 5 = 7$. Since both sides equal 7, the answer is correct.

Exercise Set 2.5

Directions: Solve for the variable by isolating the unknown. Check your answer.

1) $8x - 18 = 30$

2) $7x + 15 = 50$

3) $3x = 20 - 2x$

4) $24 + 4x = 72$

5) $5x + 4x = 63$

6) $9x = x + 64$

7) $\frac{x}{6} + 12 = 19$

8) $32 - 5x = 3x$

9) $40 = 7x - 9$

10) $18 = \frac{x}{3} + 9$

11) $15x = 54 + 6x$

12) $12 = 3x - 9$

13) $11 + \frac{x}{5} = 20$

14) $20 - x = 4x$

15) $9 + 6x = 7x - 4$

16) $\frac{x}{4} - 5 = 5$

17) $6x - 30 = 0$

18) $56 = 8x + x - 2x$

19) $x + \frac{x}{2} = 8 + x$

20) $8 + 4x = 5x - 7$

2.6 Equations with More Steps

The exercises in this section require more steps of algebra to isolate the unknown.

Example 1. $9x - 8 = 5x + 12$

- We will put the variables on the left side and the constants on the right side. This will give us a positive coefficient on the left side.
- Step 1: Add 8 to both sides: $9x - 8 + 8 = 5x + 12 + 8$. This cancels the 8 on the left side. Simplify the right side to get: $9x = 5x + 20$.
- Step 2: Subtract $5x$ from both sides: $9x - 5x = 5x + 20 - 5x$. This cancels the $5x$ on the right side. Combine like terms on the left side to get: $4x = 20$.
- Step 3: Divide by 4 on both sides: $\frac{4x}{4} = \frac{20}{4}$. This cancels the 4 on the left side. Simplify the right side to get: $x = 5$.

Check the answer. Plug 5 in for x on each side:

- $9x - 8 = 9(5) - 8 = 45 - 8 = 37$
- $5x + 12 = 5(5) + 12 = 25 + 12 = 37$

Since both sides equal 37, the answer is correct.

Example 2. $19 - 3x = 4 + 2x$

- We will put the variables on the right side and the constants on the left side. This will give us a positive coefficient on the right side.
- Step 1: Subtract 4 from both sides: $19 - 3x - 4 = 4 + 2x - 4$. This cancels the 4 on the right side. Simplify the left side to get: $15 - 3x = 2x$.
- Step 2: Add $3x$ to both sides: $15 - 3x + 3x = 2x + 3x$. This cancels the $3x$ on the left side. Combine like terms on the right side to get: $15 = 5x$.
- Step 3: Divide by 5 on both sides: $\frac{15}{5} = \frac{5x}{5}$. This cancels the 5 on the right side. Simplify the left side to get: $3 = x$. Note that $3 = x$ is equivalent to $x = 3$.

Check the answer. Plug 3 in for x on each side:

- $19 - 3x = 19 - 3(3) = 19 - 9 = 10$
- $4 + 2x = 4 + 2(3) = 4 + 6 = 10$

Since both sides equal 10, the answer is correct.

Exercise Set 2.6

Directions: Solve for the variable by isolating the unknown. Check your answer.

1) $18 + 8x = 42 + 2x$

2) $6x + 13 = 40 - 3x$

3) $7x - 36 = 4x - 12$

4) $56 - x = 5x + 8$

5) $90 - 4x = 18 + 4x$

6) $15x + 28 = 70 + 8x$

7) $6x + 3 = 12x - 3$

8) $8x + 12 - 3x = 48 - 4x$

9) $100 + 4x = 350 - 6x$

10) $x - 4 + x = 8 - x$

11) $22 + 7x - 8 = 3x + 42$

12) $8x - 3x + 5 = 8x - 19$

13) $6x - 16 + 6x = 40 + 4x$

14) $7 - 2x + 4 = 11 + 3x$

15) $22 - 8x = 9x - 22 - 6x$

16) $7x + 18 - 2x = 56 - x + 34$

2.7 Negative Coefficients or Answers

When multiplying or dividing with negative numbers, recall that the answer is negative if an odd number of minus signs are multiplying or dividing, and the answer is positive if an even number of minus signs are multiplying or dividing. For example, $(-6)(7) = -42$ whereas $(-6)(-7) = 42$. Similarly, $\frac{12}{-3} = -4$ whereas $\frac{-12}{-3} = 4$.

Sometimes it is possible to avoid a negative sign. For example, consider the equation $7 - x = x + 1$. If we put the variable terms on the left, we get $-2x = -6$, but if instead we choose to put the variable terms on the right, we get $6 = 2x$. Either way leads to the answer $x = 3$, but one way avoids the negative signs.

However, there are many situations where the solution is more efficient to deal with the negative sign. For example, consider the equation $4 - 5x = 19$. This problem can be solved in fewer steps by leaving the variable term on the left. Subtract 4 from both sides to get $-5x = 15$ and divide both sides by -5 to get $x = -3$.

Example 1. $8 - 4x = 20$
- Subtract 8 from both sides: $8 - 4x - 8 = 20 - 8$. Simplify: $-4x = 12$.
- Divide by -4 on both sides: $\frac{-4x}{-4} = \frac{12}{-4}$. Simplify: $x = -3$.

Check the answer. Plug -3 in for x on the left: $8 - 4x = 8 - 4(-3) = 8 + 12 = 20$. Since both sides equal 20, the answer is correct.

Example 2. $-2x + 9 = 24 - 5x$
- Subtract 9 from both sides and simplify: $-2x = 15 - 5x$.
- Add $5x$ to both sides and simplify: $3x = 15$.
- Divide by 3 on both sides: $\frac{3x}{3} = \frac{15}{3}$. Simplify: $x = 5$.

Check the answer. Plug 5 in for x on each side:
- $-2x + 9 = -2(5) + 9 = -10 + 9 = -1$
- $24 - 5x = 24 - 5(5) = 24 - 25 = -1$

Since both sides equal -1, the answer is correct.

Exercise Set 2.7

Directions: Solve for the variable by isolating the unknown. Check your answer.

1) $-7x - 23 = 40$

2) $50 = 8 - 6x$

3) $5x + 20 = -20$

4) $-40 = -4x - 16$

5) $5 - 3x = -2x + 1$

6) $3x + 100 = -5x + 36$

7) $8 + x = 2 - x$

8) $-7x - 45 = -20 - 2x$

9) $5 - \frac{x}{6} = 12$

10) $2x = -45 - 3x - 4x$

11) $-21 = -5x + 2x$

12) $4 = 9 - \frac{x}{7}$

13) $-111 - 4x + 30 = 5x$

14) $-17x - 24 = -9x - 72$

15) $-18 - 3x = 6 + 3x$

16) $10 - 7x = -12x - 30$

2.8 Multiply or Divide by Minus One

When an equation has several negative signs, it can be helpful to multiply both sides of the equation by minus one. For example, consider the simple equation $-x = -3$. If we multiply both sides of the equation by -1, we get $x = 3$. (We get the same answer if we *divide* both sides of the equation by -1. Whether you multiply or divide by -1 makes no difference.) Remember to multiply **every** term of the equation by -1. The sign of every term will change.

Example 1. $-3 - x = 9$
- Add 3 to both sides: $-x = 12$.
- Multiply by -1 on both sides: $x = -12$.

Check the answer. Plug -12 in for x on the left: $-3 - (-12) = -3 + 12 = 9$. Since both sides equal 9, the answer is correct.

Exercise Set 2.8

Directions: Solve for the variable by isolating the unknown. Check your answer.

1) $-x = -11$

2) $8 = 5 - x$

3) $17 = -x + 9$

4) $-4 - x = -6$

5) $-x + 1 = 10$

6) $9 - 6x = 2 - 7x$

2.9 Fractional Answers

The answers to the exercises in this section are fractions. When checking the answer, it may be necessary to add, subtract, or multiply fractions. Recall the rules for adding, subtracting, and multiplying fractions (which are reviewed in Sec. 1.7):

- To add or subtract fractions, first find a **common denominator**. For example, $\frac{5}{6} + \frac{3}{8} = \frac{20}{24} + \frac{9}{24} = \frac{29}{24}$.

- When a whole number multiplies a fraction, it just multiplies the numerator. For example, $4\frac{2}{3} = \frac{4 \cdot 2}{3} = \frac{8}{3}$. (Recall that in algebra, $4\frac{2}{3}$ will be interpreted as 4 times $\frac{2}{3}$; it is not a mixed number.)

- **Reduce** the answer if the numerator and denominator share a common factor. Divide the numerator and denominator by their **greatest common factor**. For example, $\frac{15}{20} = \frac{3}{4}$ because 15 and 20 are each divisible by 5.

Example 1. $4x - 7 = 8$

- Add 7 to both sides: $4x = 15$.

- Divide by 4 on both sides: $x = \frac{15}{4}$.

Check the answer. Plug $\frac{15}{4}$ in for x on the left: $4x - 7 = 4\left(\frac{15}{4}\right) - 7 = 15 - 7 = 8$. Since both sides equal 8, the answer is correct.

Example 2. $8x + 5 = 9 + 2x$

- Subtract 5 from both sides: $8x = 4 + 2x$.

- Subtract $2x$ from both sides: $6x = 4$.

- Divide by 6 on both sides: $x = \frac{4}{6}$. Reduce the answer: $x = \frac{2}{3}$ because $\frac{4}{6} = \frac{2}{3}$.

Check the answer. Plug $\frac{2}{3}$ in for x on each side:

- $8x + 5 = 8\left(\frac{2}{3}\right) + 5 = \frac{16}{3} + 5 = \frac{16}{3} + \frac{15}{3} = \frac{31}{3}$

- $9 + 2x = 9 + 2\left(\frac{2}{3}\right) = 9 + \frac{4}{3} = \frac{27}{3} + \frac{4}{3} = \frac{31}{3}$

Since both sides equal $\frac{31}{3}$, the answer is correct.

Exercise Set 2.9

Directions: Solve for the variable by isolating the unknown.

1) $3x + 8 = 12$

2) $9 = 5x - 8$

3) $2x + 7 = 6$

4) $6 - 8x = 20$

5) $9x - 11 = 4x - 3$

6) $2x + 18 = 5x + 8$

7) $2x - 16 = 8x - 12$

8) $5x - 9 = 11 - 3x$

2.10 Fractional Coefficients or Constants

When the variable term has been isolated, if the coefficient is a fraction, multiply both sides of the equation by the denominator and divide both sides of the equation by the numerator. For example, multiply by 3 on both sides of $\frac{2x}{3} = 10$ to get $2x = 30$ and then divide by 2 on both sides to get $x = 15$. As a check, $\frac{2x}{3} = \frac{2(15)}{3} = \frac{30}{3} = 10$. Recall the rules for arithmetic with fractions (which are reviewed in Sec. 1.7):

- To add or subtract fractions, first find a **common denominator**. This applies even if variables are involved. For example, $\frac{2x}{3} + \frac{3x}{4} = \frac{8x}{12} + \frac{9x}{12} = \frac{17x}{12}$.

- To multiply two fractions, multiply their numerators and denominators. For example, $\frac{3}{4}\frac{7}{5} = \frac{3\cdot7}{4\cdot5} = \frac{21}{20}$. (Recall that in algebra when two fractions are beside one another, this means to multiply them.)

- To divide a fraction by a whole number, multiply by the reciprocal of the whole number. For example, $\frac{1}{3}$ divided by 4 equals $\frac{1}{3}\frac{1}{4} = \frac{1}{3\cdot4} = \frac{1}{12}$.

- **Reduce** the answer if the numerator and denominator share a common factor. For example, $\frac{8}{12} = \frac{2}{3}$ because 8 and 12 are each divisible by 4.

Example 1. $4x - \frac{1}{2} = \frac{1}{3}$

- Add $\frac{1}{2}$ to both sides: $4x = \frac{1}{2} + \frac{1}{3}$.

- Make a common denominator to add the fractions: $\frac{1}{2} + \frac{1}{3} = \frac{3}{6} + \frac{2}{6} = \frac{5}{6}$. Combine this with the previous step: $4x = \frac{5}{6}$.

- Divide by 4 on both sides. Note that dividing $\frac{5}{6}$ by 4 is equivalent to multiplying $\frac{5}{6}$ by $\frac{1}{4}$. We get $x = \frac{5}{6}\frac{1}{4}$, which simplifies to $x = \frac{5}{24}$.

Check the answer. Plug $\frac{5}{24}$ in for x on the left: $4x - \frac{1}{2} = 4\left(\frac{5}{24}\right) - \frac{1}{2} = \frac{20}{24} - \frac{1}{2} = \frac{20}{24} - \frac{12}{24} = \frac{8}{24} = \frac{1}{3}$. Since both sides equal $\frac{1}{3}$, the answer is correct.

Example 2. $\frac{3x}{2} - 6 = \frac{x}{3} + 8$

- Add 6 to both sides: $\frac{3x}{2} = \frac{x}{3} + 14$.

- Subtract $\frac{x}{3}$ from both sides: $\frac{3x}{2} - \frac{x}{3} = 14$.

- Make a common denominator to subtract the fractions: $\frac{3x}{2} - \frac{x}{3} = \frac{9x}{6} - \frac{2x}{6} = \frac{7x}{6}$. Combine this with the previous step: $\frac{7x}{6} = 14$.

- Multiply by 6 on both sides: $7x = 84$.

- Divide by 7 on both sides: $x = 12$.

Check the answer. Plug 12 in for x on each side:

- $\frac{3x}{2} - 6 = \frac{3(12)}{2} - 6 = \frac{36}{2} - 6 = 18 - 6 = 12$

- $\frac{x}{3} + 8 = \frac{12}{3} + 8 = 4 + 8 = 12$

Since both sides are equal, the answer is correct.

Example 3. $2x + \frac{1}{6} = \frac{1}{2} - \frac{x}{4}$

- Subtract $\frac{1}{6}$ from both sides: $2x = \frac{1}{2} - \frac{1}{6} - \frac{x}{4}$.

- Make a common denominator to subtract the fractions: $\frac{1}{2} - \frac{1}{6} = \frac{3}{6} - \frac{1}{6} = \frac{2}{6} = \frac{1}{3}$. (Note that $\frac{2}{6}$ reduced to $\frac{1}{3}$.) Combine this with the previous step: $2x = \frac{1}{3} - \frac{x}{4}$.

- Add $\frac{x}{4}$ to both sides: $2x + \frac{x}{4} = \frac{1}{3}$.

- Make a common denominator to add the fractions: $2x + \frac{x}{4} = \frac{8x}{4} + \frac{x}{4} = \frac{9x}{4}$. Combine this with the previous step: $\frac{9x}{4} = \frac{1}{3}$.

- Multiply by 4 on both sides: $9x = \frac{4}{3}$.

- Divide by 9 on both sides: $x = \frac{4}{27}$.

Check the answer. Plug $\frac{4}{27}$ in for x on each side:

- $2x + \frac{1}{6} = 2\left(\frac{4}{27}\right) + \frac{1}{6} = \frac{8}{27} + \frac{1}{6} = \frac{16}{54} + \frac{9}{54} = \frac{25}{54}$

- $\frac{1}{2} - \frac{x}{4} = \frac{1}{2} - \frac{4}{27}\frac{1}{4} = \frac{1}{2} - \frac{4}{108} = \frac{54}{108} - \frac{4}{108} = \frac{50}{108} = \frac{25}{54}$

Since both sides are equal, the answer is correct.

Exercise Set 2.10

Directions: Solve for the variable by isolating the unknown.

1) $3x + \dfrac{1}{12} = \dfrac{3}{4}$

2) $\dfrac{2x}{3} - 5 = 7$

3) $\dfrac{3}{4} = \dfrac{1}{6} + \dfrac{5x}{3}$

4) $\dfrac{5}{6} - \dfrac{9x}{2} = \dfrac{8}{3}$

5) $8x - \frac{1}{3} = 2x + \frac{5}{2}$

6) $\frac{x}{5} + 4 = \frac{x}{8} + 10$

7) $3x + \frac{7}{6} = \frac{3x}{2} - \frac{9}{4}$

8) $\frac{5}{4} - \frac{3x}{16} = \frac{5x}{8} - \frac{7}{12}$

2.11 Special Solutions

Until now, we've only solved equations that have one unique answer. It's important to be aware that not every equation that can be written down has one unique answer. The examples below illustrate equations that have special solutions.

- $x = x + 2$ has **no solution** because there isn't a value of x that can make this equation true. One way to see this is to subtract x from both sides to get $0 = 2$, which is definitely false. Another way to see it is to consider that no number can equal itself plus two.

- $x = x$ can be solved by **all real numbers**. Any value of x equals itself. No matter what number you plug in for x, it will be true that $x = x$. For example, $7 = 7$.

- In Chapter 6, we'll explore equations where the variable is squared, like $x^2 = 9$. Equations of this form often have **multiple solutions**. For example, $x = 3$ and $x = -3$ both solve $x^2 = 9$ because $3^2 = 9$ and $(-3)^2 = (-3)(-3) = 9$.

Example 1. $x - x = 5$
Combine like terms to get $0 = 5$. Since this is false, there is no solution.

Example 2. $2x + x = 3x$
Combine like terms to get $3x = 3x$. Divide both sides by 3 to get $x = x$. Since x always equals itself, this is solved by all real numbers.

Exercise Set 2.11

Directions: Determine whether each equation has one unique solution, no solution, multiple solutions, or is solved by all real numbers.

1) $6x - 2x = 4x$

2) $5 + x = x + 7$

3) $3x + 2 - 3x = 0$

4) $8 + 4x - 8 = 4x$

5) $11 + x - 4 = x + 7$

6) $\frac{2x}{3} = \frac{8x}{12}$

2.12 Isolating Symbols in Formulas

Consider the formula for the area of a triangle, which is $A = \frac{1}{2}bh$. If you know the base and height of a triangle, this formula lets you solve for the area. Suppose that you know the area of the triangle and the base of the triangle, but wish to solve for the height. In that case, you could apply algebra, treating the height as the variable and the other symbols as constants. We could isolate the symbol h as follows. First multiply both sides of the equation by 2 to get $2A = bh$, and then divide both sides of the equation by the base to get $\frac{2A}{b} = h$. Now if you know the area and base, the formula tells you to multiply the area by 2 and divide by the base to find the height.

Example 1. Solve for Q in the formula $C = \frac{Q}{V}$.
Multiply both sides by V to get $CV = Q$, which is equivalent to $Q = CV$.

Example 2. Solve for W in the formula $P = 2L + 2W$.
Subtract $2L$ from both sides to get $P - 2L = 2W$.
Divide both sides by 2 to get $\frac{P}{2} - L = W$, which is equivalent to $W = \frac{P}{2} - L$. Note that we divided the entire left-hand side of the equation by 2. It would also be okay to write $\frac{P-2L}{2} = W$ (see Chapters 4-5).

Exercise Set 2.12

Directions: Solve for the indicated symbol by isolating it.

1) Solve for R in $V = IR$

2) Solve for m in $d = \frac{m}{V}$

3) Solve for t in $v = \frac{d}{t}$

4) Solve for K in $C = K - 273$

5) Solve for θ in $\varphi = 90 - \theta$

6) Solve for T in $PV = nRT$

7) Solve for C in $F = \frac{9}{5}C + 32$

8) Solve for m in $y = mx + b$

9) Solve for b in $A = \frac{b(h_1 + h_2)}{2}$

10) Solve for y in $3x + 2y = 6z$

2.13 Word Problems

The word problems in this section can be solved by following these steps:

- Read the problem carefully. Identify the given information.
- Determine what the problem is asking you to solve for. Call this x. Write out what x represents. For example, "x = Ryan's age now."
- If it seems like there is a second unknown, note that you will be able to express it in terms of the first unknown. For example, if x represents Ryan's age now and you know that Julie is 8 years older than Ryan, then Julie's age is $x + 8$.
- Write an equation that relates the given information to the unknown(s). Try to reason this out based on what is stated in the problem. Some common signal words are listed below. Signal words can help, but since the English language allows the same idea to be communicated many different ways, beware that signal words are not foolproof.
- Solve the equation by isolating the unknown. Check the answer by plugging it back into the equation. Also check if the answer seems reasonable.

Common addition keywords include total, sum, together, combined, more than, gained, increased by, raised to, etc.

- If an unknown number increases by 5, write $x + 5$.
- For a number that is 4 more than x, write $x + 4$.

Common subtraction keywords include difference, between, less than, left over, fewer, minus, decreased by, lost, smaller than, taken away, after, etc.

- For 8 less than a number, write $x - 8$.
- If an event occurred 7 years ago, write $x - 7$.

Common multiplication keywords include times, product, twice, double, triple, each, of (like one-half of the students), increased by a factor of, decreased by a factor of, etc.

- For 3 times a number, write $3x$.
- If a number doubles, write $2x$.
- For the product of x and 9, write $9x$.

It may help to review Sec. 1.4.

Common division keywords include out of, per, average, etc.

- To make half as much, write $\frac{x}{2}$.
- If a quantity is divided into 6 equals parts, write $\frac{x}{6}$.

Common equal sign language includes is, was, makes, will be, equals, etc.

Example 1. After Rita spent $8 on her dinner, she had $17 left over. How much money did Rita have before she ate dinner?

- $x =$ how much money Rita had before she ate dinner.
- Rita started with x dollars. She spent $8. She has $17 left: $x - 8 = 17$.
- Add 8 to both sides to get $x = 25$: Rita had $25 before she ate dinner.

Example 2. Jeff is twice as old as Kelly. The sum of their ages is 24. How old is Kelly?

- $x =$ Kelly's age.
- $2x =$ Jeff's age (since Jeff is twice as old as Kelly).
- The sum of their ages is: $x + 2x = 24$.
- Combine like terms on the left side to get $3x = 24$.
- Divide by 3 on both sides to get $x = 8$: Kelly is 8 years old.
- Check: Jeff is $2x = 2(8) = 16$ and the sum of their ages is $8 + 16 = 24$.

Example 3. After Violet received 12 bracelets on her birthday, she had four times as many bracelets as she had originally. How many bracelets does she have now?

- $x =$ the number of bracelets that Violet had originally (note: this isn't what we are solving for, but once we find x, we will be able to determine the answer).
- $4x =$ the number of bracelets that Violet has now.
- Violet started with x. She received 12 more. She has $4x$ now: $x + 12 = 4x$.
- Subtract x from both sides to get $12 = 3x$.
- Divide by 3 on both sides to get $4 = x$.
- The answer is $4x = 4(4) = 16$ (this is how many bracelets Violet has now).
- Check: Violet had $x = 4$ originally and has $x + 12 = 4 + 12 = 16$ now, which is 4 times what she had originally.

Exercise Set 2.13

Directions: Use algebra to solve each word problem.

1) Alice had $75 originally. She spent $28 on a pair of shoes and $19 on a t-shirt. How much money does she have left over?

2) Three consecutive odd numbers have a sum of 63. What are the numbers?

3) Susan is three times as old as Dave. Their ages in years have a sum of 64? What are their ages?

4) After Melissa gave one-third of her beads away, she still had 24 beads left. How many beads did Melissa have in the beginning?

5) Two numbers have a difference of 25 and a sum of 47. What are the numbers?

6) Doug has $10. After he bought 4 apples, he had $7 left over. How much does it cost to buy one apple?

7) One number is 8 times another number. The numbers have a difference of 77. What are the numbers?

8) At a theater where there is no sales tax, an adult ticket costs twice as much as a child's ticket. Wendy bought 2 adult tickets and 5 children's tickets for $63. How much does each kind of ticket cost?

3 EXPONENTS

3.1 Product Rule for Exponents

A whole number exponent represents repeated multiplication. For example, x^8 is a concise way of stating $xxxxxxxx$. In the expression x^8, there are 8 x's multiplying one another. The number of x's multiplying one another in the quantity x^m is equal to m. Similarly, the number of x's multiplying one another in the quantity x^n is equal to n. It follows that if we multiply x^m and x^n together, the total number of x's multiplying one another will equal $m + n$. This rule is represented by the following formula:

$$x^m x^n = x^{m+n}$$

For example, if $m = 5$ and $n = 3$, this formula becomes $x^5 x^3 = x^{5+3} = x^8$. It is easy to verify this, since $x^5 = xxxxx$, $x^3 = xxx$, and $x^5 x^3 = (xxxxx)(xxx) = xxxxxxxx = x^8$. The total number of x's multiplying one another is 8. This rule also works if the base is a number. For example, $2^3 2^4 = 8(16) = 128 = 2^7 = 2^{3+4}$.

Note that $x^1 = x$ (there is just one x). For example, $x^2 x = x^2 x^1 = x^{2+1} = x^3$.

Example 1. $x^6 x^3 = x^{6+3} = x^9$ **Example 2.** $x^4 x^2 x = x^4 x^2 x^1 = x^{4+2+1} = x^7$

Exercise Set 3.1

Directions: Simplify each expression.

1) $x^3 x^2 =$

2) $x^7 x^3 =$

3) $x^8 x^5 =$

4) $x^6 x^6 =$

5) $x^4 x =$

6) $xx =$

7) $x^9 x^6 =$

8) $x^4 x^4 x^4 =$

9) $x^7 x^5 x^2 =$

10) $x^8 x^7 x =$

11) $xxxx =$

12) $x^9 x^8 x^7 x^6 =$

3.2 Quotient Rule for Exponents

The number of x's multiplying one another in the quantity x^m is equal to m. Similarly, the number of x's multiplying one another in the quantity x^n is equal to n. It follows that if we divide x^m by x^n, the total number of x's multiplying one another will equal $m - n$. This rule is represented by the following formula (which assumes that $x \neq 0$):

$$\frac{x^m}{x^n} = x^{m-n}$$

For example, if $m = 7$ and $n = 4$, this formula becomes $\frac{x^7}{x^4} = x^{7-4} = x^3$. It is easy to verify this, since $x^7 = xxxxxxx$, $x^4 = xxxx$, and $\frac{x^7}{x^4} = \frac{xxxxxxx}{xxxx} = xxx = x^3$. This rule also works if the base is a number. For example, $\frac{2^5}{2^3} = \frac{32}{8} = 4 = 2^2 = 2^{5-3}$.

Recall that $x^1 = x$. For example, $\frac{x^4}{x} = \frac{x^4}{x^1} = x^{4-1} = x^3$.

Example 1. $\frac{x^5}{x^4} = x^{5-4} = x^1 = x$ **Example 2.** $\frac{x^6 x^4}{x^3} = \frac{x^{6+4}}{x^3} = \frac{x^{10}}{x^3} = x^{10-3} = x^7$

Exercise Set 3.2

Directions: Simplify each expression, assuming that $x \neq 0$.

1) $\dfrac{x^6}{x^2} =$ 2) $\dfrac{x^8}{x^5} =$

3) $\dfrac{x^{12}}{x^7} =$ 4) $\dfrac{x^5}{x} =$

5) $\dfrac{x^4 x^3}{x^2} =$ 6) $\dfrac{x^9}{x^6 x^2} =$

7) $\dfrac{x^7 x}{x^6} =$ 8) $\dfrac{x^{14}}{x^6 x} =$

9) $\dfrac{x^9 x^8}{x^6 x^4} =$ 10) $\dfrac{x^9 x}{x^3 x^3} =$

3.3 An Exponent of Zero

In the equation $\frac{x^m}{x^n} = x^{m-n}$, if we set $m = n$, we get $\frac{x^n}{x^n} = x^{n-n}$ which simplifies to $1 = x^0$. This is the reason that any nonzero number raised to the power of zero equals one.

$$x^0 = 1 \text{ (if } x \neq 0)$$

For example, $8^0 = 1$ and $\left(-\frac{1}{3}\right)^0 = 1$. As long as the number is nonzero, if it has an exponent of zero, the answer is one. Even $\left(\sqrt{2}\right)^0 = 1$.

Note that 0^0 is **indeterminate**. When x equals zero, we run into trouble with $\frac{x^n}{x^n}$, since division by zero is a problem. Since $\frac{0}{0}$ is indeterminate, it follows that 0^0 is similarly indeterminate.

(Let's review a couple of special cases of arithmetic. The problem $\frac{1}{0}$ is **undefined**, while the problem $\frac{0}{0}$ is **indeterminate**. To understand why, consider the problem $\frac{12}{3}$, which asks the question, "Which number times 3 is equal to 12?" The answer is 4. Now try this with $\frac{1}{0}$, which asks, "Which number times 0 is equal to 1?" This is undefined since no matter what you multiply by 0, it will *never* equal 1. Anything times zero equals zero. Now try this with $\frac{0}{0}$, which asks, "Which number times 0 is equal to 0?" This is indeterminate because there are an infinite number of possible answers. Any number times zero will equal zero.)

Example 1. $7^0 = 1$ **Example 2.** 0^0 is indeterminate   **Example 3.** $\frac{1}{0}$ is undefined

Exercise Set 3.3

Directions: Determine the best answer to each exponent problem.

1) $1^0 =$

2) $\left(\frac{1}{0}\right)^2$

3) $(-1)^0 =$

4) $0^{6(2)-4(3)} =$

3.4 Negative Exponents

In the equation $\frac{x^m}{x^n} = x^{m-n}$, if we set $m = 0$, we get $\frac{x^0}{x^n} = x^{0-n}$. Since $x^0 = 1$, the left side simplifies to $\frac{1}{x^n}$. Since $0 - n = -n$, the right side simplifies to x^{-n}. Putting these together, we get $\frac{1}{x^n} = x^{-n}$. We may swap the left and right sides in accordance with the reflexive property (Sec. 1.6) to get:

$$x^{-n} = \frac{1}{x^n}$$

This tells us that a negative exponent is equivalent to applying the absolute value of the exponent to the reciprocal of the base. (Recall that we did this using numbers in Sec. 1.11.) For example, $x^{-4} = \frac{1}{x^4}$. Note that an exponent of negative one is equivalent to a **reciprocal**: $x^{-1} = \frac{1}{x}$. A reciprocal is also referred to as a **multiplicative inverse** since $x^{-1}x = x^{-1}x^1 = x^{-1+1} = x^0 = 1$.

Note that the power rules discussed in the previous sections apply whether or not the exponents are positive, zero, or negative. For example, $x^6x^{-4} = x^{6+(-4)} = x^{6-4} = x^2$. It may help to review Sec. 1.10 (regarding arithmetic with negative numbers).

Example 1. $x^9x^{-6} = x^{9+(-6)} = x^3$ **Example 2.** $x^2x^{-7} = x^{2+(-7)} = x^{-5} = \frac{1}{x^5}$

Example 3. $\frac{x^4}{x^{-3}} = x^{4-(-3)} = x^{4+3} = x^7$ **Example 4.** $\frac{x^{-6}}{x} = \frac{x^{-6}}{x^1} = x^{-6-1} = x^{-7} = \frac{1}{x^7}$

Exercise Set 3.4

Directions: Simplify each expression, assuming that $x \neq 0$.

1) $x^9x^{-5} =$ 2) $x^{-8}x^4 =$

3) $x^7x^{-7} =$ 4) $x^{-5}x^{-5} =$

5) $xx^{-4} =$

6) $x^{-8}x^{-6} =$

7) $x^{12}x^{-4}x^{-3} =$

8) $x^{7}x^{6}x^{-5} =$

9) $x^{-5}x^{-3}x^{-2} =$

10) $x^{8}x^{-7}x =$

11) $\dfrac{x^2}{x^{-4}} =$

12) $\dfrac{x^{-9}}{x^7} =$

13) $\dfrac{x^{-6}}{x^{-6}} =$

14) $\dfrac{x^8}{x^{-8}} =$

15) $\dfrac{x^{-3}}{x^{-4}} =$

16) $\dfrac{x^{-2}}{x} =$

17) $\dfrac{x^8 x^{-3}}{x^{-7}} =$

18) $\dfrac{x^{-6}}{x^7 x^{-5}} =$

19) $\dfrac{x^{-9}x}{x^{-8}x^{-5}} =$

20) $\dfrac{x^{-7}x^{-4}}{x^{-9}x^{-3}} =$

3.5 Power of a Power

Consider the quantity $(x^m)^n$. The number of x's multiplying one another in x^m is equal to m, and the number of x^m's multiplying one another in $(x^m)^n$ is equal to n, which means that there are a total of mn (meaning m times n) x's multiplying one another in $(x^m)^n$. We can express this general rule as:

$$(x^m)^n = x^{mn}$$

If you didn't quite follow the previous paragraph, don't worry. We'll do it again now using numbers. Consider the quantity $(x^3)^4$. First note that $x^3 = xxx$ (this has 3 x's multiplying one another). Now note that $(x^3)^4 = x^3 x^3 x^3 x^3$ (this has 4 x^3's multiplying one another). When we substitute $x^3 = xxx$ into $(x^3)^4 = x^3 x^3 x^3 x^3$, we get a total of 12 x's multiplying one another: $(x^3)^4 = xxxxxxxxxxxx = 10^{12}$. The easy to way to do this is to apply the general rule $(x^m)^n = x^{mn}$ with $m = 3$ and $n = 4$: $(x^3)^4 = x^{3(4)} = x^{12}$. The main idea is to multiply the exponents of $(x^3)^4$ to make x^{12}.

Compare the rules $x^m x^n = x^{m+n}$ where the exponents are added to the rule $(x^m)^n = x^{mn}$ where the exponents are multiplied. It's important to remember both rules and to be able to keep them straight. **Tip:** You can tell them apart with a simple example. For example, with $m = 3$ and $n = 2$, it's easy to see that $x^3 x^2 = (xxx)(xx) = x^5$, which requires adding $m + n$, which reminds you that $x^m x^n = x^{m+n}$ adds exponents, and it's easy to see that $(x^3)^2 = x^3 x^3 = (xxx)(xxx) = x^6$, which requires multiplying mn, which reminds you that $(x^m)^n = x^{mn}$ multiplies exponents.

Example 1. $(x^5)^4 = x^{5(4)} = x^{20}$ **Example 2.** $(x^{-3})^2 = x^{-3(2)} = x^{-6} = \frac{1}{x^6}$

Exercise Set 3.5

Directions: Simplify each expression, assuming that $x \neq 0$.

1) $(x^6)^4 =$

2) $(x^3)^3 =$

3) $(x^7)^{-3} =$

4) $(x^{-6})^{-8} =$

5) $(x^{-9})^6 =$

6) $(x^7)^0 =$

7) $(x^{-5})^{-5} =$

8) $(x^8)^2 =$

9) $(x^{-7})^1 =$

10) $(x^{-2})^8 =$

11) $(x^{-3})^{-8} =$

12) $(x^4)^{-1} =$

13) $(x^7)^6 =$

14) $(x)^{-8} =$

15) $(x^0)^{-2} =$

16) $(x^{-9})^{-9} =$

3.6 Roots

In the equation $(x^m)^n = x^{mn}$, if we let $m = \frac{1}{n}$, the equation becomes $\left(x^{1/n}\right)^n = x$ since $mn = \left(\frac{1}{n}\right)n = 1$ such that $x^{mn} = x^1 = x$ on the right side.

Recall the definition of a root from Sec. 1.11. The nth root of x, denoted $\sqrt[n]{x}$, asks the question, "Which number raised to the nth power equals x?" This means that $\left(\sqrt[n]{x}\right)^n = x$. In the quantity $\left(\sqrt[n]{x}\right)^n$, the number of $\sqrt[n]{x}$'s multiplying one another equals n. For example, for $n = 3$, we get $\left(\sqrt[3]{x}\right)^3 = \sqrt[3]{x}\sqrt[3]{x}\sqrt[3]{x} = x$. This is easy to check with numbers. For example, for $x = 8$ we get $\left(\sqrt[3]{8}\right)^3 = \sqrt[3]{8}\sqrt[3]{8}\sqrt[3]{8} = 2(2)(2) = 8$.

Compare the equations from the previous two paragraphs: $\left(x^{1/n}\right)^n = x$ and $\left(\sqrt[n]{x}\right)^n = x$. Since $\left(x^{1/n}\right)^n$ equals x and since $\left(\sqrt[n]{x}\right)^n$ also equals x, it follows that $\left(x^{1/n}\right)^n$ equals $\left(\sqrt[n]{x}\right)^n$. (This is the transitive property discussed in Sec. 1.6.) Compare $\left(x^{1/n}\right)^n$ and $\left(\sqrt[n]{x}\right)^n$, which we just reasoned must be equal. By comparison, we can see that:
$$x^{1/n} = \sqrt[n]{x}$$
This shows that $x^{1/n}$ represents the nth root of x. For example, when $n = 2$, we get:
$$x^{1/2} = \sqrt{x}$$
An exponent of one-half is equivalent to a square root. More generally, an exponent of $1/n$ is equivalent to the nth root of x. For example, $27^{1/3} = 3$ because $27^{1/3} = \sqrt[3]{27}$ and $\sqrt[3]{27} = 3$.

As mentioned in Sec. 1.11, we will only take the positive root unless a $\pm$ sign or a minus sign is written before the root. For example, $4^{1/2} = 2$, even though both 2 and -2 equal 4 when squared. (When solving for a variable raised to an exponent, in that context we will include both positive and negative roots. See Sec. 3.10 and Chapter 6.)

Example 1. (A) What does $x^{1/4}$ mean? $x^{1/4} = \sqrt[4]{x}$

What does it equal when $x = 16$? $16^{1/4} = \sqrt[4]{16} = 2$ because $2^4 = 16$

(B) For these values of n and x, show that $\left(x^{1/n}\right)^n = x$. Plug in $n = 4$ and $x = 16$ to get

$\left(16^{1/4}\right)^4 = 2^4 = 16$.

Example 2. What does $x^{-1/3}$ mean? $x^{-1/3} = \frac{1}{x^{1/3}} = \frac{1}{\sqrt[3]{x}}$

What does it equal when $x = 8$? $8^{-1/3} = \frac{1}{8^{1/3}} = \frac{1}{\sqrt[3]{8}} = \frac{1}{2}$ because $2^3 = 8$.

Exercise Set 3.6

Directions: Interpret each expression and plug in the specified value.

1) What does $x^{1/2}$ mean? What does it equal when $x = 49$?

2) What does $x^{1/3}$ mean? What does it equal when $x = 64$?

3) What does $x^{-1/2}$ mean? What does it equal when $x = 81$?

4) What does $x^{1/5}$ mean? What does it equal when $x = 32$?

5) What does $x^{-1/4}$ mean? What does it equal when $x = 625$?

3.7 Fractional Exponents

In the equation $(x^m)^n = x^{mn}$, if we let $m = \frac{1}{p}$, the equation becomes $\left(x^{1/p}\right)^n = x^{n/p}$ since $mn = \left(\frac{1}{p}\right)n = \frac{n}{p}$ such that $x^{mn} = x^{n/p}$ on the right side. According to Sec. 3.6, $x^{1/p} = \sqrt[p]{x}$. If we substitute $x^{1/p} = \sqrt[p]{x}$ into $\left(x^{1/p}\right)^n = x^{n/p}$, we get $\left(\sqrt[p]{x}\right)^n = x^{n/p}$. We can use the reflexive property (Sec. 1.6) to swap the left and right sides:

$$x^{n/p} = \left(\sqrt[p]{x}\right)^n$$

This equation shows that we may interpret a fractional exponent as follows. One way to evaluate a fractional exponent $x^{n/p}$ is to first take the pth root of x and then raise that to the power of n. When we do this, we're interpreting the denominator of the exponent as a root and the numerator of the exponent as a power. That's exactly what we did in Sec. 1.11.

If we let $n = \frac{1}{q}$ in the equation $(x^m)^n = x^{mn}$, we get $(x^m)^{1/q} = x^{m/q}$, which similarly becomes $\sqrt[q]{x^m} = x^{m/q}$. This shows that we may use the numerator of the exponent as a power first and then use the denominator of the exponent as a root. It doesn't matter whether we treat the power or the root first. For example, $27^{2/3} = \left(\sqrt[3]{27}\right)^2 = 3^2 = 9$ is equivalent to $27^{2/3} = \sqrt[3]{27^2} = \sqrt[3]{729} = 9$. The exponent 2/3 involves a square and a cube root. When we took the cube root first and then squared the answer, we got 9, and when we applied the square first and then took the cube root, we still got 9.

The previous rules regarding exponents still apply when the exponents are fractions, as the examples below illustrate.

- $x^{1/2}x^{1/3} = x^{1/2+1/3} = x^{5/6}$ because $\frac{1}{2} + \frac{1}{3} = \frac{3}{6} + \frac{2}{6} = \frac{5}{6}$. (Recall from Sec. 1.7 that the way to add or subtract fractions is to find a common denominator.)
- $\frac{x^{1/6}}{x^{3/4}} = x^{1/6-3/4} = x^{-7/12} = \frac{1}{x^{7/12}}$ because $\frac{1}{6} - \frac{3}{4} = \frac{2}{12} - \frac{9}{12} = -\frac{7}{12}$.
- $\left(x^{1/4}\right)^{5/2} = x^{(1/4)(5/2)} = x^{5/8}$ because $\frac{1}{4}\frac{5}{2} = \frac{1(5)}{4(2)} = \frac{5}{8}$.

Example 1. $x^{3/2}x^{4/5} = x^{3/2+4/5} = x^{23/10}$ because $\frac{3}{2} + \frac{4}{5} = \frac{15}{10} + \frac{8}{10} = \frac{23}{10}$

Example 2. $\frac{x^{5/8}}{x^{1/3}} = x^{5/8-1/3} = x^{7/24}$ because $\frac{5}{8} - \frac{1}{3} = \frac{15}{24} - \frac{8}{24} = \frac{7}{24}$

Example 3. $\left(x^{8/7}\right)^{5/3} = x^{(8/7)(5/3)} = x^{40/21}$ because $\frac{8}{7}\frac{5}{3} = \frac{8(5)}{7(3)} = \frac{40}{21}$

Example 4. $x^{-4/7} = \frac{1}{x^{4/7}}$

Exercise Set 3.7

Directions: Simplify each expression, assuming that $x \neq 0$.

1) $x^{4/5}x^{2/3} =$

2) $x^{5/6}x^{-3/8} =$

3) $x^{-5/6}x^{1/2} =$

4) $x^{-5/2}x^{-4/3} =$

5) $x^{3/4}x^{2/3}x =$

6) $x^{3/8}x^{-1/4}x^{-1/2} =$

7) $\dfrac{x^{5/6}}{x^{1/3}} =$

8) $\dfrac{x^{5/3}}{x^{-3/4}} =$

9) $\dfrac{x^{-7/2}}{x^{9/4}} =$

10) $\dfrac{x^{3/4}x^{3/4}}{x^{3/8}} =$

11) $\dfrac{x^{-4/5}}{x^{2/5}x^{1/2}} =$

12) $\dfrac{x^{2/3}x^{1/2}}{x^{3/4}x^{-1/3}} =$

13) $\left(x^{3/4}\right)^{6} =$

14) $\left(x^{5/6}\right)^{-3/2} =$

15) $\left(x^{-4/9}\right)^{3/2} =$

16) $\left(x^{-3}\right)^{-5/6} =$

3.8 Exponent of a Product or Quotient

When a product of quantities is enclosed in parentheses and the parentheses have an exponent like $(x^m y^n)^p$, the exponent multiplies the exponent of each quantity:

$$(x^m y^n)^p = x^{mp} y^{np}$$

For example, $(x^2 y^3)^4 = x^{2(4)} y^{3(4)} = x^8 y^{12}$. A longer way to look at this is $(x^2 y^3)^4 = (x^2 y^3)(x^2 y^3)(x^2 y^3)(x^2 y^3) = x^8 y^{12}$. Similarly, if a quotient of quantities is enclosed in parentheses and the parentheses have an exponent, we get:

$$\left(\frac{x^m}{y^n}\right)^p = \frac{x^{mp}}{y^{np}}$$

For example, $\left(\frac{x^5}{y^2}\right)^6 = \frac{x^{5(6)}}{y^{2(6)}} = \frac{x^{30}}{y^{12}}$.

Example 1. $(3x^2)^5 = (3^1 x^2)^5 = 3^{1(5)} x^{2(5)} = 3^5 x^{10} = 243 x^{10}$

Example 2. $\left(\frac{x^4}{7}\right)^2 = \left(\frac{x^4}{7^1}\right)^2 = \frac{x^{4(2)}}{7^{1(2)}} = \frac{x^8}{7^2} = \frac{x^8}{49}$ **Example 3.** $\left(\frac{x^3}{4}\right)^{-2} = \left(\frac{4}{x^3}\right)^2 = \frac{4^2}{x^{3(2)}} = \frac{16}{x^6}$

Exercise Set 3.8

Directions: Simplify each expression, assuming that $x \neq 0$.

1) $(9x^3)^2 =$

2) $(-5x^8)^3 =$

3) $(-4x^7)^4 =$

4) $(2x)^{-5} =$

5) $(6x^7)^0 =$

6) $(8x^{-4})^{-1} =$

7) $(64x^5)^{1/2} =$

8) $\left(27x^{3/4}\right)^{2/3} =$

9) $\left(\dfrac{x^4}{5}\right)^3 =$

10) $\left(\dfrac{3}{x^2}\right)^4 =$

11) $\left(-\dfrac{x^9}{4}\right)^5 =$

12) $\left(-\dfrac{x^7}{6}\right)^2 =$

13) $\left(\dfrac{x^6}{5}\right)^{-2} =$

14) $\left(\dfrac{x^8}{7}\right)^{-1} =$

15) $\left(\dfrac{x^8}{256}\right)^{3/4} =$

16) $\left(\dfrac{x^{10}}{32}\right)^{4/5} =$

3.9 Scientific Notation

In scientific notation, a power of ten is used to position the decimal point immediately following the first digit of a number, like 3.78×10^4 or 9.6×10^{-3}. The times symbol ($\times$) is common for scientific notation. Since it is wedged between numbers, it shouldn't cause any confusion with a variable (x). To express a number in scientific notation, follow these steps:

- Count how many digits the decimal point needs to move in order to follow the first digit of the number. For example, in 6294 (which is equivalent to 6294.0) the decimal point needs to move 3 places to the left to form 6.294, while in 0.08136 the decimal point needs to move 2 places to the right to form 8.136.
- If the number is greater than 1, multiply by a positive power of 10 where the power equals the answer to the previous step. Multiplying by a positive power of 10 allows us to move the decimal point to the left. For example, $6294 = 6.294 \times 10^3$.
- If the number is less than 1, multiply by a negative power of 10 where the power equals the negative of the answer to the first step. Multiplying by a negative power of 10 allows us to move the decimal point to the right. For example, $0.08136 = 8.136 \times 10^{-2}$.

Example 1. $27{,}693 = 2.7693 \times 10^4$ (moved left 4 places)

Example 2. $0.0092 = 9.2 \times 10^{-3}$ (moved right 3 places)

Example 3. $5900 = 5.900 \times 10^3 = 5.9 \times 10^3$ (moved left 3 places)
Note: 5.9 is equivalent to 5.900. Trailing decimal zeros have no effect mathematically. (However, in a science or engineering class, a trailing decimal zero may be important if it is a significant figure.)

Exercise Set 3.9

Directions: Express each number using scientific notation.

1) 518 =

2) 0.094 =

3) 0.00675 =

4) 639,415 =

5) 1694 =

6) 0.00001487 =

7) 0.62 =

8) 81,300 =

9) 9,583,000 =

10) 0.000256 =

3.10 Simple Equations with Exponents

In the equations of this section, the variable has an exponent and the equation is simple enough that it can be solved by isolating the unknown (Sec. 2.4). Not all equations with exponents of variables can be solved this way (such as those of Chapter 6).

First isolate the variable with the exponent. Then raise both sides of the equation to the reciprocal of the exponent. For example, take the cube root of both sides of $x^3 = 8$ to get $\sqrt[3]{x^3} = \sqrt[3]{8}$. The left side simplifies to x since $\sqrt[3]{x^3} = (x^3)^{1/3} = x^1 = x$. As another example, raise both sides of $x^{1/4} = 3$ to the power of 4 to get $\left(x^{1/4}\right)^4 = 3^4$. The left side simplifies to $x^{(1/4)4} = x^1 = x$.

When taking an even root of both sides of an equation, include a $\pm$ sign to allow for all possible solutions. For example, for $x^4 = 16$, when we take the fourth root of both sides, since this is an even root, we will write $x = \pm 16^{1/4}$, which gives two answers: $x = \pm 2$. There are two answers because $(-2)^4 = 16$ and $2^4 = 16$. For an odd root, there will only be one answer. For example, for $x^3 = 27$, the only answer is $x = 27^{1/3}$ which simplifies to $x = 3$. It may help to review Sec. 1.11 and Sec.'s 3.6-3.7.

Example 1. $x^3 - 7 = 20$
- Add 7 to both sides: $x^3 = 27$.
- Take the cube root of both sides: $x = 27^{1/3}$. Simplify: $x = 3$.

Check the answer. Plug 3 in for x on the left: $x^3 - 7 = 3^3 - 7 = 27 - 7 = 20$. Since both sides equal 20, the answer is correct.

Example 2. $2x^4 = 512$
- Divide by 2 on both sides: $x^4 = 256$.
- Take the fourth root of both sides: $x = \pm 256^{1/4}$. Simplify: $x = \pm 4$.

Check the answers. Plug ± 4 in for x on the left: $2x^4 = 2(\pm 4)^4 = 2(256) = 512$. Since both sides equal 512, the answers are correct.

Example 3. $11 + 3x^{1/3} = 26$

- Subtract 11 from both sides: $3x^{1/3} = 15$.
- Divide by 3 on both sides: $x^{1/3} = 5$.
- Cube both sides: $x = 5^3$. Simplify: $x = 125$.

Check the answer. Plug 125 in for x on the left: $11 + 3x^{1/3} = 11 + 3(125)^{1/3} = 11 + 3(5) = 11 + 15 = 26$. Since both sides equal 26, the answer is correct.

Exercise Set 3.10

Directions: Solve for the variable by isolating the unknown. Check your answer.

1) $x^5 + 18 = 50$

2) $72 = x^4 - 9$

3) $5x^{1/2} = 20$

4) $7 = 2x^{1/3} + 1$

5) $\dfrac{x^3}{5} = -25$

6) $4 + 6x^2 = 100$

7) $3x^4 - 1 = 2$

8) $900 = -4x^3 + 36$

9) $\dfrac{x^{1/2}}{4} = 3$

10) $5x^3 = x^3 + 500$

11) $5x^2 - 48 = 50 + 3x^2$

12) $\dfrac{x^{-2}}{4} = 9$

13) $2x^{3/4} = 54$

14) $3x^{-2/3} = 48$

3.11 Special Solutions Involving Exponents

An equation with a variable that has an exponent may have a single unique answer, multiple solutions, may be indeterminate, may be undefined, or may have no solution.

- $x^2 = 9$ has **multiple solutions** since $x = -3$ and $x = 3$ both equal 9 when they are squared.
- $x^2 = x^2 + 1$ has **no solution** because when x^2 is subtracted from both sides, we get $0 = 1$ which can never be true.
- $x^0 = 1$ is **indeterminate** since it is satisfied by all nonzero numbers. Any nonzero number raised to the power of zero equals one (Sec. 3.3).
- $x = (1 - 1)^{-1}$ is **undefined** since the right side simplifies to $0^{-1} = \frac{1}{0}$ and because $\frac{1}{0}$ is undefined (Sec. 3.3).

Example 1. $3x^2 = 48$

Divide by 3 on both sides to get $x^2 = 16$. Square root both sides to get two solutions: $x = \pm 4$.

Example 2. $x^2 = -1$

Square root both sides to get $x = (-1)^{1/2}$ which is equivalent to $x = \sqrt{-1}$. This has no real solution (there is only an imaginary solution).

Exercise Set 3.11

Directions: Determine whether each equation has one unique solution, has no solution, has multiple solutions, is indeterminate, or is undefined.

1) $\dfrac{x^5}{x^5} = 1$

2) $6x^2 + 8 = 4x^2 + 2x^2$

3) $x^2 - 4 = 0$

4) $x = \left[\dfrac{8(2)-4^2}{8(2)+4^2}\right]^{2(3)-6}$

5) $x^2 + 4 = 0$

6) $x^{-1} = 0$

3.12 Word Problems Involving Exponents

The word problems in this section involve exponents. It may help to review Sec. 2.13 (which has tips for solving word problems) and Sec. 1.4.

Common root and power keywords include squared, cubed, square root, cube root, raised to the power, exponent, etc.

- If an unknown number is squared, write x^2.
- For the cube root of an unknown number, write $x^{1/3}$.

For other kinds of keywords, review Sec.'s 2.13 and 1.4.

Example 1. A positive number raised to the fourth power is 6 more than 75. What is the number?

- x = the positive number.
- x^4 = the positive number raised to the fourth power.
- According to the problem: $x^4 - 6 = 75$.
- Add 6 to both sides to get $x^4 = 81$.
- Take the fourth root of both sides to get $x = 81^{1/4} = \pm 3$.
- The **positive** number is 3. (The problem asked for the positive answer.)
- Check: $x^4 - 6 = 3^4 - 6 = 81 - 6 = 75$.

Example 2. The square root of Cory's age (in years) is 5. How old is Cory?

- x = Cory's age.
- $\sqrt{x}$ = the square root of Cory's age.
- According to the problem: $\sqrt{x} = 5$.
- Square both sides to get $x = 5^2 = 25$.
- Cory is 25 years old.
- Check: $\sqrt{x} = \sqrt{25} = 5$.

Exercise Set 3.12

Directions: Use algebra to solve each word problem.

1) Sarah is buying carpet for a square room that measures 12 feet wide. How many square feet of carpet does she need to purchase?

2) On a board game, Steve's token is 35 steps away from the finish line. Steve rolled a 3 and a cube, which allow him to advance his token 3 cubed steps. How far from the finish line will his token be after he moves it?

3) Five times the square root of a number equals eighty. What is the number?

4) When a positive number is raised to the power of two-thirds, the result is 20 less than the square of 6. What is the number?

4 THE DISTRIBUTIVE PROPERTY

4.1 Distributing with Variables

The distributive property is $a(b + c) = ab + ac$. You can verify this formula by trying it out for a variety of numbers. For example, if $a = 5$, $b = 4$, and $c = 6$, the left side is $a(b + c) = 5(4 + 6) = 5(10) = 50$ and the right side is also $ab + ac = 5(4) + 5(6) = 20 + 30 = 50$. Both sides will be equal no matter which numbers you try. Test it out!

When applying the distributive property to expressions that have variables, you need to remember the rules for exponents from Chapter 3. For example, to distribute the $3x^4$ in the expression $3x^4(5x^3 - 7x^2)$, we will need to apply the rule $x^m x^n = x^{m+n}$. If you let $a = 3x^4$, $b = 5x^3$, and $c = -7x^2$, according to the distributive property, we get the following:
$$3x^4(5x^3 - 7x^2) = 3x^4(5x^3) + 3x^4(-7x^2) = 15x^7 - 21x^6$$
Note that $x^4 x^3 = x^{4+3} = x^7$ and $x^4 x^2 = x^{4+2} = x^6$.

Example 1. $4(3x + 5) = 4(3x) + 4(5) = 12x + 20$

Example 2. $2x(4x - 3) = 2x(4x) + 2x(-3) = 8x^2 - 6x$

Example 3. $6x^3(3x^2 - x + 5) = 6x^3(3x^2) + 6x^3(-x) + 6x^3(5) = 18x^5 - 6x^4 + 30x^3$

Exercise Set 4.1

Directions: Apply the distributive property to each expression.

1) $7(9x - 8) =$

2) $x(x - 1) =$

3) $6x(4x + 7) =$

4) $5x^2(4x^2 - 8x) =$

5) $8x^3(6x^5 + 4x^3) =$

6) $9x(x^2 - 6x + 5) =$

7) $4x^2(2x^4 + 3x^2 - 6) =$

8) $3x^5(-8x^6 - 7x^4 + x^2) =$

9) $x(x^3 - x^2 + x - 1) =$

10) $2x^{11}(3x^{12} - 8x^7 + 6x^2) =$

11) $5x^{-2}(3x^2 + 7x) =$

12) $9x^4(6x^{-2} + 4x^{-8}) =$

13) $2x^{-3}(4x^{-2} - 6x^{-4}) =$

14) $8x^{-4}(3x^6 + 2x^4 - 4x^2) =$

15) $3x^{2/3}\left(2x^{3/4} - x^{1/3}\right) =$

4.2 Distributing Minus Signs

If the expression before the parentheses is negative, that minus sign gets distributed to each term in parentheses. Remember that if two negative numbers multiply one another, the product is positive (Sec. 1.10).

Example 1. $-3(2x + 4) = -3(2x) - 3(4) = -6x - 12$

Example 2. $-(x - 1) = -(x) - (-1) = -x + 1$

Example 3. $-2x^2(x^2 + 3x - 5) = -2x^2(x^2) - 2x^2(3x) - 2x^2(-5) = -2x^4 - 6x^3 + 10x^2$

Exercise Set 4.2

Directions: Apply the distributive property to each expression.

1) $-6(4x - 8) =$

2) $-x(x^2 - 2x) =$

3) $-(-x - 5) =$

4) $-7x(-4x^2 + 9) =$

5) $-x(-3x^2 - 2x + 4) =$

6) $-8x^2(7x^5 + 5x^3 - 3x) =$

4.3 The FOIL Method

The FOIL method is a common way to apply the distributive method to multiplication of the following form:

$$(a + b)(c + d) = ac + ad + bc + bd$$

The word FOIL is an abbreviation. If you can remember this abbreviation, it will help you multiply expressions like the one shown above.

- F stands for "first." The first term, ac, comes from multiplying the first terms of $(a + b)(c + d)$. Note that a and c come first in each expression.
- O stands for "outside." The second term, ad, comes from multiplying the outside terms of $(a + b)(c + d)$. Note that a and d are on the outside.
- I stands for "inside." The third term, bc, comes from multiplying the inside terms of $(a + b)(c + d)$. Note that b and c are on the inside.
- L stands for "last." The last term, bd, comes from multiplying the last terms of $(a + b)(c + d)$. Note that b and d come last in each expression.

The FOIL method applies the distributive property. We first distribute a to $(c + d)$ to get $ac + ad$ and next distribute b to $(c + d)$ to get $bc + bd$.

You can verify the formula above by trying it out for a variety of numbers. For example, if $a = 2$, $b = 3$, $c = 4$, and $d = 5$ the left side is $(a + b)(c + d) = (2 + 3)(4 + 5) = 5(9) = 45$ and the right side is also $ac + ad + bc + bd = 2(4) + 2(5) + 3(4) + 3(5) = 8 + 10 + 12 + 15 = 18 + 27 = 45$. Both sides will be equal no matter which numbers you try. Test it out!

Example 1. $(x + 4)(x - 3) = x(x) + x(-3) + 4(x) + 4(-3) = x^2 - 3x + 4x - 12 = x^2 + x - 12$ In the last step, we combined like terms (Sec. 2.2): $-3x + 4x = x$.

Example 2. $(3x - 2)(2x^2 + 4) = 3x(2x^2) + 3x(4) - 2(2x^2) - 2(4) = 6x^3 + 12x - 4x^2 - 8 = 6x^3 - 4x^2 + 12x - 8$ In the last step, we reordered the terms in order to express our answer in decreasing powers of the variable.

Exercise Set 4.3

Directions: Apply the FOIL method to expand each expression.

1) $(x + 7)(x + 4) =$

2) $(x + 8)(x - 6) =$

3) $(x - 1)(x + 5) =$

4) $(-x + 9)(x + 3) =$

5) $(x - 4)(x - 6) =$

6) $(-x - 2)(x + 8) =$

7) $(x - 5)(-x - 7) =$

8) $(9 - x)(6 - x) =$

9) $(6x - 4)(3x + 7) =$

10) $(5x + 2)(3x + 4) =$

11) $(2x^2 + 6)(x^2 - 5) =$

12) $(3x^2 - 4x)(2x^2 + 6) =$

13) $(5x^6 + 7x^4)(8x^3 - 6x) =$

14) $(5x + 3x^{-1})(4x^{-3} - 2x^{-4}) =$

15) $\left(x^{3/4} - 2x^{1/2}\right)\left(3x^{1/2} + 5x^{1/4}\right) =$

4.4 The Square of the Sum

The expression $(u + v)^2$ is equivalent to $(u + v)(u + v)$. If the FOIL method is applied to this expression, we get $(u + v)^2 = (u + v)(u + v) = u(u) + u(v) + v(u) + v(v) = u^2 + uv + vu + v^2 = u^2 + 2uv + v^2$. If we ignore the middle steps, we get:
$$(u + v)^2 = u^2 + 2uv + v^2$$
It is handy to remember this formula. You can apply it whenever two terms are added together (even if there is a minus sign, like Example 1) and the sum is squared.

To apply this formula, first identify u and v. For example, $u = 3x$ and $v = -2$ in the expression $(3x - 2)^2$, which leads to $(3x - 2)^2 = (3x)^2 + 2(3x)(-2) + (-2)^2$, which simplifies to $9x^2 - 12x + 4$. Note that $(3x)^2 = 3^2 x^2 = 9x^2$ (Sec. 3.8).

Example 1. $(x - 3)^2 = x^2 + 2(x)(-3) + (-3)^2 = x^2 - 6x + 9$ since $u = x$ and $v = -3$.

Example 2. $(2x + 4)^2 = (2x)^2 + 2(2x)(4) + 4^2 = 4x^2 + 16x + 16$ since $u = 2x$ and $v = 4$.

Example 3. $(4x^3 + x)(4x^3 + x) = (4x^3)^2 + 2(4x^3)(x) + x^2 = 16x^6 + 8x^4 + x^2$ since $u = 4x^3$ and $v = x$.

Exercise Set 4.4

Directions: Apply the square of the sum formula to expand each expression.

1) $(x + 8)^2 =$

2) $(x - 7)(x - 7) =$

3) $(4x + 9)^2 =$

4) $(6x - 5)(6x - 5) =$

5) $(-x + 6)^2 =$

6) $(-x - 4)^2 =$

7) $(1 - x)^2 =$

8) $(5x^2 + 8)^2 =$

9) $(6x^2 - 3x)^2 =$

10) $(9x^5 + 4x^3)(9x^5 + 4x^3) =$

4.5 The Difference of Squares

Consider the expression $(u + v)(u - v)$, which has one plus sign and one minus sign. If we apply the FOIL method, we get $(u + v)(u - v) = u(u) + u(-v) + v(u) + v(-v) = u^2 - uv + vu - v^2 = u^2 - v^2$. Observe that the cross terms (uv and $-vu$) cancel out. If we ignore the middle steps, we get:

$$(u + v)(u - v) = u^2 - v^2$$

It is handy to remember this formula. You can apply it whenever an expression has the form $(u + v)(u - v)$. Note the different signs in this expression.

To apply this formula, first identify u and v. For example, $u = 4x$ and $v = 5$ in the expression $(4x + 5)(4x - 5)$, which leads to $(4x + 5)(4x - 5) = (4x)^2 - 5^2$, which simplifies to $16x^2 - 25$. Note that $(4x)^2 = 4^2x^2 = 16x^2$ (Sec. 3.8).

Example 1. $(x + 2)(x - 2) = x^2 - 2^2 = x^2 - 4$ since $u = x$ and $v = 2$.

Example 2. $(3x^2 + 4)(3x^2 - 4) = (3x^2)^2 - 4^2 = 9x^4 - 16$ since $u = 3x^2$ and $v = 4$.

Exercise Set 4.5

Directions: Apply the difference of squares formula to expand each expression.

1) $(x + 5)(x - 5) =$

2) $(x - 4)(x + 4) =$

3) $(x + 1)(x - 1) =$

4) $(-x + 3)(-x - 3) =$

5) $(6 + x)(6 - x) =$

6) $(2x + 7)(2x - 7) =$

7) $(4x - 9)(4x + 9) =$

8) $(x^2 + x)(x^2 - x) =$

9) $(3x^5 + 4x^3)(3x^5 - 4x^3) =$

10) $\left(x^{3/2} + x^{-1}\right)\left(x^{3/2} - x^{-1}\right) =$

4.6 Beyond FOIL

The FOIL concept may be applied even if there are three or more terms enclosed in one of the parentheses. The main idea is that each term in the first expression multiplies every term in the second expression. For example,

$$(x - 2)(x^2 - 3x + 1) = x(x^2) + x(-3x) + x(1) - 2(x^2) - 2(-3x) - 2(1)$$
$$= x^3 - 3x^2 + x - 2x^2 + 6x - 2 = x^3 - 5x^2 + 7x - 2$$

In the last step, we combined like terms: $-3x^2 - 2x^2 = -5x^2$ and $x + 6x = 7x$. The x and the -2 are each distributed to $(x^2 - 3x + 1)$ in the previous expansion. That is,

$$(x - 2)(x^2 - 3x + 1) = x(x^2 - 3x + 1) - 2(x^2 - 3x + 1)$$

Example 1. $(x + 3)(x^2 + 2x - 4) = x(x^2) + x(2x) + x(-4) + 3(x^2) + 3(2x) + 3(-4)$
$= x^3 + 2x^2 - 4x + 3x^2 + 6x - 12 = x^3 + 5x^2 + 2x - 12$ In the last step, we combined like terms (Sec. 2.2): $2x^2 + 3x^2 = 5x^2$ and $-4x + 6x = 2x$.

Example 2. $(x^2 + 3x - 2)(x^2 - 4x + 5) = x^2(x^2) + x^2(-4x) + x^2(5) + 3x(x^2) + 3x(-4x)$
$+ 3x(5) - 2(x^2) - 2(-4x) - 2(5) = x^4 - 4x^3 + 5x^2 + 3x^3 - 12x^2 + 15x - 2x^2 + 8x - 10$
$= x^4 - x^3 - 9x^2 + 23x - 10$ In the last step, combined like terms (Sec. 2.2): $-4x^3 + 3x^3$
$= -x^3, 5x^2 - 12x^2 - 2x^2 = -9x^2$, and $15x + 8x = 23x$.

Exercise Set 4.6

Directions: Expand each expression.

1) $(x + 4)(x^2 - 6x + 7) =$

2) $(2x - 1)(3x^2 + 4x - 5) =$

3) $(-x + 2)(x^2 - 8x - 9) =$

4) $(4x - 6)(x^2 + 7x - 3) =$

5) $(3x + 8)(2x^2 - 5x + 9) =$

6) $(x^2 - 4x - 7)(x^2 + 3x - 7) =$

7) $(2x^2 - 3x + 6)(3x^2 - 4x - 5) =$

4.7 Expansion of Powers

If two or more terms are in parentheses and the parentheses have an exponent that is an integer greater than two, like $(x - 4)^3$, you can expand this by multiplying one step at a time, like the example below.

$$(x - 4)^3 = (x - 4)(x - 4)(x - 4) = [x(x) + x(-4) - 4(x) - 4(-4)](x - 4)$$
$$= (x^2 - 4x - 4x + 16)(x - 4) = (x^2 - 8x + 16)(x - 4)$$
$$= x^2(x) + x^2(-4) - 8x(x) - 8x(-4) + 16(x) + 16(-4)$$
$$= x^3 - 4x^2 - 8x^2 + 32x + 16x - 64 = x^3 - 12x^2 + 48x - 64$$

Example 1. $(x + 3)^3 = (x + 3)(x + 3)(x + 3) = [x(x) + x(3) + 3(x) + 3(3)](x + 3)$
$= (x^2 + 3x + 3x + 9)(x + 3) = (x^2 + 6x + 9)(x + 3) = x^2(x) + x^2(3) + 6x(x) +$
$6x(3) + 9(x) + 9(3) = x^3 + 3x^2 + 6x^2 + 18x + 9x + 27 = x^3 + 9x^2 + 27x + 27$

Example 2. $(3x - 2)^3 = (3x - 2)(3x - 2)(3x - 2)$
$= [3x(3x) + 3x(-2) - 2(3x) - 2(-2)](3x - 2) = (9x^2 - 6x - 6x + 4)(3x - 2)$
$= (9x^2 - 12x + 4)(3x - 2) = 9x^2(3x) + 9x^2(-2) - 12x(3x) - 12x(-2) + 4(3x) + 4(-2)$
$= 27x^3 - 18x^2 - 36x^2 + 24x + 12x - 8 = 27x^3 - 54x^2 + 36x - 8$

Exercise Set 4.7

Directions: Expand each expression.

1) $(x - 5)^3 =$

2) $(6x + 3)^3 =$

3) $(x - 2)^4 =$

4) $(x^2 + 5x - 2)^3 =$

4.8 The Binomial Expansion

Consider the expression $(u + v)^n$, where n is a whole number.

- $(u + v)^1 = u + v$
- $(u + v)^2 = (u + v)(u + v) = u^2 + uv + uv + v^2 = u^2 + 2uv + v^2$
- $(u + v)^3 = (u^2 + 2uv + v^2)(u + v) = u^3 + u^2v + 2u^2v + 2uv^2 + uv^2 + v^3$
  $= u^3 + 3u^2v + 3uv^2 + v^3$
- $(u + v)^4 = (u^3 + 3u^2v + 3uv^2 + v^3)(u + v)$
  $= u^4 + u^3v + 3u^3v + 3u^2v^2 + 3u^2v^2 + 3uv^3 + uv^3 + v^4$
  $= u^4 + 4u^3v + 6u^2v^2 + 4uv^3 + v^4$
- $(u + v)^5 = (u^4 + 4u^3v + 6u^2v^2 + 4uv^3 + v^4)(u + v)$
  $= u^5 + u^4v + 4u^4v + 4u^3v^2 + 6u^3v^2 + 6u^2v^3 + 4u^2v^3 + 4u^2v^4 + uv^4 + v^5)$
  $= u^5 + 5u^4v + 10u^3v^2 + 10u^2v^3 + 5uv^4 + v^5$

When we arrange these expressions in a triangle (called **Pascal's triangle**), a pattern can be seen. Each expression below is referred to as a **binomial expansion**.

$$u + v$$
$$u^2 + 2uv + v^2$$
$$u^3 + 3u^2v + 3uv^2 + v^3$$
$$u^4 + 4u^3v + 6u^2v^2 + 4uv^3 + v^4$$
$$u^5 + 5u^4v + 10u^3v^2 + 10u^2v^3 + 5uv^4 + v^5$$
$$u^6 + 6u^5v + 15u^4v^2 + 20u^3v^3 + 15u^2v^4 + 6uv^5 + v^6$$
$$u^7 + 7u^6v + 21u^5v^2 + 35u^4v^3 + 35u^3v^4 + 21u^2v^5 + 7uv^6 + v^7$$
$$u^8 + 8u^7v + 28u^6v^2 + 56u^5v^3 + 70u^4v^4 + 56u^3v^5 + 28u^2v^6 + 8uv^7 + v^8$$

Which patterns do you see in the triangle above?

- $(u + v)^n$ begins with u^n and ends with v^n
- The second term is $nu^{n-1}v$ and the second-to-last term is nuv^{n-1}. For example, these terms are $7u^6v$ and $7uv^6$ for $(u + v)^7$.
- Each coefficient equals the sum of the two coefficients above it. For example, for $(u + v)^7$, note that $21 = 6 + 15$ and $35 = 20 + 15$, and for $(u + v)^8$, note that $28 = 7 + 21$, $56 = 21 + 35$, and $70 = 35 + 35$.
- The exponents add up to n. For example, u^5v^2 and u^4u^3 give $5 + 2 = 4 + 3 = 7$.

The coefficients of the binomial expansion can be determined using the formula that follows. The exclamation mark (!) indicates a **factorial**, which means to multiply by successively smaller integers until reaching one. For example, $4! = 4(3)(2)(1) = 24$. Note that $1! = 1$ and $0! = 1$ also. (Why does $0! = 1$? One reason behind this is so that the formula $n! = n(n-1)!$ will hold. For example, $5! = 5(4!) = 5(24) = 120$. When we apply this to $n = 1$, we get $1! = 1(0!) = 1$, which works provided that $0! = 1$.)

$$(u + v)^n = u^n + nu^{n-1}v + \frac{n(n-1)}{2!}u^{n-2}v^2 + \frac{n(n-1)(n-2)}{3!}u^{n-3}v^3 + \cdots + v^n$$

For example, for $n = 7$, since $\frac{7(6)}{2!} = \frac{42}{2} = 21$, $\frac{7(6)(5)}{3!} = \frac{210}{6} = 35$, $\frac{7(6)(5)(4)}{4!} = \frac{840}{24} = 35$, $\frac{7(6)(5)(4)(3)}{5!} = \frac{2520}{120} = 21$, etc., we get:

$$(u + v)^7 = u^7 + 7u^6v + 21u^5v^2 + 35u^4v^3 + 35u^3v^4 + 21u^2v^5 + 7uv^6 + v^7$$

To apply the formula for the binomial expansion, first identify u and v. For example, $u = 2x$ and $v = -3$ in the expression $(2x - 3)^4$, which leads to:

$$(2x - 3)^4 = u^4 + 4u^3v + 6u^2v^2 + 4uv^3 + v^4$$
$$= (2x)^4 + 4(2x)^3(-3) + 6(2x)^2(-3)^2 + 4(2x)(-3)^3 + (-3)^4$$
$$= 16x^4 + 4(8x^3)(-3) + 6(4x^2)(9) + 4(2x)(-27) + 81$$
$$= 16x^4 - 96x^3 + 216x^2 - 216x + 81$$

Recall the rule $(x^m y^n)^p = x^{mp}y^{np}$ from Sec. 3.8. For example, $(2x)^3 = 2^3x^3 = 8x^3$.

Example 1. $(x - 4)^3 = u^3 + 3u^2v + 3uv^2 + v^3 = x^3 + 3x^2(-4) + 3x(-4)^2 + (-4)^3$
$= x^3 - 12x^2 + 3(16)x - 64 = x^3 - 12x^2 + 48x - 64$ since $u = x$ and $v = -4$.

Example 2. $(2x + 5)^5 = u^5 + 5u^4v + 10u^3v^2 + 10u^2v^3 + 5uv^4 + v^5$
$= (2x)^5 + 5(2x)^4(5) + 10(2x)^3(5)^2 + 10(2x)^2(5)^3 + 5(2x)(5)^4 + 5^5$
$= 32x^5 + 5(16x^4)(5) + 10(8x^3)(25) + 10(4x^2)(125) + 5(2x)(625) + 3125$
$= 32x^5 + 400x^4 + 2000x^3 + 5000x^2 + 6250x + 3125$ since $u = 2x$ and $v = 5$.

Exercise Set 4.8

Directions: Apply the binomial expansion formula to expand each expression.

1) $(x + 9)^4 =$

2) $(x - 3)^5 =$

3) $(2 + x)^6 =$

4) $(-x + 4)^7 =$

5) $(3x + 2)^4 =$

6) $(2x - 4)^3 =$

7) $(x + 10)^9 =$

4.9 Distributing Fractions

The distributive property applies even if the numbers (or expressions) are fractions. For example, we may apply $a(b+c) = ab + ac$ to $\frac{x}{2}\left(\frac{x}{3} - 1\right) = \frac{x}{2}\frac{x}{3} - \frac{x}{2}1 = \frac{x^2}{6} - \frac{x}{2}$.

Example 1. $\frac{1}{2}(6x - 4) = \frac{1}{2}6x - \frac{1}{2}4 = 3x - 2$

Example 2. $\frac{x}{3}\left(\frac{x}{4} - \frac{2}{3}\right) = \frac{x}{3}\frac{x}{4} - \frac{x}{3}\frac{2}{3} = \frac{x^2}{12} - \frac{2x}{9}$

Exercise Set 4.9

Directions: Apply the distributive property to each expression.

1) $\frac{3}{4}(8x + 12) =$

2) $24x\left(\frac{x}{6} - \frac{2}{3}\right) =$

3) $\frac{x}{2}(2x^2 + 4x - 14) =$

4) $\frac{2}{3}\left(\frac{x}{5} + \frac{9}{8}\right) =$

5) $\frac{5}{x}(4x^3 - 3x^2 + 2x) =$

6) $\frac{3x}{8}\left(\frac{4x}{5} - \frac{2}{7}\right) =$

4.10 Distributing Across Fractions

If the numerator or denominator of a fraction includes a sum of terms, like $\frac{3x+2}{4}$ or $\frac{x-3}{x+3}$, when another number or expression multiplies this fraction, the distributive property still applies. For example, $5\frac{3x+2}{4} = \frac{5}{1}\frac{3x+2}{4} = \frac{5(3x+2)}{1(4)} = \frac{5(3x)+5(2)}{4} = \frac{15x+10}{4}$ and $\frac{x}{2}\frac{x-3}{x+3} = \frac{x(x-3)}{2(x+3)} = \frac{x(x)-x(3)}{2(x)+2(3)} = \frac{x^2-3x}{2x+6}$.

It's important to note what is and what isn't allowed regarding fractions. For example:

- $\frac{x+5}{2} = \frac{x}{2} + \frac{5}{2}$ is allowed. Think of it as $\frac{x+5}{2} = \frac{1}{2}(x+5) = \frac{x}{2} + \frac{5}{2}$. This works for any numbers. For example, if $x = 15$, the left side is $\frac{x+5}{2} = \frac{15+5}{2} = \frac{20}{2} = 10$ and the right side is $\frac{x}{2} + \frac{5}{2} = \frac{15}{2} + \frac{5}{2} = \frac{20}{2} = 10$.

- $\frac{3}{x+2}$ **doesn't** split into $\frac{3}{x} + \frac{3}{2}$. For example, if $x = 4$, the left side is $\frac{3}{4+2} = \frac{3}{6} = \frac{1}{2}$ and the right side is $\frac{3}{x} + \frac{3}{2} = \frac{3}{4} + \frac{3}{2} = \frac{3}{4} + \frac{6}{4} = \frac{9}{4}$, which isn't $\frac{1}{2}$.

- $\frac{6x+12}{3x+4}$ **doesn't** split into $\frac{6x}{3x} + \frac{12}{4}$. For example, if $x = 1$, the left side is $\frac{6x+12}{3x+4} = \frac{18}{7}$ and the right side is $\frac{6x}{3x} + \frac{12}{4} = \frac{6}{3} + \frac{12}{4} = 2 + 3 = 5$, which isn't $\frac{18}{7}$.

Example 1. $6\frac{4x-3}{5x} = \frac{6}{1}\frac{4x-3}{5x} = \frac{6(4x-3)}{1(5x)} = \frac{6(4x)-6(3)}{5x} = \frac{24x-18}{5x}$

Example 2. $\frac{2}{3}\frac{x^2+1}{7} = \frac{2(x^2+1)}{3(7)} = \frac{2(x^2)+2(1)}{21} = \frac{2x^2+2}{21}$

Example 3. $\frac{3}{x}\frac{x+4}{x-5} = \frac{3(x+4)}{x(x-5)} = \frac{3(x)+3(4)}{x(x)-x(5)} = \frac{3x+12}{x^2-5x}$

Example 4. $\frac{x+2}{x+3}\frac{x+1}{x+4} = \frac{(x+2)(x+1)}{(x+3)(x+4)} = \frac{x(x)+x(1)+2(x)+2(1)}{x(x)+x(4)+3(x)+3(4)} = \frac{x^2+x+2x+2}{x^2+4x+3x+12} = \frac{x^2+3x+2}{x^2+7x+12}$

Exercise Set 4.10

Directions: Apply the distributive property to each expression.

1)

$$9\,\frac{7x - 9}{5} =$$

2)

$$6x\,\frac{8x - 1}{7} =$$

3)

$$\frac{1}{5}\,\frac{x^2}{3x^2 + 4} =$$

4)

$$\frac{x}{3}\,\frac{4x + 8}{9} =$$

5)

$$\frac{x^3}{5}\,\frac{2x^2 - 3x + 6}{7x - 4} =$$

6)

$$\frac{x + 3}{x - 4}\,\frac{x + 3}{x + 4} =$$

7)

$$\frac{5x^2 - 2}{6x^2 + 3}\,\frac{3x + 4}{2x^2 - 3} =$$

4.11 Distributing into Roots and Powers

Consider the expression $2(3x + 1) = 2(3x) + 2(1) = 6x + 2$. If we square both sides of this equation (and ignore the middle), we get $2^2(3x + 1)^2 = (6x + 2)^2$, which we may also write as $4(3x + 1)^2 = (6x + 2)^2$. Now suppose that we had begun with the expression $4(3x + 1)^2$. We can apply the distributive property to write this as $(6x + 2)^2$ as follows. First, square root 4 to get $\sqrt{4} = 2$ and then distribute the 2 to get $2(3x + 1) = 6x + 2$. That is, in order to distribute the 4 into $(3x + 1)^2$, we must first take the square root of 4 to account for the square outside of the parentheses.

More generally, we may write:
$$a(b + c)^d = \left(a^{1/d}b + a^{1/d}c\right)^d$$
In the previous example, $4(3x + 1)^2$, $a = 4$, $b = 3x$, $c = 1$, and $d = 2$, such that:
$$4(3x + 1)^2 = \left(4^{1/2}3x + 4^{1/2}1\right)^2 = (6x + 2)^2$$

Here is a simple way to look at it. Do the opposite to the coefficient to what is being done to the parentheses. For example:

- In $4(3x + 1)^2$, the parentheses are squared, so we square root the 4 to bring it inside: $4(3x + 1)^2 = \left[4^{1/2}(3x + 1)\right]^2 = [2(3x + 1)]^2 = (6x + 2)^2$.
- In $27(2x - 3)^3$, the parentheses are cubed, so we cube root the 27 to bring it inside: $27(2x - 3)^3 = \left[27^{1/3}(2x - 3)\right]^3 = [3(2x - 3)]^3 = (6x - 9)^3$.
- In $5\sqrt{x + 7} = 5(x + 7)^{1/2}$, the parentheses are square rooted, so we square the 5 to bring it inside: $5(x + 7)^{1/2} = [5^2(x + 7)]^{1/2} = [25(x + 7)]^{1/2} = (25x + 175)^{1/2}$ $= \sqrt{25x + 175}$.

This applies the properties $(x^m y^n)^p = x^{mp}y^{np}$ (Sec. 3.8) and $x^n = \left(x^{1/n}\right)^n = x^1 = x$.

Example 1. $1000(5x - 2)^3 = \left[1000^{1/3}(5x - 2)\right]^3 = [10(5x - 2)]^3 = (50x - 20)^3$

Example 2. $2(4x + 1)^{1/3} = [2^3(4x + 1)]^{1/3} = [8(4x + 1)]^{1/3} = (32x + 8)^{1/3}$

Exercise Set 4.11

Directions: Distribute the coefficient into the root or power similar to the examples.

1) $25(7x - 8)^2 =$

2) $64(6x + 5)^3 =$

3) $9(5x - 9)^{1/2} =$

4) $x^{12}(6x^3 - 9x)^2 =$

5) $81x^8(x^2 - 3x + 4)^4 =$

6) $36x^2 \left(\frac{x}{3} - \frac{2}{x} \right)^2 =$

7) $x^6\sqrt{8x^2 - 4} =$

8) $6x(2x + 8)^{-1} =$

9) $4x^4(3x - 5)^{-2} =$

10) $x^{-3}(x^{-4} + x^{-1})^{-1} =$

4.12 Distributing in Equations

The equations in this section involve the distributive property.

Example 1. $3(x - 2) = x + 4$
- Distribute on the left side: $3x - 6 = x + 4$.
- Add 6 to both sides: $3x = x + 10$.
- Subtract x from both sides: $2x = 10$.
- Divide by 2 on both sides: $x = \frac{10}{2}$. Simplify: $x = 5$.

Check the answer. Plug 5 in for x on each side:
- $3(x - 2) = 3(5 - 2) = 3(3) = 9$
- $x + 4 = 5 + 4 = 9$

Since both sides equal 9, the answer is correct.

Exercise Set 4.12

Directions: Solve for the variable by isolating the unknown. Check your answer.

1) $5x - 12 = 2(x + 3)$

2) $6(2x - 5) = 16x - 50$

3) $-2(3x - 4) = x - 13$

4) $4(6x - 12) = 3(5x + 8)$

5 FACTORING EXPRESSIONS

5.1 Factoring Variables

When the distributive property, $a(b + c) = ab + ac$, is applied in reverse, we call it **factoring**: $ab + ac = a(b + c)$. For example, consider the example of the distributive property $5x^2(2x + 3) = 10x^3 + 15x^2$. If we write this in reverse, we say that we have factored $5x^2$ out of $10x^3 + 15x^2$.

$$10x^3 + 15x^2 = 5x^2(2x + 3)$$

Factoring means to pull out an expression that is common to each term. The terms $10x^3$ and $15x^2$ each have a factor of $5x^2$. Specifically, $10x^3$ equals $5x^2$ times $2x$ and $15x^2$ equals $5x^2$ times 3.

We generally factor out the greatest common expression, meaning the largest power of x and the greatest factor of the coefficients that is common to each term. For example, for $18x^9$ and $24x^7$ the largest power of x that is common to both expressions is x^7 and the greatest factor that is common to the coefficients is 6. We can factor $18x^9 + 24x^7$ as $18x^9 + 24x^7 = 6x^7(3x^2 + 4)$.

For a single variable, factor out the greatest common expression as follows:
- Identify the greatest factor that is common to every coefficient. For example, the greatest common factor of 21 and 28 is 7 since $21 = 3(7)$ and $28 = 4(7)$.
- Identify the smallest power of the variable. For example, the smallest power of x for x^5 and x^3 is x^3.
- Combine the answers to the previous bullet points. For example, $21x^5 + 28x^3 = 7x^3(3x^2 + 4)$.
- Check your answer by distributing: $7x^3(3x^2 + 4) = 21x^5 + 28x^3$.

Example 1. $15x^3 + 25x^2 = 5x^2(3x + 5)$

Example 2. $36x^5 - 18x^3 = 9x^3(4x^2 - 2)$

Example 3. $12x^7 - 24x^5 + 4x^4 = 4x^4(3x^3 - 6x + 1)$

Exercise Set 5.1

Directions: Factor out the greatest common expression.

1) $10x^2 + 14x =$

2) $6x^3 - 9x^2 =$

3) $80x^6 + 60x^4 =$

4) $18x^9 + 36x^8 =$

5) $5x^5 - 3x^3 =$

6) $32x^2 - 24 =$

7) $9x^4 + 15x^3 =$

8) $16x^7 - 20x^4 =$

9) $32x^8 + 12x^3 =$

10) $90x^{15} - 108x^9 =$

11) $30x^3 - 60x^2 + 45 =$

12) $72x^7 - 96x^6 + 144x^5 =$

5.2 Factoring Minus Signs

It is possible to factor out a negative coefficient. This is convenient when every term is negative, or even when the first term is negative. Remember that if two negative numbers multiply one another, the product is positive (Sec. 1.10).

Example 1. $-2x^3 - 6x^2 = -2x^2(x + 3)$ since $-2x^2(x + 3) = -2x^2(x) - 2x^2(3)$

Example 2. $-10x^4 + 25x^2 = -5x^2(2x^2 - 5)$ since $-5x^2(2x^2 - 5) = -5x^2(2x^2) - 5x^2(-5)$

Exercise Set 5.2

Directions: Factor out the greatest common expression along with a minus sign.

1) $-8x^2 - 12x =$

2) $-15x^5 - 9x^3 =$

3) $-2x^4 + 1 =$

4) $-36x^8 + 72x^4 =$

5) $-12x^3 - 16x^2 + 8x =$

6) $-24x^2 + 18x - 21 =$

7) $-28x^{15} - 42x^{12} - 35x^9 =$

8) $-54x^{5/4} + 30x^{3/4} + 12x^{1/4} =$

5.3 Factoring a Difference of Squares

Recall the difference of squares formula from Sec. 4.5:
$$(u + v)(u - v) = u^2 - v^2$$
When two squared expressions are subtracted, this formula may be applied in reverse to factor the expression.
$$u^2 - v^2 = (u + v)(u - v)$$
To apply this formula, first identify u and v. For example, in $9x^2 - 4$, we can identify $u = \sqrt{9x^2} = \sqrt{9}\sqrt{x^2} = 3x$ and $v = \sqrt{4} = 2$, such that $9x^2 - 4 = (3x + 2)(3x - 2)$.

Example 1. $x^2 - 9 = x^2 - 3^2 = (x + 3)(x - 3)$

Example 2. $4x^2 - 49 = (2x)^2 - 7^2 = (2x + 7)(2x - 7)$

Exercise Set 5.3

Directions: Apply the difference of squares formula to factor each expression.

1) $x^2 - 25 =$

2) $16x^2 - 81 =$

3) $36x^2 - 1 =$

4) $225x^6 - 100 =$

5) $49x^8 - 144 =$

6) $-64x^2 + 121 =$

5.4 Factoring a Square of the Sum

Recall the square of the sum formula from Sec. 4.4:
$$(u + v)^2 = u^2 + 2uv + v^2$$
When an expression with three terms has this same form, this formula may be applied in reverse to factor the expression.
$$u^2 + 2uv + v^2 = (u + v)^2$$
To apply this formula, first identify u and v. For example, in $4x^2 + 12x + 9$, we can identify $u = \sqrt{4x^2} = \sqrt{4}\sqrt{x^2} = 2x$ and $v = \sqrt{9} = 3$, such that $4x^2 + 12x + 9 = (2x + 3)^2$. Note that $u^2 = (2x)^2 = 4x^2$, $v^2 = 3^2 = 9$, and $2uv = 2(2x)(3) = 12x$. All 3 terms match.

Example 1. $25x^2 + 20x + 4 = (5x)^2 + 2(5x)(2) + 2^2 = (5x + 2)^2$

Example 2. $9x^2 - 24x + 16 = (3x)^2 + 2(3x)(-4) + 4^2 = (3x - 4)^2$ Note the sign of the middle term. You can check the answer using the FOIL method (Sec. 4.3).

Exercise Set 5.4

Directions: Apply the of square of the sum formula to factor each expression.

1) $x^2 + 14x + 49 =$

2) $x^2 - 2x + 1 =$

3) $4x^2 + 20x + 25 =$

4) $16x^2 + 64x + 64 =$

5) $49x^2 - 126x + 81 =$

6) $100x^2 - 100x + 25 =$

7) $36x^2 - 144x + 144 =$

8) $64x^2 + 96x + 36 =$

9) $9x^2 + 54x + 81 =$

10) $49x^2 - 112x + 64 =$

11) $169x^2 - 442x + 289 =$

12) $225x^2 + 750x + 625 =$

13) $\frac{x^2}{4} + \frac{3x}{4} + \frac{9}{16} =$

14) $x^4 + 6x^2 + 9 =$

15) $49x^6 - 28x^3 + 4 =$

16) $4x^{18} + 20x^9 + 25 =$

17) $x^{4/3} + 6x^{2/3} + 9 =$

5.5 Factoring a Quadratic

The term **quadratic** refers to an expression or equation with three terms: one with a squared variable (x^2), one with a variable with no power shown (x), and a constant term. Chapter 6 is devoted to the quadratic equation, including the quadratic formula. Since the current chapter is focused on factoring, in this section we will focus only on how to factor a quadratic expression (for the simplest cases where the coefficients and constants are integers). We will learn more general techniques for dealing with quadratic equations in Chapter 6.

Beware that factoring the quadratic isn't as straightforward or as clear-cut as most students would like. Factoring the quadratic is a technique. There are guidelines for how to go about it, but it takes a little trial and error. The better your logic and reasoning skills, the less trial and error it will take. The most important point to keep in mind is this: Most students who are persistent in trying to learn how to factor the quadratic and who keep practicing the technique eventually get the hang of it. Aim for this. When you finally reach this stage, you will be fluent in this method and the solutions will start to come easily. In the meantime, if you find it a little frustrating or it seems like there is too much guessing, keep working at it, knowing that most students who keep working at it eventually get the hang of it. You can do it!

A quadratic expression has the form $ax^2 + bx + c$ where a, b, and c are constants. This is a quadratic "expression" because it doesn't have an equal sign (Sec. 1.2). Don't call it a quadratic "equation." (In Chapter 6, we'll explore quadratic "equations," which *do* have equal signs.)

When a quadratic expression is factored, it is rewritten in the form $(dx + e)(fx + g)$, where $(dx + e)(fx + g)$ is equal to $ax^2 + bx + c$. The idea is to find the new constants d, e, f, and g given the original constants a, b, and c. Here is an example with numbers: $3x^2 - 2x - 8$ can be factored as $(3x + 4)(x - 2)$ because if we apply the FOIL method (Sec. 4.3) to $(3x + 4)(x - 2)$ we get $(3x + 4)(x - 2) = 3x^2 - 6x + 4x - 8 = 3x^2 - 2x - 8$.

We wish to begin with an expression of the form $3x^2 - 2x - 8$ and rewrite it in the form $(3x + 4)(x - 2)$. If the new form equals the original form, we will say that the quadratic expression has been factored. The question is how to determine the new form. That is the focus of this section. As you read the examples and discussion in this section, try your best to understand the reasoning. It's okay (and normal) if it doesn't seem 100% clear the first time you read through it. Try to understand it as well as you can the first time through. If needed, you can read it again another time.

Consider the quadratic expression $x^2 + 5x + 6$. This expression is a little simpler, so it is a good place to start. One thing that makes this simpler is that the coefficient of x^2 is one (because $1x^2 = x^2$). This is simpler for problems (like those of this section) where the coefficients and constants are integers. When we FOIL an expression of the form $(dx + e)(fx + g)$, we get $dfx^2 + dgx + efx + eg$. If d, e, f, and g are integers, the only way that the coefficient of x^2 can be 1 is if $d = f = 1$. If we set $d = f = 1$, we get an expression of the form $(x + e)(x + g)$. For this example, we only need to find two constants (e and g) instead of four. If we FOIL $(x + e)(x + g)$, we get $x^2 + gx + ex + eg$. By comparing $x^2 + 5x + 6$ with $x^2 + gx + ex + eg$, we see that $eg = 6$ and $g + e = 5$ in this example. Which integers can you multiply together to make 6? You should think of two possibilities: 6 times 1 equals 6, and 3 times 2 also equals 6. This means that either $e = 6$ and $g = 1$, or $e = 3$ and $g = 2$. For which of these two cases will $g + e = 5$? For $e = 6$ and $g = 1$ we get $6 + 1 = 7$, while for $e = 3$ and $g = 2$ we get $3 + 2 = 5$. The correct answer is $e = 3$ and $g = 2$ because this makes $g + e = 5$. This means that $x^2 + 5x + 6$ factors as $(x + 3)(x + 2)$. One nice thing about factoring a quadratic expression is that it's easy to check the answer. Apply the FOIL method (Sec. 4.3). If the "FOILed" expression equals the quadratic expression, it is correct.
$$(x + 3)(x + 2) = x^3 + 2x + 3x + 6 = x^2 + 5x + 6$$
Since $(x + 3)(x + 2)$ equals $x^2 + 5x + 6$, we have correctly factored this expression.

That was a long paragraph. We were trying to do two things at once: Use an example to help illustrate what it means to factor the quadratic, and thoroughly describe the reasoning involved in factoring it. One main idea behind the technique is that we are basically trying to apply the FOIL method backwards.

Now let's look at an example that isn't quite as simple: $8x^2 + 22x + 15$. What makes this less simple is the coefficient 8 before x^2. Imagine applying the FOIL method in reverse. When we FOIL $(dx + e)(fx + g)$, we get $dfx^2 + dgx + efx + eg$, which we wish to equal $8x^2 + 22x + 15$. When we do that, d and f are factors of 8 (since $df = 8$) and e and g are factors of 15 (since $eg = 15$). Therefore, the way to go about this is to think about the possible factors of 8 and 15:

- 8 can be factored as 8 times 1 or as 4 times 2.
- 15 can be factored as 15 times 1 or as 5 times 3.

Only one combination of these factors will make the coefficient of x in the middle term equal 22. The question is, which one? Note that there are many possibilities, such as $(8x + 5)(x + 3)$, $(8x + 3)(x + 5)$, $(8x + 15)(x + 1)$, $(8x + 1)(x + 15)$, $(4x + 5)(2x + 3)$, $(4x + 3)(2x + 5)$, $(8x + 15)(x + 1)$, $(8x + 1)(x + 15)$, and many others. In each case, the first terms multiply together to make $8x^2$ since $(8x)(x) = 8x^2$ and $(4x)(2x) = 8x^2$ and the last terms multiply together to make 15 since $(5)(3) = 15$ and $(15)(1) = 15$. However, only one of these combinations (which are not all even listed above) will have the middle term equal $22x$. Since $dgx + efx = 22x$ in this example, we need $dg + ef$ to add up to 22. We already know that either d and f equal 8 and 1 or they equal 4 and 2 (in any order) since $df = 8$, and that either e and g equal 15 and 1 or they equal 5 and 3 (in any order) since $eg = 15$.

- d and f are either 8 and 1 or 4 and 2 (in any order).
- e and g are either 15 and 1 or 5 and 3 (in any order).

In order for $dg + ef = 22$, we need one pair of numbers from the first bullet point and one pair of numbers from the second bullet point where the two products dg and ef add up to 22. Note that dg involves one number from each bullet point, as does ef. Here are the possibilities:

- d and f could be 8 and 1 (in any order) and e and g could be 15 and 1 (in any order). Here we can make $dg + ef = (8)(1) + (15)(1) = 8 + 15 = 23$, which is close, but 23 doesn't equal 22 so it's not close enough. The other option, $(8)(15) + (1)(1) = 120 + 1 = 121$, is way off.
- d and f could be 8 and 1 (in any order) and e and g could be 5 and 3 (in any order). The best we can do here is $(8)(3) + (5)(1) = 24 + 5 = 29$, which isn't 22. The alternative, $(8)(5) + (3)(1) = 40 + 3 = 43$ is even worse.

- d and f could be 4 and 2 (in any order) and e and g could be 15 and 1 (in any order). Nothing comes close. We either get $(4)(15) + (1)(2) = 60 + 2 = 62$ or $(4)(1) + (15)(2) = 4 + 30 = 34$.
- d and f could be 4 and 2 (in any order) and e and g could be 5 and 3 (in any order). Finally, we find the solution $(4)(3) + (5)(2) = 12 + 10 = 22$ that works. We need $d = 4, e = 5, f = 2$, and $g = 3$. That way, we get $dg + ef = (4)(3) + (5)(2) = 12 + 10 = 22$.

If we use $d = 4, e = 5, f = 2$, and $g = 3$, which we finally discovered in the last bullet point, we find that $(4x + 5)(2x + 3)$ factors $8x^2 + 22x + 15$. Again, it's really easy to *check* the answer; just apply the FOIL method:
$$(4x + 5)(2x + 3) = 8x^2 + 12x + 10x + 15 = 8x^2 + 22x + 15$$

We spent a page and a half discussing the previous solution. Most of that went into trying to help you understand the problem, the solution, and all of the reasoning that went into it. Also, we used four bullet points to work out all of the combinations. Now we need to learn how to solve such problems more efficiently, so they can be solved in a couple of lines rather than over an entire page.

Look at the previous four bullet points. We didn't really need to consider all of those combinations. This is where reasoning skills and logic can help you solve the problem quicker. We were looking at the pairs 8 and 1 or 4 and 2 and the pairs 15 and 1 or 5 and 3. The two numbers from one of the first pairs multiply the two numbers from one of the last pairs, and we wanted these products to add up to 22. Here is how some reasoning could have helped us narrow down the possibilities:

- We could exclude 8 and 1 pretty quickly. The only number 8 can multiply and not exceed 22 is 1, and when we subtract $(8)(1) = 8$ from 22, we get $22 - 8 = 14$. We can't make 14 from 1 and 15.
- We could also exclude 15 and 1 pretty quickly. Similarly, in this example 15 could only multiply the 1, and then $22 - 15 = 7$ can't be made from 1 and 8.
- This leaves the pair 4 and 2 and the pair 5 and 3. We can make 22 from these pairs: $4(3) + 2(5) = 12 + 10 = 22$.

With reasoning and practice, you can learn to do this efficiently.

Following is one strategy that may be used to factor a quadratic expression:

- List the possible factors of the coefficient of x^2 and the possible factors of the constant term. For example, for $6x^2 + 25x + 21$, the possible factors of 6 are 6 and 1 or 3 and 2 and the possible factors of 21 are 21 and 1 or 7 and 3.
- Organize these possibilities like the example below. This is to help you visualize applying the FOIL method to different possible combinations.

$$(\quad x + \quad)(\quad x + \quad)$$

6	21	1	1
3	7	2	3
2	3	3	7
1	1	6	21

- It's important to visualize this table properly. Columns 1 and 3 go together, and columns 2 and 4 go together. For example, the 6 on the top left has to go with the 1 at the top of the third column, as highlighted below. When thinking about this 6 and 1 pair, this pair can go with any of the four pairs from columns 2 and 4, such as the 3 and 7 highlighted below.

$$(\quad x + \quad)(\quad x + \quad)$$

6	21	1	1
3	7	2	3
2	3	3	7
1	1	6	21

- The table above has 16 possible combinations. Any of the 4 pairs from columns 1 and 3 (which are 6-1, 3-2, 2-3, and 1-6) can go along with any of the 4 pairs from columns 2 and 4 (which are 21-1, 7-3, 3-7, and 1-21).
- Our goal is to find two pairs (one from columns 1 and 3, the other from columns 2 and 4), so that the product of the numbers from columns 1 and 4 plus the product of the numbers from columns 2 and 3 equals the coefficient of $25x$.
- For example, for the two pairs highlighted above, when we multiply the 6 from column 1 with the 7 from column 4 and multiply the 3 from column 2 with the 1 from column 3, we get $6(7) + 3(1) = 42 + 3 = 45$, which isn't 25.
- We can rule out the 21 and 1 from columns 2 and 4 from being one of the pairs because the 21 is clearly too large for the sum to come out to 25 in this case (it would take a pair of 4 and 1 to go with the 21 and 1 in order to get the two products to add up to 25).

- The two pairs highlighted below will work. One pair is the 6 and 1 from columns 1 and 3. The second pair is the 7 and 3 from columns 2 and 4. Multiply column 1 times column 4 and multiply column 2 times column 3. We get $(6)(3) + (7)(1) = 18 + 7 = 25$.

$$(\; x + \quad)(\; x + \quad)$$

6	21	1	1
3	7	2	3
2	3	3	7
1	1	6	21

- Therefore, the answer is $(6x + 7)(1x + 3)$, but since $1x = x$, we will write this as $(6x + 7)(x + 3)$.
- Check the answer by using the FOIL method:

$$(6x + 7)(x + 3) = 6x^2 + 18x + 7x + 21 = 6x^2 + 25x + 21$$

- Since $(6x + 7)(x + 3)$ equals $6x^2 + 25x + 21$, this factoring is correct.

Example 1. Factor $x^2 + 8x + 12$

The coefficient of x^2 factors as $(1)(1)$. The 12 factors as $(12)(1)$, $(6)(2)$, or $(4)(3)$. If the coefficient of x^2 is 1, we don't need to list 1 and 12 in addition to 12 and 1, we don't need to list 2 and 6 in addition to 6 and 2, and similarly for 3 and 4 and 4 and 3.

- Make a table showing the factors.
- Columns 1 and 3 have the factors of 1 (from $1x^2 = x^2$).
- Columns 2 and 4 have the possible factors of 12.

$$(\; x + \quad)(\; x + \quad)$$

1	12	1	1
	6		2
	4		3

- Our goal is to find a pair from columns 1 and 3 (there is only 1 pair here) and a pair from columns 2 and 4, so that column 1 times column 4 plus column 2 times column 3 equals the coefficient of $8x$.
- The 6 and 2 from columns 2 and 4 work because $(1)(2) + (6)(1) = 2 + 6 = 8$.
- $(1)(2)$ is the "outside" and $(6)(1)$ is the "inside" of the FOIL method.

The solution is therefore $(x + 6)(x + 2)$. Recall that $1x = 1$.

Check: $(x + 6)(x + 2) = x^2 + 2x + 6x + 12 = x^2 + 8x + 12$

Example 2. Factor $3x^2 + 17x + 10$ Since the coefficient of x^2 isn't 1, here order does matter. We need to consider both 1 and 10 and 10 and 1, both 2 and 5 and 5 and 2, etc.

- The factors of 3 are 3 and 1.
- The factors of 10 are 10 and 1 or 5 and 2.

$$(\;\; x + \;\;\;\;)(\;\; x + \;\;\;\;)$$

3	10	1	1
1	5	3	2
	2		5
	1		10

Since $(3)(5) + (2)(1) = 15 + 2 = 17$, the solution is $(3x + 2)(x + 5)$.

Check: $(3x + 2)(x + 5) = 3x^2 + 15x + 2x + 10 = 3x^2 + 17x + 10$

Example 3. Factor $10x^2 + 42x + 8$

- The factors of 10 are 10 and 1 or 5 and 2.
- The factors of 8 are 8 and 1 or 4 and 2.

$$(\;\; x + \;\;\;\;)(\;\; x + \;\;\;\;)$$

10	8	1	1
5	4	2	2
2	2	5	4
1	1	10	8

Since $(2)(1) + (8)(5) = 2 + 40 = 42$, the solution is $(2x + 8)(5x + 1)$.

Check: $(2x + 8)(5x + 1) = 10x^2 + 2x + 40x + 8 = 10x^2 + 42x + 8$

Example 4. Factor $14x^2 + 27x + 9$

- The factors of 14 are 14 and 1 or 7 and 2.
- The factors of 9 are 9 and 1 or 3 and 3.

$$(\;\; x + \;\;\;\;)(\;\; x + \;\;\;\;)$$

14	9	1	1
7	3	2	3
2	1	7	9
1		14	

Since $(7)(3) + (3)(2) = 21 + 6 = 27$, the solution is $(7x + 3)(2x + 3)$.

Check: $(7x + 3)(2x + 3) = 14x^2 + 21x + 6x + 9 = 14x^2 + 27x + 9$

Exercise Set 5.5

Directions: Factor each quadratic expression like the examples.

1) $x^2 + 9x + 20 =$

2) $x^2 + 10x + 24 =$

3) $5x^2 + 37x + 14 =$

4) $6x^2 + 31x + 35 =$

5) $10x^2 + 53x + 63 =$

6) $32x^2 + 12x + 1 =$

7) $36x^2 + 65x + 25 =$

8) $27x^2 + 195x + 100 =$

9) $7x^2 + 22x + 16 =$

10) $16x^2 + 40x + 9 =$

11) $20x^2 + 82x + 42 =$

12) $24x^2 + 22x + 5 =$

13) $36x^2 + 71x + 33 =$

14) $9x^2 + 20x + 4 =$

15) $24x^2 + 73x + 24 =$

16) $15x^2 + 58x + 32 =$

5.6 Factoring a Quadratic with Negative Signs

If a quadratic expression has a positive constant term, a negative x term, and a positive x^2 term, like $x^2 - 7x + 12$, then each expression will have a minus sign if it is factored. For example, $x^2 - 7x + 12$ factors as $(x - 3)(x - 4)$ because $(x - 3)(x - 4) = x^2 - 4x - 3x + 12 = x^2 - 7x + 12$.

If a quadratic expression has a negative constant term and positive x^2 term (regardless of the sign of the x term), like $x^2 - 4x - 5$ or $x^2 + 4x - 5$, then the two expressions will have opposite signs when it is factored. For example:

- $x^2 - 4x - 5$ factors as $(x - 5)(x + 1)$ because $(x - 5)(x + 1) = x^2 + x - 5x - 5$
  $= x^2 - 4x - 5$.
- $x^2 + 4x - 5$ factors as $(x + 5)(x - 1)$ because $(x + 5)(x - 1) = x^2 - x + 5x - 5$
  $= x^2 + 4x - 5$.

Observe that the x term is negative in one example above and positive in the other, whereas the constant term is negative in both examples. The expression $(x - 5)(x + 1)$ expands to $x^2 - 4x - 5$ because $-5x$ is more negative than x is positive (in $x - 5x$), whereas the expression $(x + 5)(x - 1)$ expands to $x^2 + 4x - 5$ because $5x$ is more positive than $-x$ is negative (in $-x + 5x$). When factoring a quadratic expression that includes negative signs, it is necessary to reason the signs of the factored expressions.

Example 1. Factor $x^2 - 5x + 6$

- If only the x term is negative, both expressions are negative when it is factored.
- If the coefficient of x^2 is 1, we don't need to consider 1 and 6 in addition to 6 and 1 or 2 and 3 in addition to 3 and 2.

$$(\; x + \quad)(\; x + \quad)$$

1	−6	1	−1
	−3		−2

Since $(1)(-2) + (-3)(1) = -2 - 3 = -5$, the solution is $(x - 3)(x - 2)$.

Check: $(x - 3)(x - 2) = x^2 - 2x - 3x + 6 = x^2 - 5x + 6$

It may help to review the rules for multiplying with negative numbers (Sec. 1.10).

Example 2. Factor $2x^2 + 5x - 3$

- If the constant term is negative (and the x^2 term is positive), one expression is positive and one expression will be negative when it is factored.
- The factors of 2 are 2 and 1.
- The factors of 3 are 3 and 1. Since the constant term is negative, we need to consider -3 and 1 as well as 3 and -1 (plus the opposite order).

$$(\quad x + \quad)(\quad x + \quad)$$

2	3	1	−1
1	1	2	−3
	−3		1
	−1		3

Since $(2)(3) + (-1)(1) = 6 - 1 = 5$, the solution is $(2x - 1)(x + 3)$.

Check: $(2x - 1)(x + 3) = 2x^2 + 6x - x - 3 = 2x^2 + 5x - 3$

Example 3. Factor $2x^2 - 5x - 3$

- If the constant term is negative (and the x^2 term is positive), one expression is positive and one expression will be negative when it is factored.
- The factors of 2 are 2 and 1.
- The factors of 3 are 3 and 1. Since the constant term is negative, we need to consider -3 and 1 as well as 3 and -1 (plus the opposite order).

$$(\quad x + \quad)(\quad x + \quad)$$

2	3	1	−1
1	1	2	−3
	−3		1
	−1		3

Since $(2)(-3) + (1)(1) = -6 + 1 = -5$, the solution is $(2x + 1)(x - 3)$.

Check: $(2x + 1)(x - 3) = 2x^2 - 6x + x - 3 = 2x^2 - 5x - 3$

It is instructive to compare Examples 2 and 3, which only differ in the minus signs.

Exercise Set 5.6

Directions: Factor each quadratic expression like the examples.

1) $x^2 - 8x + 7 =$

2) $x^2 + 10x - 11 =$

3) $3x^2 - 2x - 5 =$

4) $9x^2 - 21x + 10 =$

5) $6x^2 + 11x - 35 =$

6) $10x^2 - 39x - 27 =$

7) $32x^2 - 44x + 9 =$

8) $24x^2 + 58x - 5 =$

Want more practice? Solve the problems of Sec. 6.5 on a sheet of scratch paper.

5.7 Completing the Square

Consider the expression $x^2 + 6x + 5$. It is almost a square of the sum, but not quite. Compare the previous expression with $(x+3)^2 = (x+3)(x+3) = x^2 + 3x + 3x + 9 = x^2 + 6x + 9$. If $x^2 + 6x + 5$ had been $x^2 + 6x + 9$ instead of $x^2 + 6x + 5$, we would have been able to factor it as $(x+3)(x+3)$, but since there is a 5 instead of a 9 in $x^2 + 6x + 5$, we can't quite do it.

Actually, there *is* a way. We can rewrite 5 as $9 - 4$ because $5 = 9 - 4$:
$$x^2 + 6x + 5 = x^2 + 6x + 9 - 4 = (x+3)^2 - 4$$
This technique is called **completing the square** and has applications in various branches of mathematics, science, and engineering. The strategy is:

- Determine which expression squared could make the coefficients of x^2 and x. For example, for $4x^2 - 12x + 17$, the coefficients of x^2 and x are 4 and -12. We can make these coefficients from $(2x-3)^2 = (2x-3)(2x-3) = 4x^2 - 6x - 6x + 9 = 4x^2 - 12x + 9$. The only problem is that the constant equals 9 instead of 17.

- **Handy formula**: There is a handy formula for the previous step. If you identify the constants a, b, and c as the coefficients of the quadratic expression (with the x^2 term first, the x term second, and the constant term last), the expression will be $\left(\sqrt{a}\, x + \frac{b}{2\sqrt{a}}\right)^2$. For example, for $4x^2 - 12x + 17$, we get $a = 4, b = -12$, and $c = 17$, for which $\left(\sqrt{a}\, x + \frac{b}{2\sqrt{a}}\right)^2 = \left(\sqrt{4}\, x + \frac{-12}{2\sqrt{4}}\right)^2 = \left(2x - \frac{6}{\sqrt{4}}\right)^2 = \left(2x - \frac{6}{2}\right)^2 = (2x-3)^2$.

- Rewrite the constant term using addition or subtraction to make the constant term needed for the squared expression from the previous step. For example, when we expand $(2x-3)^2$, we get $(2x-3)^2 = 4x^2 - 12x + 9$. We can rewrite the 17 of $4x^2 - 12x + 17$ in the form $17 = 9 + 8$. This gives:
$$x^2 - 12x + 17 = x^2 - 12 + 9 + 8 = (2x-3)^2 + 8$$

Check the answer by using the FOIL method (or the square of the sum formula):
$$(2x-3)^2 + 8 = (2x-3)(2x-3) + 8 = 4x^2 - 12x + 9 + 8 = 4x^2 - 12x + 17$$

For curious students: Where does the formula $\left(\sqrt{a}\, x + \frac{b}{2\sqrt{a}}\right)^2$ come from? It comes from the square of the sum formula (Sec. 5.4): $(u + v)^2 = u^2 + 2uv + v^2$. Compare this with $ax^2 + bx + c$. For the coefficients of x^2 and x to match, we need $u^2 = ax^2$ and $2uv = bx$. Square root both sides of $u^2 = ax^2$ to get $u = \sqrt{a}\, x$ and divide both sides of $2uv = bx$ by $2u$ to get $v = \frac{bx}{2u} = \frac{bx}{2\sqrt{a}\, x} = \frac{b}{2\sqrt{a}}$. Plug $u = \sqrt{a}\, x$ and $v = \frac{b}{2\sqrt{a}}$ into $(u + v)^2$ to get the formula $\left(\sqrt{a}\, x + \frac{b}{2\sqrt{a}}\right)^2$.

Example 1. Complete the square for $x^2 + 8x + 4$.

$$a = 1, b = 8, \left(\sqrt{a}\, x + \frac{b}{2\sqrt{a}}\right)^2 = \left(\sqrt{1}\, x + \frac{8}{2\sqrt{1}}\right)^2 = (x + 4)^2 = x^2 + 8x + 16$$

Now rewrite 4 as $4 = 16 - 12$:

$$x^2 + 8x + 4 = x^2 + 8x + 16 - 12 = (x + 4)^2 - 12$$

Check: $(x + 4)^2 - 12 = (x + 4)(x + 4) - 12 = x^2 + 8x + 16 - 12 = x^2 + 8x + 4$

Example 2. Complete the square for $25x^2 - 30x + 20$.

$$a = 25, b = -30, \left(\sqrt{a}\, x + \frac{b}{2\sqrt{a}}\right)^2 = \left(\sqrt{25}\, x + \frac{-30}{2\sqrt{25}}\right)^2$$

$$= \left(5x - \frac{15}{5}\right)^2 = (5x - 3)^2 = 25x^2 - 30x + 9$$

Now rewrite 20 as $20 = 9 + 11$:

$$25x^2 - 30x + 20 = 25x^2 - 30x + 9 + 11 = (5x - 3)^2 + 11$$

Check: $(5x - 3)^2 + 11 = (5x - 3)(5x - 3) + 11 = 25x^2 - 30x + 9 + 11 = 25x^2 - 30x + 20$

Example 3. Complete the square for $4x^2 + 28x - 11$.

$$a = 4, b = 28, \left(\sqrt{a}\, x + \frac{b}{2\sqrt{a}}\right)^2 = \left(\sqrt{4}\, x + \frac{28}{2\sqrt{4}}\right)^2$$

$$= \left(2x + \frac{14}{2}\right)^2 = (2x + 7)^2 = 4x^2 + 28x + 49$$

Now rewrite -11 as $-11 = 49 - 60$:

$$4x^2 + 28x - 11 = 4x^2 + 28x + 49 - 60 = (2x + 7)^2 - 60$$

Check: $(2x + 7)^2 - 60 = (2x + 7)(2x + 7) - 60 = 4x^2 + 28x + 49 - 60 = 4x^2 + 28x - 11$

Exercise Set 5.7

Directions: Complete the square for each expression like the examples.

1) $x^2 + 18x + 4 =$

2) $x^2 - 12x + 60 =$

3) $36x^2 + 36x - 36 =$

4) $64x^2 - 80x + 16 =$

5) $81x^2 + 108x + 54 =$

6) $100x^2 - 20x - 1 =$

7) $49x^2 - 126x + 49 =$

8) $144x^2 + 72x =$

5.8 Factoring with Fractions

When two or more terms are added or subtracted in the numerator or denominator of a fraction, the factoring technique that we learned in Sec.'s 5.1-5.2 may be used. For example, $\frac{4x^2+6x}{3x+9} = \frac{2x(2x+3)}{3(x+3)} = \frac{2x}{3}\frac{2x+3}{x+3}$.

Example 1. $\frac{12x^2+9x}{5} = \frac{3x(4x+3)}{5}$ Alternate answers: $\frac{3x}{5}(4x+3)$ or $3x\frac{4x+3}{5}$

Example 2. $\frac{8}{5x^5-20x^3} = \frac{8}{5x^3(x^2-4)}$ Alternate answers: $\frac{8}{5x^3}\frac{1}{x^2-4}$ or $\frac{1}{5x^3}\frac{8}{x^2-4}$

Example 3. $\frac{6x^2-10x}{9x+6} = \frac{2x(3x-5)}{3(x+2)}$ Alternate answer: $\frac{2x}{3}\frac{3x-5}{x+2}$

Exercise Set 5.8

Directions: Factor the greatest common expression where terms are added or subtracted.

1)

$$\frac{14x^5 - 42x^2}{3} =$$

2)

$$\frac{5x}{24x + 30} =$$

3)

$$\frac{36x^8 + 63x^5}{32x^7 - 40x^3} =$$

4)

$$\frac{48x^5 - 60x^4 + 96x^3}{36x^2} =$$

5.9 Factoring Expressions out of Roots and Powers

If two or more terms are added or subtracted and this expression is raised to a power, you may factor a common expression out of the power by raising it to the same power, like the example below.

$$(8x^3 + 20x^2)^3 = [(4x^2)(2x + 5)]^3 = (4x^2)^3(2x + 5)^3 = 64x^6(2x + 5)^3$$

First, we factored $4x^2$ out of $8x^3 + 20x^2$ to write $8x^3 + 20x^2 = 4x^2(2x + 5)$. Then we applied the rule $(x^m)^n = x^{mn}$ from Sec. 3.5 to write $[(4x^2)(2x + 5)]^3 = (4x^2)^3(2x + 5)^3$. Finally, we applied the rule $(x^m y^n)^p = x^{mp} y^{np}$ from Sec. 3.8 to write $(4x^2)^3 = 4^3(x^2)^3 = 64x^6$.

Example 1. $(30x^2 + 25x)^2 = [(5x)(6x + 5)]^2 = (5x)^2(6x + 5)^2 = 25x^2(6x + 5)^2$

Example 2. $(8x^7 - 16x^3)^{1/3} = [(8x^3)(x^4 - 2)]^{1/3} = (8x^3)^{1/3}(x^4 - 2)^{1/3} = 2x(x^4 - 2)^{1/3}$
Note: $(8x^3)^{1/3} = 2x$ because $(2x)^3 = 2^3 x^3 = 8x^3$

Example 3. $\sqrt{12x + 20} = \sqrt{4(3x + 5)} = \sqrt{4}\sqrt{3x + 5} = 2\sqrt{3x + 5}$

Exercise Set 5.9

Directions: Factor out the greatest common expression like the examples.

1) $(6x + 9)^2 =$

2) $(3x^4 - 4x^2)^3 =$

3) $(16x^3 + 20x)^5 =$

4) $(12x^4 - 21x^3)^6 =$

5) $(2x^9 + 6x^5)^4 =$

6) $(x^{12} + x^9 - x^7)^8 =$

7) $(24x^3 - 60x - 48)^2 =$

8) $\sqrt{18x^2 - 27} =$

9) $\sqrt{25x^4 + 75x^3 + 50x^2} =$

10) $(24x^8 + 40x^6)^{2/3} =$

11) $(9x^5 - 12x^3)^{-1} =$

12) $(14x^7 + 10x^6)^{-3} =$

5.10 Factoring Perfect Squares out of Square Roots

It is customary to factor perfect squares out of square roots when the number inside of the radical is a multiple of a perfect square. A number that is the square of a whole number is considered a **perfect square**. Numbers that are perfect squares include 1, 4, 9, 16, 25, 36, 49, 64, 81, 100, 121, 144, etc. For example, 49 is a perfect square since 49 is equal to 7^2 (since $7^2 = 7 \cdot 7 = 49$).

As an example, consider the irrational number $\sqrt{12}$. Since 12 is a multiple of 4 (since $12 = 3 \cdot 4$) and since 4 is a perfect square (since $2^2 = 2 \cdot 2 = 4$), we can factor the 4 out of the radical as follows:

$$\sqrt{12} = \sqrt{4(3)} = \sqrt{4}\sqrt{3} = 2\sqrt{3}$$

Note that $\sqrt{4} = 2$ because $2^2 = 4$. We applied the rule that $\sqrt{ab} = \sqrt{a}\sqrt{b}$. To see that $\sqrt{ab} = \sqrt{a}\sqrt{b}$, recall the rule $(x^m y^n)^p = x^{mp} y^{np}$ from Sec. 3.8. Set $m = n = 1$ to get $(x^1 y^1)^p = x^{1p} y^{1p}$, which simplifies to $(xy)^p = x^p y^p$. Now set $p = 1/2$ to get $(xy)^{1/2} = x^{1/2} y^{1/2}$. Then recall that $x^{1/2} = \sqrt{x}$ from Sec. 3.7 to get $\sqrt{xy} = \sqrt{x}\sqrt{y}$. Finally, if you replace x with a and y with b, $\sqrt{xy} = \sqrt{x}\sqrt{y}$ will become $\sqrt{ab} = \sqrt{a}\sqrt{b}$.

The main idea is to think of which numbers that are perfect squares can evenly multiply into the value under the radical. For example, since $63 = 9(7)$ and since 9 is a perfect square (since $3^2 = 9$), we can factor $\sqrt{63}$ as $\sqrt{63} = \sqrt{9(7)} = \sqrt{9}\sqrt{7} = 3\sqrt{7}$.

Example 1. $\sqrt{18} = \sqrt{9(2)} = \sqrt{9}\sqrt{2} = 3\sqrt{2}$

Example 2. $\sqrt{75} = \sqrt{25(3)} = \sqrt{25}\sqrt{3} = 5\sqrt{3}$

Example 3. $\sqrt{80} = \sqrt{16(5)} = \sqrt{16}\sqrt{5} = 4\sqrt{5}$

Example 4. $\sqrt{28} = \sqrt{4(7)} = \sqrt{4}\sqrt{7} = 2\sqrt{7}$

Exercise Set 5.10

Directions: Factor any perfect squares out of the square root.

1) $\sqrt{8} =$

2) $\sqrt{27} =$

3) $\sqrt{20} =$

4) $\sqrt{108} =$

5) $\sqrt{98} =$

6) $\sqrt{45} =$

7) $\sqrt{24} =$

8) $\sqrt{90} =$

9) $\sqrt{700} =$

10) $\sqrt{288} =$

11) $\sqrt{192} =$

12) $\sqrt{150} =$

5.11 Factoring in Equations

If the product of two numbers equals zero, at least one of the numbers must be zero. Algebraically, this means that if $xy = 0$, then it must either be the case that $x = 0$ or that $y = 0$. Either way, the product will be zero.

This idea is relevant to factoring. For example, consider the equation $3x^3 - 9x^2 = 0$. If we factor the left side, we get $3x^2(x - 3) = 0$. There are two possible solutions to this problem. One solution is that $3x^2 = 0$, which simplifies to $x = 0$, and the second solution is $x - 3 = 0$, which simplifies to $x = 3$. That is, the only way that $3x^2(x - 3)$ can equal zero is if $3x^2$ equals zero or if $x - 3$ equals zero. We can see that the two solutions $x = 0$ and $x = 3$ both solve the original equation by plugging each one into the original equation. When $x = 0$, we get $3x^3 - 9x^2 = 3(0)^3 - 9(0)^2 = 0 - 0 = 0$. When $x = 3$, we get $3(3)^3 - 9(3)^2 = 3(27) - 9(9) = 81 - 81 = 0$.

It may seem tempting to divide both sides of the equation $3x^3 - 9x^2 = 0$ by x^2, but there is a subtle problem with doing that. The problem is that if $x = 0$ (which we just saw is indeed a valid solution), dividing by x^2 would involve dividing by zero. Since division by zero is a problem (for example, $\frac{1}{0}$ is undefined and $\frac{0}{0}$ is indeterminate, as discussed in Sec. 3.3), we must be careful to avoid dividing by zero. There is another problem with dividing by x^2. If we divide both sides of $3x^3 - 9x^2 = 0$ by x^2, we get $3x - 9 = 0$, which simplifies to $x = 3$. Although $x = 3$ is a valid solution, it isn't the *only* solution. This technique loses the alternate solution, $x = 0$, which is also valid. The technique of factoring that we applied in the previous paragraph found both of the solutions.

Note that zero is special. If $xy = 0$, we know for sure that either $x = 0$ or $y = 0$. If the right hand side isn't zero, we can't make a similar argument. For example, consider the equation $uv = 4$. You might want to make a similar argument that $u = 1$ and that $v = 4$ because $1(4) = 4$, but this argument doesn't work because there are an infinite number of values of u and v for which the product is 4. For example, if $u = \frac{1}{2}$ and $v = 8$, we get $\frac{1}{2}(8) = \frac{8}{2} = 4$. For the equation $uv = 4$, there is no way to determine the values of u and v without additional information. In contrast, for the equation $xy = 0$, we know for sure that either x or y equals zero. Therefore, for the equation $x(x - 3) = 0$, we can reason that either $x = 0$ or that $x = 3$, but for the equation $x(x - 3) = 4$, we can't make the similar argument. In the second case, we can distribute on the left side to get $x^2 - 3x = 4$, subtract 4 from both sides to get $x^2 - 3x - 4 = 0$, and then apply the factoring technique from Sec. 5.6. This gives $(x - 4)(x + 1) = 0$. Now that we have zero on the right side (instead of 4), we may apply our argument to determine that either $x - 4 = 0$ (for which $x = 4$) or $x + 1 = 0$ (for which $x = -1$). Problems of this latter type will be explored in Chapter 6.

Example 1. $9x^2 - 36x = 0$

- Factor the left side: $3x(3x - 12) = 0$.
- There are two possible solutions: either $3x = 0$ or $3x - 12 = 0$.
- Add 12 to both sides of the second equation: either $3x = 0$ or $3x = 12$.
- Divide by 3 on both sides in each equation: $x = 0$ or $x = 4$.

Check the answers. Plug each value of x into the original equation:

- $9x^2 - 36x = 9(0)^2 - 36(0) = 0 - 0 = 0$
- $9x^2 - 36x = 9(4)^2 - 36(4) = 9(16) - 144 = 144 - 144 = 0$

Since both sides equal 0 in each case, the answers are correct.

Exercise Set 5.11

Directions: Solve for the variable by isolating the unknown. Check your answer.

1) $8x^2 + 32x = 0$

2) $12x^5 - 72x^4 = 0$

3) $x^9 - 7x^8 = 0$

4) $10x^6 = 20x^5$

6 QUADRATIC EQUATIONS

6.1 Vocabulary Related to the Quadratic

A **polynomial** refers to terms of the form ax^n that are added together (or subtracted), where the exponents are nonnegative integers. For example, $4x^3 + 6x^2 - 7x + 9$ is a polynomial because each term has the form ax^n with a nonnegative exponent. The last term, 9, is a constant term (corresponding to $n = 0$ because $x^0 = 1$). The other constants (the $4, 6$, and -7) are called **coefficients**. It is conventional to order the terms from the highest power to the lowest power.

The **degree** of the polynomial refers to the largest exponent. For example, $7x^5 + 3x^2$ is degree five.

A polynomial with degree one is **linear** and has the form $mx + b$. This is the equation for a straight line.

A polynomial with degree two is **quadratic**. For example, $5x^2 + 4x + 9$ is quadratic. It has three terms: a **quadratic** term ($5x^2$), a **linear** term ($4x$), and a **constant** term (9). When a quadratic appears in the form $ax^2 + bx + c$ (with three terms in this order), it is said to be in **standard form**.

The **discriminant** of a quadratic equals $b^2 - 4ac$. This expression helps to determine the nature of the answers to a quadratic equation, such as how many solutions there are and whether they are real or complex (Sec. 6.8).

6.2 The Simplest Quadratics

A quadratic equation is very easy to solve if it is missing the linear term (meaning that there isn't a term proportional to the first power of x) like $5x^2 = 80$ or if it is missing the constant term like $2x^2 = 6x$.

- If the quadratic equation is missing the linear term, isolate the quadratic term, divide both sides by its coefficient, and square root both sides. Include a $\pm$ sign for two possible answers. For example, for $5x^2 - 80 = 0$, add 80 to both sides to get $5x^2 = 80$, divide by 5 on both sides to get $x^2 = 16$, and square root both sides: $x = \pm 4$. The reason for the $\pm$ sign is that $x = -4$ and $x = 4$ both solve the original equation: $5(-4)^2 - 80 = 5(16) - 80 = 80 - 80 = 0$ in addition to $5(4)^2 = 5(16) - 80 = 80 - 80 = 0$.

- If the quadratic equation is missing the constant term, bring all of the terms to the same side of the equation and factor out the variable (like we did in Sec. 5.11). There will be two possible answers. For example, for $2x^2 = 6x$, subtract $6x$ from both sides to get $2x^2 - 6x = 0$ and factor out $2x$ to get $2x(x - 3) = 0$. As discussed in Sec. 5.11, either $2x = 0$ or $x - 3 = 0$. The solution to $2x = 0$ is $x = 0$ and the solution to $x - 3 = 0$ is $x = 3$. The two answers to $2x^2 = 6x$ are $x = 0$ and $x = 3$. Check: $2x^2 = 2(0)^2 = 0$ agrees with $6x = 6(0) = 0$ and $2x^2 = 2(3)^2 = 2(9) = 18$ agrees with $6x = 6(3) = 18$.

Example 1. $2x^2 - 18 = 0$

- Add 18 to both sides: $2x^2 = 18$.
- Divide by 2 on both sides: $x^2 = 9$.
- Square root both sides: $x = \pm 3$.

Check the answers. Plug each value of x into the original equation: $2x^2 - 18 = 2(\pm 3)^2 - 18 = 2(9) - 18 = 18 - 18 = 0$. Since both sides equal 0, the answers are correct.

Example 2. $5x^2 - 15x = 0$

- Factor the left side: $5x(x - 3) = 0$.
- There are two possible solutions: either $5x = 0$ or $x - 3 = 0$.
- Divide by 5 in the first equation. Add 3 in the second equation: $x = 0$ or $x = 3$.

Check the answers. Plug each value of x into the original equation:

- $5x^2 - 15x = 5(0)^2 - 15(0) = 0 - 0 = 0$
- $5x^2 - 15x = 5(3)^2 - 15(3) = 5(9) - 45 = 45 - 45 = 0$

Since both sides equal 0 in each case, the answers are correct.

Exercise Set 6.2

Directions: Solve for the variable in each equation.

1) $3x^2 = 48$

2) $2x^2 - 3x = 0$

3) $3x^2 = 4x$

4) $9x^2 - 225 = 0$

5) $x^2 + 36 = 100$

6) $x^2 + 5x = 0$

7) $\frac{x^2}{2} = 3x$

8) $2x^2 - 48 = 50$

6.3 The Quadratic Formula

If a quadratic equation is written in the **standard form** of $ax^2 + bx + c = 0$, it can be solved by applying the following equation, which is called the **quadratic formula**:

$$x = \frac{-b \pm \sqrt{b^2 - 4ac}}{2a}$$

The $\pm$ sign represents that there are generally two solutions to a quadratic equation: $x = \frac{-b+\sqrt{b^2-4ac}}{2a}$ and $x = \frac{-b-\sqrt{b^2-4ac}}{2a}$. Given a quadratic equation with numbers, such as $2x^2 - 3x - 2 = 0$, by comparison with the general equation $ax^2 + bx + c = 0$, we can identify the constants. For example, for $2x^2 - 3x - 2 = 0$, the constants are $a = 2$, $b = -3$, and $c = -2$. We can then plug these numbers into the quadratic formula to determine the solutions:

$$x = \frac{-b \pm \sqrt{b^2 - 4ac}}{2a} = \frac{-(-3) \pm \sqrt{(-3)^2 - 4(2)(-2)}}{2(2)}$$

$$x = \frac{3 \pm \sqrt{9 + 16}}{4} = \frac{3 \pm \sqrt{25}}{4} = \frac{3 \pm 5}{4}$$

$$x = \frac{3 + 5}{4} = \frac{8}{4} = 2 \quad \text{or} \quad x = \frac{3 - 5}{4} = -\frac{2}{4} = -\frac{1}{2}$$

It's important to keep track of the signs carefully. (It may help to review Sec. 1.10.) For example, $-(-3) = 3$, $(-3)^2 = (-3)(-3) = 9$, and $9 - 4(2)(-2) = 9 + 16 = 25$. Check the answers by plugging them into the original equation:

- $2x^2 - 3x - 2 = 2(2)^2 - 3(2) - 2 = 2(4) - 6 - 2 = 8 - 8 = 0$
- $2x^2 - 3x - 2 = 2\left(-\frac{1}{2}\right)^2 - 3\left(-\frac{1}{2}\right) - 2 = 2\left(\frac{1}{4}\right) + \frac{3}{2} - 2 = \frac{1}{2} + \frac{3}{2} - 2 = 2 - 2 = 0$

For curious students: Where does the quadratic equation come from? There is more than one way to derive it. One way to solve $ax^2 + bx + c = 0$ is to assume that the answer has both rational and irrational parts, meaning that x has the following form: $x = d + \sqrt{e}$, where d and e are rational, but where $\sqrt{e}$ may be irrational. Plug this in for x in the equation $ax^2 + bx + c = 0$:

$$a\left(d + \sqrt{e}\right)^2 + b\left(d + \sqrt{e}\right) + c = 0$$

Apply the FOIL method. Note that $\sqrt{e}\sqrt{e} = \left(\sqrt{e}\right)^2 = e$.

$$a\left(d^2 + 2d\sqrt{e} + e\right) + b\left(d + \sqrt{e}\right) + c = 0$$

Distribute like we learned in Chapter 4.

$$ad^2 + 2ad\sqrt{e} + ae + bd + b\sqrt{e} + c = 0$$

For the left side to be zero, the rational and irrational parts must separately be zero:

$$ad^2 + ae + bd + c = 0 \quad \text{and} \quad 2ad\sqrt{e} + b\sqrt{e} = 0$$

- In the right equation, factor out the $\sqrt{e}$ to get $(2ad + b)\sqrt{e} = 0$. As in Sec. 5.11, either $2ad + b = 0$ or $e = 0$. In the first case, $2ad = -b$, for which $d = -\dfrac{b}{2a}$.

- Plug $d = -\dfrac{b}{2a}$ into the left equation: $a\left(-\dfrac{b}{2a}\right)^2 + ae + b\left(-\dfrac{b}{2a}\right) + c = 0$. Simplify this: $\dfrac{ab^2}{4a^2} + ae - \dfrac{b^2}{2a} + c = 0$ becomes $\dfrac{b^2}{4a} + ae - \dfrac{b^2}{2a} + c = 0$. Note that $\dfrac{b^2}{4a} - \dfrac{b^2}{2a} = \dfrac{b^2}{4a} - \dfrac{2b^2}{4a} = \dfrac{b^2 - 2b^2}{4a} = -\dfrac{b^2}{4a}$ such that $ae - \dfrac{b^2}{4a} + c = 0$. This becomes $ae = \dfrac{b^2}{4a} - c$.

  Divide by a on both sides to get $e = \dfrac{b^2}{4a^2} - \dfrac{c}{a}$.

Plug $d = -\dfrac{b}{2a}$ and $e = \dfrac{b^2}{4a^2} - \dfrac{c}{a}$ into the equation $x = d + \sqrt{e}$ (from the previous page):

$$x = d + \sqrt{e} = -\frac{b}{2a} + \sqrt{\frac{b^2}{4a^2} - \frac{c}{a}}$$

Make a common denominator inside of the radical:

$$x = -\frac{b}{2a} + \sqrt{\frac{b^2}{4a^2} - \frac{4ac}{4a^2}} = -\frac{b}{2a} + \sqrt{\frac{b^2 - 4ac}{4a^2}}$$

Apply the rule $\sqrt{\dfrac{u}{v}} = \dfrac{\sqrt{u}}{\sqrt{v}}$. Note that $\sqrt{4a^2} = \pm 2a$ because $(\pm 2a)^2 = 2^2 a^2 = 4a^2$.

$$x = -\frac{b}{2a} \pm \frac{\sqrt{b^2 - 4ac}}{2a}$$

$$x = \frac{-b \pm \sqrt{b^2 - 4ac}}{2a}$$

A more formal and rigorous way to derive the quadratic formula is to apply the factoring technique from Sec. 5.5, while working exclusively with variables. The method that we applied here is less formal, but helps to see how the formula can come about.

This section includes a few "starter" problems broken down into steps to help acquaint you with the quadratic formula. Sec.'s 6.5-6.7 provide much more practice.

Following is the strategy for using the quadratic formula:
- If the given quadratic equation isn't already in standard form, put the equation in standard form (Sec. 6.4).
- Once the equation is in standard form, identify the constants a, b, and c by comparing the equation with the general form $ax^2 + bx + c = 0$.
- Plug the values of a, b, and c into the quadratic formula and simplify.
- Check the answers by plugging them into the original equation.

Example 1. $3x^2 - 14x - 24 = 0$
- Identify the constants by comparing $ax^2 + bx + c = 0$ to $3x^2 - 14x - 24 = 0$:
$$a = 3 \quad , \quad b = -14 \quad , \quad c = -24$$
- Plug these values into the quadratic formula:
$$x = \frac{-b \pm \sqrt{b^2 - 4ac}}{2a} = \frac{-(-14) \pm \sqrt{(-14)^2 - 4(3)(-24)}}{2(3)}$$
- Follow the order of operations (Sec. 1.5). Take care with the minus signs. For example, $-(-14) = 14$, $(-14)^2 = (-14)(-14) = 196$, and $-4(3)(-24) = 288$.
$$x = \frac{14 \pm \sqrt{196 + 288}}{6} = \frac{14 \pm \sqrt{484}}{6} = \frac{14 \pm 22}{6}$$
- Work out the arithmetic for two separate answers: one for each sign ($\pm$).
$$x = \frac{14 + 22}{6} = \frac{36}{6} = 6 \quad \text{or} \quad x = \frac{14 - 22}{6} = -\frac{8}{6} = -\frac{4}{3}$$
- The two answers are $x = 6$ and $x = -\frac{4}{3}$.
- Check each answer by plugging it into the original equation:
$$3x^2 - 14x - 24 = 3(6)^2 - 14(6) - 24 = 3(36) - 84 - 24 = 108 - 108 = 0$$
$$3x^2 - 14x - 24 = 3\left(-\frac{4}{3}\right)^2 - 14\left(-\frac{4}{3}\right) - 24 = 3\left(\frac{16}{9}\right) + \frac{56}{3} - 24$$
$$= \frac{16}{3} + \frac{56}{3} - 24 = \frac{72}{3} - 24 = 24 - 24 = 0$$

Exercise Set 6.3

Directions: Use the quadratic formula to solve for the variable in each equation.

1) $x^2 - 7x + 12 = 0$

(A) Determine the constants a, b, and c.

(B) Plug these values into the quadratic formula.

(C) Simplify and determine all of the answers.

2) $2x^2 - 8x - 10 = 0$

(A) Determine the constants a, b, and c.

(B) Plug these values into the quadratic formula.

(C) Simplify and determine all of the answers.

3) $4x^2 + 7x - 15 = 0$

(A) Determine the constants a, b, and c.

(B) Plug these values into the quadratic formula.

(C) Simplify and determine all of the answers.

4) $6x^2 - x - 40 = 0$

(A) Determine the constants a, b, and c.

(B) Plug these values into the quadratic formula.

(C) Simplify and determine all of the answers.

6.4 Standard Form of the Quadratic

If a quadratic equation is in standard form, which is $ax^2 + bx + c = 0$, we can identify the constants a, b, and c, and plug these constants into the quadratic formula in order to solve for the variable. However, when you come across a quadratic equation, it isn't always in standard form. In this section, we will explore how to put a given quadratic equation into standard form. We won't solve the equations in this section; we'll just practice rewriting quadratic equations in standard form. (Sec.'s 6.5-6.7 focus on using the quadratic formula.)

Consider the equation $-2x + 8x^2 = 3x^2 + 9$. This is a quadratic equation, but it isn't in standard form for a few reasons: it has 4 terms (instead of 3), not all of the terms are on the same side of the equation, and the terms aren't in the right order. We can put this equation in standard form by:

- bringing all of the terms to the same side of the equation (it doesn't matter if we put them all on the right side or all on the left side)
- combining like terms (Sec. 2.2)
- putting the terms in the standard order (with the quadratic term first, the linear term second, and the constant term last)

Subtract $3x^2$ and subtract 9 from both sides of the equation: $-2x + 8x^2 - 3x^2 - 9 = 0$. Combine like terms ($8x^2$ and $-3x^2$ are like terms): $-2x + 5x^2 - 9 = 0$. Reorder the terms: $5x^2 - 2x - 9 = 0$. The equation $5x^2 - 2x - 9 = 0$ is in standard form.

Example 1. Rewrite $3 - 4x = 2x^2$ in standard form.

- Subtract $2x^2$ from both sides: $3 - 4x - 2x^2 = 0$.
- Reorder the terms: $-2x^2 - 4x + 3 = 0$.

Example 2. Rewrite $7 + 5x^2 + 6x = 4x^2 + 9$ in standard form.

- Subtract $4x^2$ and 9 from both sides: $7 + 5x^2 + 6x - 4x^2 - 9 = 0$.
- Combine like terms: $-2 + x^2 + 6x = 0$.
- Reorder the terms: $x^2 + 6x - 2 = 0$.

Exercise Set 6.4

Directions: Rewrite each equation in standard form. You don't need to solve the equation.

1) $9x + 5 - 4x^2 = 0$

2) $24 - 5x + 2x^2 - 3x - 18 + x^2 = 0$

3) $8x^2 + 3 = 6x^2 - 12x$

4) $5x^2 - 7 = 6x$

5) $4x - 9x^2 = 6x - 8$

6) $11x - 9 + 3x^2 = 5x + 9 - 4x^2$

7) $3 + 8x = 7 - 2x^2$

8) $5x^2 - 4x = 7 - 4x$

9) $x^2 - 9x + 5 = 2x^2 - 3x + 11$

10) $(x + 3)(x + 6) = 25$

6.5 Quadratics with Integer Solutions

The quadratic equations in this section have answers that are integers. (This makes these equations easy enough that you could factor them like we did in Sec.'s 5.5-5.6, except that the purpose of this section is to practice using the quadratic formula.) If the given equation isn't in standard form, $ax^2 + bx + c = 0$, first put the equation in standard form like we did in Sec. 6.4 before using the quadratic formula.

Example 1. $2x^2 + 6x - 20 = 0$

$$a = 2 \quad , \quad b = 6 \quad , \quad c = -20$$

(Note: When all of the constants share a common factor, it sometimes helps to divide both sides of the equation by that factor. Here, we could have divided both sides by 2.)

$$x = \frac{-b \pm \sqrt{b^2 - 4ac}}{2a} = \frac{-6 \pm \sqrt{6^2 - 4(2)(-20)}}{2(2)}$$

$$x = \frac{-6 \pm \sqrt{36 + 160}}{4} = \frac{-6 \pm \sqrt{196}}{4} = \frac{-6 \pm 14}{4}$$

$$x = \frac{-6 + 14}{4} = \frac{8}{4} = 2 \quad \text{or} \quad x = \frac{-6 - 14}{4} = \frac{-20}{4} = -5$$

Check: $2x^2 + 6x - 20 = 2(2)^2 + 6(2) - 20 = 2(4) + 12 - 20 = 8 - 8 = 0$

and $2x^2 + 6x - 20 = 2(-5)^2 + 6(-5) - 20 = 2(25) - 30 - 20 = 50 - 50 = 0$

Example 2. $18 + x^2 - 9x = 0$

First rewrite the equation in standard form. Reorder the terms:

$$x^2 - 9x + 18 = 0$$

$$a = 1 \quad , \quad b = -9 \quad , \quad c = 18$$

$$x = \frac{-b \pm \sqrt{b^2 - 4ac}}{2a} = \frac{-(-9) \pm \sqrt{(-9)^2 - 4(1)(18)}}{2(1)}$$

$$x = \frac{9 \pm \sqrt{81 - 72}}{2} = \frac{9 \pm \sqrt{9}}{2} = \frac{9 \pm 3}{2}$$

$$x = \frac{9 + 3}{2} = \frac{12}{2} = 6 \quad \text{or} \quad x = \frac{9 - 3}{2} = \frac{6}{2} = 3$$

Check: $18 + x^2 - 9x = 18 + 6^2 - 9(6) = 18 + 36 - 54 = 54 - 54 = 0$

and $18 + x^2 - 9x = 18 + 3^2 - 9(3) = 18 + 9 - 27 = 27 - 27 = 0$

Exercise Set 6.5

Directions: Use the quadratic formula to solve for the variable in each equation.

1) $x^2 - 8x + 12 = 0$

2) $2x^2 + 8x - 10 = 0$

3) $16x + 30 + 2x^2 = 0$

4) $14 - x^2 - 5x = 0$

5) $4x^2 - 16x = -16$

6) $30 - x = x^2$

7) $0 = x^2 + 11x + 18$

8) $9x = -3x^2 - 6$

9) $7x^2 - 10x = 5x^2 + 100$

10) $2x^2 + 21x + 7 = 9 + 5x^2 - 2$

11) $10x + 24 = x^2$

12) $5x^2 + 10x - 90 = 30 - 15x$

13) $3x^2 - 4x = 75 - 4x$

14) $7x^2 + 5x + 26 = 6x^2 + 8x + 80$

15) $\frac{x^2}{2} = 24 - x$

16) $(2x - 8)(x - 6) = 6$

6.6 Quadratics with Fractional Answers

The quadratic equations in this section have answers that are fractions.

Example 1. $15x^2 - 11x + 2 = 0$

$$a = 15 \quad , \quad b = -11 \quad , \quad c = 2$$

$$x = \frac{-b \pm \sqrt{b^2 - 4ac}}{2a} = \frac{-(-11) \pm \sqrt{(-11)^2 - 4(15)(2)}}{2(15)}$$

$$x = \frac{11 \pm \sqrt{121 - 120}}{30} = \frac{11 \pm \sqrt{1}}{30} = \frac{11 \pm 1}{30}$$

$$x = \frac{11 + 1}{30} = \frac{12}{30} = \frac{12 \div 6}{30 \div 6} = \frac{2}{5} \quad \text{or} \quad x = \frac{11 - 1}{30} = \frac{10}{30} = \frac{1}{3}$$

Check: $15x^2 - 11x + 2 = 15\left(\frac{2}{5}\right)^2 - 11\left(\frac{2}{5}\right) + 2 = \frac{15(4)}{25} - \frac{22}{5} + 2 = \frac{60}{25} - \frac{110}{25} + \frac{50}{25} = 0$

and $15x^2 - 11x + 2 = 15\left(\frac{1}{3}\right)^2 - 11\left(\frac{1}{3}\right) + 2 = \frac{15(1)}{9} - \frac{11}{3} + 2 = \frac{15}{9} - \frac{33}{9} + \frac{18}{9} = 0$

Exercise Set 6.6

Directions: Use the quadratic formula to solve for the variable in each equation.

1) $8x^2 - 10x + 3 = 0$

2) $12x^2 - 5x - 2 = 0$

3) $21x^2 + 47x + 20 = 0$

4) $45x^2 = 32 - 52x$

5) $24x^2 - 2x = 15$

6) $30 = 89x - 24x^2$

7) $8x^2 + 3x = 5 - 6x^2$

8) $27x^2 + 18x = -16 - 24x$

9) $10x^2 - 17x - 54 = 12x + 18$

10) $24 - 9x^2 - 29x = 12x^2 + 45 + 29x$

6.7 Quadratics with Irrational Answers

The quadratic equations in this section have irrational numbers. If the square root contains any perfect squares, factor them out of the square root. It may help to review Sec. 5.10. For example, $\sqrt{99} = \sqrt{9(11)} = \sqrt{9}\sqrt{11} = 3\sqrt{11}$.

Example 1. $2x^2 + 6x + 3 = 0$

$$a = 2 \quad , \quad b = 6 \quad , \quad c = 3$$

$$x = \frac{-b \pm \sqrt{b^2 - 4ac}}{2a} = \frac{-6 \pm \sqrt{6^2 - 4(2)(3)}}{2(2)} = \frac{-6 \pm \sqrt{36 - 24}}{4} = \frac{-6 \pm \sqrt{12}}{4}$$

Since 12 is evenly divisible by 4 and since 4 is a perfect square (because $2^2 = 4$), we will factor this perfect square out of the square root: $\sqrt{12} = \sqrt{4(3)} = \sqrt{4}\sqrt{3} = 2\sqrt{3}$.

$$x = \frac{-6 \pm 2\sqrt{3}}{4}$$

We can reduce this fraction by factoring the 2 out of the numerator. This 2 will cancel a 2 from the denominator.

$$x = \frac{2\left(-3 \pm \sqrt{3}\right)}{2(2)} = \frac{-3 \pm \sqrt{3}}{2}$$

The two answers are $x = \frac{-3+\sqrt{3}}{2}$ and $x = \frac{-3-\sqrt{3}}{2}$. It is simpler just to write $x = \frac{-3\pm\sqrt{3}}{2}$.

Check: The easiest way to check these answers is with a calculator.

$$x = \frac{-3 + \sqrt{3}}{2} \approx -0.634 \quad \text{or} \quad x = \frac{-3 - \sqrt{3}}{2} \approx -2.37$$

Check: $2x^2 + 6x + 3 \approx 2(-0.634)^2 + 6(-0.634) + 3 \approx 0.804 - 3.80 + 3 = 0.004 \approx 0$ and $2x^2 + 6x + 3 \approx 2(-2.37)^2 + 6(-2.37) + 3 \approx 11.2 - 14.2 + 3 = 0$.

(The first answer is slightly off due to rounding values to three digits. If you keep additional digits throughout the calculation, the values will match more closely.)

Exercise Set 6.7

Directions: Use the quadratic formula to solve for the variable in each equation.

1) $x^2 + 5x + 2 = 0$

2) $3x^2 + 6x - 5 = 0$

3) $4x^2 + 4 = 9x$

4) $5x^2 - 2x = 2$

5) $2x + 6 = 3x^2$

6) $7 = 8x - 2x^2$

7) $9x^2 - 10x = 5 + 5x^2$

8) $5x^2 + 15x = 8x - 1$

6.8 Using the Discriminant

The **discriminant** is the part of the quadratic inside of the square root: $b^2 - 4ac$. The discriminant is useful because it determines the nature of the answers to a quadratic equation:

- If the discriminant is positive, there are two distinct real answers.
- If the discriminant is zero, there is one distinct real answer.
- If the discriminant is negative, there are two distinct complex answers.

Why? The quadratic takes the square root of the discriminant. Look at the quadratic formula below. You'll see that $b^2 - 4ac$ is inside of the square root. The square root of a negative number isn't real. For example, try to find a number that you can square to make the number -9. You won't be able to find a real number that does this. If you try -3, it doesn't work because $(-3)^2 = (-3)(-3) = 9$; the minus sign gets squared. Therefore, if the discriminant is negative, the answers are complex (meaning that they aren't real); they include imaginary numbers.

$$x = \frac{-b \pm \sqrt{b^2 - 4ac}}{2a}$$

If the discriminant is zero, there is only one distinct answer because ± 0 is the same regardless of the sign. For example, $\frac{6 \pm 0}{2}$ gives $\frac{6+0}{2} = \frac{6}{2} = 3$ and $\frac{6-0}{2} = \frac{6}{2} = 3$.

Example 1. Calculate the discriminant for $5x^2 + 8x + 3 = 0$.

$$a = 5 \quad , \quad b = 8 \quad , \quad c = 3 \quad , \quad b^2 - 4ac = 8^2 - 4(5)(3) = 64 - 60 = 4$$

Since the discriminant, which equals 4, is positive, there are two real answers.

Example 2. Calculate the discriminant for $3x^2 + 6x + 3 = 0$.

$$a = 3 \quad , \quad b = 6 \quad , \quad c = 3 \quad , \quad b^2 - 4ac = 6^2 - 4(3)(3) = 36 - 36 = 0$$

Since the discriminant equals zero, there is one real answer.

Example 3. Calculate the discriminant for $x^2 + 4x + 5 = 0$.

$$a = 1 \quad , \quad b = 4 \quad , \quad c = 5 \quad , \quad b^2 - 4ac = 4^2 - 4(1)(5) = 16 - 20 = -4$$

Since the discriminant, which equals -4, is negative, there are two complex answers.

Exercise Set 6.8

Directions: Calculate the discriminant and determine the nature of the answers (without actually calculating the answers).

1) $x^2 + 2x + 3 = 0$

2) $2x^2 - 5x + 3 = 0$

3) $5x^2 - 7x - 3 = 0$

4) $-4x^2 + 8x - 4 = 0$

5) $\frac{x^2}{2} - \frac{3x}{4} + \frac{3}{8} = 0$

6) $4x^2 + 6 = 10x$

7) $12x - 9x^2 = 4$

8) $25 - 15x = 2x^2$

9) $5x^2 + 15x = 4x - 6$

10) $8x^2 + 11 = 12x + 6$

6.9 Factoring the Quadratic

When the quadratic equation is relatively easy to factor, it may be easier to solve the quadratic equation by factoring than by using the quadratic formula. Recall that we learned how to factor a quadratic in Sec.'s 5.5-5.6. We will apply that factoring method in this section to solve quadratic equations.

For example, consider the equation $3x^2 + 10x - 8 = 0$. We can factor this equation as $(3x - 2)(x + 4) = 0$ using the strategy from Sec. 5.5. If the product of two numbers is zero, at least one of the numbers must be zero. Since $(3x - 2)(x + 4) = 0$, either $3x - 2 = 0$ or $x + 4 = 0$. For the case $3x - 2 = 0$, add 2 to both sides to get $3x = 2$ and divide by 3 on both sides to get $x = \frac{2}{3}$. For the case $x + 4 = 0$, subtract 4 from both sides to get $x = -4$. We found the two solutions to $3x^2 + 10x - 8 = 0$ by factoring. One solution is $x = \frac{2}{3}$ and the other is $x = -4$. We can check these answers by plugging them into the original equation:

- $3\left(\frac{2}{3}\right)^2 + 10\left(\frac{2}{3}\right) - 8 = 3\left(\frac{4}{9}\right) + \frac{20}{3} - 8 = \frac{12}{9} + \frac{60}{9} - \frac{72}{9} = 0$
- $3(-4)^2 + 10(-4) - 8 = 3(16) - 40 - 8 = 48 - 48 = 0$

In order to factor a quadratic equation, first you need to put the equation in standard form (Sec. 6.4). This ensures that one side of the equation is zero. This is important because zero is a special number. When a product of two numbers equals zero, we know that one of the numbers must be zero. However, when a product of two numbers is nonzero, we don't have enough information to determine either number. Compare:
- If $xy = 0$, it must be the case that either $x = 0$ or $y = 0$.
- If $xy = 6$, there is no way to determine x or y. You might suggest that x could be 3 and y could be 2 because $3(2) = 6$, but they could also be 6 and 1 because $(6)(1) = 6$, they could be -2 and -3 because $(-2)(-3) = 6$, and they could even be $\frac{1}{4}$ and 24 because $\left(\frac{1}{4}\right)(24) = 6$. The possibilities are endless.

Tip: It may help to review Sec.'s 5.5 and 5.6 before proceeding.

Example 1. $x^2 - 9x + 20 = 0$

- If only the x term is negative, both expressions are negative when it is factored.
- The factors of 20 are 20 and 1, 10 and 2, or 5 and 4.

$$(\ x + \ )(\ x + \ )$$

1	-20	1	-1
	-10		-2
	-5		-4

Since $(1)(-4) + (-5)(1) = -4 - 5 = -9$, the solution is $(x - 5)(x - 4)$. Note that $(1)(-4)$ is the "outside" product and $(-5)(1)$ is the "inside" product of the FOIL method (Sec. 4.3) when we expand $(x - 5)(x - 4) = x^2 - 4x - 5x + 20 = x^2 - 9x + 20$. Since $(x - 5)(x - 4) = 0$, one of these two factors must be zero:

$$x - 5 = 0 \quad \text{or} \quad x - 4 = 0$$
$$x = 5 \quad \text{or} \quad x = 4$$

Check the answers by plugging them into the original equation:

- $x^2 - 9x + 20 = 5^2 - 9(5) + 20 = 25 - 45 + 20 = 0$
- $x^2 - 9x + 20 = 4^2 - 9(4) + 20 = 16 - 36 + 20 = 0$

Example 2. $5x^2 + 2x - 3 = 0$

- If the constant term is negative (and the x^2 term is positive), one expression is positive and one expression will be negative when it is factored.
- The factors of 5 are 5 and 1. The factors of -3 are 3 and -1 or -3 and 1.

$$(\ x + \ )(\ x + \ )$$

5	3	1	-1
1	1	5	-3
	-3		1
	-1		3

Since $(5)(1) + (-3)(1) = 5 - 3 = 2$, the solution is $(5x - 3)(x + 1)$.

$$5x - 3 = 0 \quad \text{or} \quad x + 1 = 0$$
$$5x = 3 \quad \text{or} \quad x = -1$$
$$x = \frac{3}{5} \quad \text{or} \quad x = -1$$

Check the answers by plugging them into the original equation:

- $5x^2 + 2x - 3 = 5\left(\frac{3}{5}\right)^2 + 2\left(\frac{3}{5}\right) - 3 = 5\left(\frac{9}{25}\right) + \frac{6}{5} - 3 = \frac{45}{25} + \frac{30}{25} - \frac{75}{25} = 0$
- $5x^2 + 2x - 3 = 5(-1)^2 + 2(-1) - 3 = 5(1) - 2 - 3 = 5 - 5 = 0$

Exercise Set 6.9

Directions: Solve for the variable by factoring the quadratic equation.

1) $x^2 + 8x + 7 = 0$

2) $2x^2 - 7x - 15 = 0$

3) $7x^2 - 6 = 11x$

4) $x^2 - 12x = -36$

5) $6x^2 = 5x + 4$

6) $26x - 35 = 3x^2$

7) $4x^2 + 45x = 8x - 40$

8) $6x^2 + 12 = 56x - 3x^2$

6.10 Quadratics in Word Problems

The word problems in this section involve quadratic equations. It may help to review Sec. 2.13 (which has tips for solving word problems), Sec. 1.4, and Sec. 3.12.

Example 1. A number squared is 35 more than twice the number. What is the number?

- x = the number and x^2 = the number squared.
- According to the problem: $x^2 = 2x + 35$.
- Put the equation in standard form (Sec. 6.4): $x^2 - 2x - 35 = 0$.

$$a = 1 \quad , \quad b = -2 \quad , \quad c = -35$$

$$x = \frac{-b \pm \sqrt{b^2 - 4ac}}{2a} = \frac{-(-2) \pm \sqrt{(-2)^2 - 4(1)(-35)}}{2(1)}$$

$$x = \frac{2 \pm \sqrt{4 + 140}}{2} = \frac{2 \pm \sqrt{144}}{2} = \frac{2 \pm 12}{2}$$

$$x = \frac{2 + 12}{2} = \frac{14}{2} = 7 \quad \text{or} \quad x = \frac{2 - 12}{2} = \frac{-10}{2} = -5$$

Check: $x^2 = 7^2 = 49$ agrees with $2x + 35 = 2(7) + 35 = 14 + 35 = 49$
and $x^2 = (-5)^2 = 25$ agrees with $2x + 36 = 2(-5) + 35 = -10 + 35 = 25$

Exercise Set 6.10

Directions: Use algebra to solve each word problem.

1) A number squared is 24 less than ten times the number. What could the number be?

2) The product of two consecutive even numbers is 360. What are the numbers?

3) Two numbers have a difference of 17 and a product of 200. What are the numbers?

7 VARIABLES IN THE DENOMINATOR

7.1 Reciprocate Both Sides

If there is a variable in a denominator and the equation has a simple structure like $\frac{1}{x} = \frac{a}{b}$, the variable can be solved for by taking the **reciprocal** of both sides. Recall that the **reciprocal** of a fraction has the numerator and denominator swapped (Sec. 1.7). For example, the reciprocal of $\frac{3}{4}$ is equal to $\frac{4}{3}$. Also, the reciprocal of an integer equals one divided by the integer. For example, the reciprocal of 5 is equal to $\frac{1}{5}$.

- If we take the reciprocal of both sides of $\frac{1}{x} = \frac{a}{b}$, the equation becomes $\frac{x}{1} = \frac{b}{a}$. Since $\frac{x}{1} = x$, this reduces to $x = \frac{b}{a}$.

- If we take the reciprocal of both sides of $\frac{1}{x} = c$, the equation becomes $\frac{x}{1} = \frac{1}{c}$, which reduces to $x = \frac{1}{c}$.

- If we take the reciprocal of both sides of $\frac{1}{x} = \frac{1}{d}$, the equation becomes $\frac{x}{1} = \frac{d}{1}$, which reduces to $x = d$.

For equations where the variable is in the denominator, but where the equation has more than two terms, note that you will need to isolate the variable term before you reciprocate both sides. For example, if $\frac{1}{x} - \frac{1}{6} = \frac{1}{3}$, you would need to combine like terms before taking the reciprocal of both sides, or use the method presented in Sec. 7.6.

Example 1. $\frac{1}{x} = \frac{2}{3}$ Take the reciprocal of both sides: $x = \frac{3}{2}$. Note that $\frac{x}{1} = x$.

Example 2. $\frac{1}{x} = 4$ Take the reciprocal of both sides: $x = \frac{1}{4}$.

Example 3. $\frac{1}{x} = \frac{1}{7}$ Take the reciprocal of both sides: $x = 7$.

Exercise Set 7.1

Directions: Solve for the variable in each equation.

1) $\frac{1}{x} = \frac{3}{5}$

2) $\frac{1}{x} = 6$

3) $\frac{1}{x} = \frac{1}{9}$

4) $\frac{1}{x} = \frac{8}{5}$

5) $\frac{5}{6} = \frac{1}{x}$

6) $2 = \frac{1}{x}$

7) $\frac{1}{x} = -\frac{2}{7}$

8) $\frac{1}{x} = -\frac{1}{8}$

9) $-5 = \frac{1}{x}$

10) $\frac{1}{10} = \frac{1}{x}$

11) $-\frac{1}{x} = \frac{9}{4}$

12) $-\frac{1}{x} = -6$

7.2 Cross Multiplying

If two fractions are set equal to one another, like $\frac{3}{x} = \frac{4}{7}$, a simple way to proceed is to **cross multiply**. This means to multiply along the diagonals, as illustrated below.

$$3(7) = 4x$$

In the example above, we multiplied 3 and 7 and we multiplied 4 and x, which gave us the equation $3(7) = 4x$. Cross multiplying basically combines two multiplications into a single step. We could have achieved the same result as follows:

- Multiply by x on both sides of $\frac{3}{x} = \frac{4}{7}$ to get $3 = \frac{4x}{7}$.

- Multiply by 7 on both sides of $3 = \frac{4x}{7}$ to get $3(7) = 4x$.

To cross multiply, look for the following diagonals:

- Multiply the numerator of the first fraction by the denominator of the second fraction. In the example above, we multiplied 3 by 7.

- Multiply the numerator of the second fraction by the denominator of the first fraction. In the example above, we multiplied 4 by x.

- Form a new equation using these two products: $3(7) = 4x$.

After we cross multiply, we can solve for the variable. In this example, the equation simplifies to $21 = 4x$. Divide by 4 on both sides to get $\frac{21}{4} = x$.

Example 1. $\frac{2}{x} = \frac{3}{5}$ Cross multiply: $2(5) = 3x$. Simplify: $10 = 3x$.

Divide by 3 on both sides: $\frac{10}{3} = x$.

Example 2. $\frac{8}{7} = \frac{4}{x}$ Cross multiply: $8x = 4(7)$. Simplify: $8x = 28$.

Divide by 8 on both sides: $x = \frac{28}{8}$. Reduce the answer: $x = \frac{7}{2}$ because $\frac{28}{8} = \frac{28/4}{8/4} = \frac{7}{2}$.

Exercise Set 7.2

Directions: Solve for the variable by cross multiplying.

1) $\dfrac{5}{x} = \dfrac{2}{3}$

2) $\dfrac{6}{x} = \dfrac{3}{4}$

3) $\dfrac{4}{x} = \dfrac{1}{7}$

4) $\dfrac{3}{x} = \dfrac{4}{3}$

5) $\dfrac{12}{x} = \dfrac{3}{5}$

6) $\dfrac{7}{x} = 3$

7) $\dfrac{8}{x} = -\dfrac{6}{5}$

8) $\dfrac{2}{3} = \dfrac{10}{x}$

9) $\frac{6}{x} = -2$

10) $-\frac{4}{x} = -\frac{1}{8}$

11) $-\frac{24}{x} = \frac{18}{5}$

12) $\frac{1}{x} = 6$

13) $\frac{25}{x} = \frac{1}{3}$

14) $\frac{22}{5} = \frac{11}{x}$

15) $-\frac{7}{9} = -\frac{3}{x}$

16) $-\frac{36}{x} = \frac{3}{4}$

7.3 Cross Multiplying with Powers

The equations in this section involve cross multiplying and also involve power rules. It may help to review Chapter 3 (especially, Sec.'s 3.1-3.8). For example, consider the equation $\frac{x^2}{3} = \frac{9}{x}$. If we cross multiply, we get $x^2(x) = 3(9)$, which simplifies to $x^3 = 27$ according to the rule $x^m x^n = x^{m+n}$ (in this case, $m = 2$ and $n = 1$ since $x^1 = 1$). Now take the cube root of both sides, like we did in Sec. 3.10, to get $x = \sqrt[3]{27} = 3$. We can check the answer by plugging it into the original equation: $\frac{x^2}{3} = \frac{3^2}{3} = \frac{9}{3} = 3$ agrees with $\frac{9}{x} = \frac{9}{3} = 3$.

Example 1. $\frac{x^3}{4} = \frac{8}{x^2}$ Cross multiply: $x^3(x^2) = 4(8)$. Simplify: $x^5 = 32$.

Take the fifth root of both sides: $x = \sqrt[5]{32} = 2$.

Check the answer. Plug 2 in for x in the original equation: $\frac{x^3}{4} = \frac{2^3}{4} = \frac{8}{4} = 2$ agrees with $\frac{8}{x^2} = \frac{8}{2^2} = \frac{8}{4} = 2$.

Example 2. $\frac{x^5}{18} = \frac{x^3}{2}$ Cross multiply: $2x^5 = 18x^3$. Subtract $18x^3$ from both sides: $2x^5 - 18x^3 = 0$. Factor like we did in Chapter 5: $2x^3(x^2 - 9) = 0$.

Either $2x^3 = 0$ or $x^2 - 9 = 0$, which means either $x = 0$ or $x = \pm\sqrt{9} = \pm3$.

Check the answer. Plug ±3 in for x in the original equation: $\frac{x^5}{18} = \frac{(\pm3)^5}{18} = \pm\frac{243}{18} = \pm\frac{27}{2}$ agrees with $\frac{x^3}{2} = \frac{(\pm3)^3}{2} = \pm\frac{27}{2}$. (Note that $\frac{243}{18} = \frac{243/9}{18/9} = \frac{27}{2}$.)

Example 3. $\frac{6}{x^{1/3}} = \frac{3}{2}$ Cross multiply: $6(2) = 3x^{1/3}$. Simplify: $12 = 3x^{1/3}$.

Divide by 3 on both sides: $4 = x^{1/3}$. Cube both sides: $4^3 = x$. Simplify: $64 = x$.

The idea behind the last step is that $\left(x^{1/3}\right)^3 = x^3$ since $(x^m)^n = x^{mn}$ and $\left(\frac{1}{3}\right)3 = \frac{3}{3} = 1$.

Check the answer. Plug 64 in for x in the original equation: $\frac{6}{x^{1/3}} = \frac{6}{64^{1/3}} = \frac{6}{\sqrt[3]{64}} = \frac{6}{4} = \frac{3}{2}$.

Exercise Set 7.3

Directions: Solve for the variable by cross multiplying.

1) $\dfrac{x^2}{5} = \dfrac{25}{x}$

2) $\dfrac{6}{x^2} = \dfrac{2}{3}$

3) $\dfrac{x^6}{4} = \dfrac{x^5}{7}$

4) $\dfrac{x}{4} = \dfrac{9}{x}$

5) $\dfrac{12}{x^2} = 3$

6) $\dfrac{x^6}{2} = \dfrac{x^8}{50}$

7) $\dfrac{x^2}{2} = \dfrac{128}{x^2}$

8) $\dfrac{54}{x} = \dfrac{x^2}{4}$

9) $\dfrac{4}{x} = -\dfrac{6}{x^2}$

10) $\dfrac{4}{x^2} = 196$

11) $\dfrac{x}{16} = -\dfrac{32}{x^2}$

12) $\dfrac{x^2}{6} = \dfrac{x^5}{6000}$

13) $-\dfrac{27}{x^2} = -\dfrac{x^2}{3}$

14) $\dfrac{18}{x^5} = \dfrac{2}{x^7}$

15) $\dfrac{x^5}{18} = \dfrac{x^7}{8}$

16) $-\dfrac{4}{x^{2/3}} = \dfrac{36}{x}$

7.4 Cross Multiply and Distribute

The equations in this section involve cross multiplying and the distributive property. It may help to review Chapter 4. For example, consider the equation $\frac{x-3}{8} = \frac{3}{4}$. When we cross multiply, we get $4(x-3) = 8(3)$. On the left side, we distribute the 4 to get $4(x) - 4(3) = 8(3)$, which simplifies to $4x - 12 = 24$. Add 12 to both sides: $4x = 36$. Divide by 4 on both sides: $x = \frac{36}{4} = 9$. Check the answer by plugging it into the original equation: $\frac{x-3}{8} = \frac{9-3}{8} = \frac{6}{8} = \frac{6/2}{8/2} = \frac{3}{4}$.

Example 1. $\frac{x+2}{4} = \frac{5}{3}$ Cross multiply: $3(x+2) = 4(5)$. Distribute: $3x + 6 = 20$.

Subtract 6 from both sides: $3x = 14$. Divide by 3 on both sides: $x = \frac{14}{3}$.

Check the answer. Plug $\frac{14}{3}$ in for x in the original equation: $\frac{x+2}{4} = \frac{\frac{14}{3}+2}{4} = \frac{\frac{14}{3}+\frac{6}{3}}{4} = \frac{20/3}{4} =$

$\frac{20/3}{4/1} = \frac{20}{3}\frac{1}{4} = \frac{20}{12} = \frac{20/4}{12/4} = \frac{5}{3}$. Recall from Sec. 1.7 that dividing by a fraction is equivalent to multiplying by its reciprocal. The reciprocal of 4 equals $\frac{1}{4}$.

Example 2. $\frac{3x-1}{x} = \frac{3x}{x+2}$ Cross multiply: $(3x-1)(x+2) = 3x^2$.

Apply the FOIL method (Sec. 4.3): $3x^2 + 6x - x - 2 = 3x^2$.

Subtract $3x^2$ from both sides: $6x - x - 2 = 0$. Simplify: $5x - 2 = 0$.

Add 2 to both sides: $5x = 2$. Divide by 5 on both sides: $x = \frac{2}{5}$.

Check the answer. Plug $\frac{2}{5}$ in for x in the original equation: $\frac{3x-1}{x} = \frac{3\left(\frac{2}{5}\right)-1}{\frac{2}{5}} = \frac{\frac{6}{5}-\frac{5}{5}}{\frac{2}{5}} = \frac{1/5}{2/5} =$

$\frac{1}{5}\frac{5}{2} = \frac{5}{10} = \frac{1}{2}$ agrees with $\frac{3x}{x+2} = \frac{3\left(\frac{2}{5}\right)}{\frac{2}{5}+2} = \frac{\frac{6}{5}}{\frac{2}{5}+\frac{10}{5}} = \frac{6/5}{12/5} = \frac{6}{5}\frac{5}{12} = \frac{30}{60} = \frac{1}{2}$. Recall from Sec. 1.7

that dividing by a fraction is equivalent to multiplying by its reciprocal. The reciprocal of $\frac{2}{5}$ equals $\frac{5}{2}$ and the reciprocal of $\frac{12}{5}$ equals $\frac{5}{12}$.

Exercise Set 7.4

Directions: Solve for the variable by cross multiplying.

1) $\dfrac{x+8}{12} = \dfrac{5}{4}$

2) $\dfrac{2}{3} = \dfrac{5x-9}{9}$

3) $\dfrac{3x-5}{6} = \dfrac{2x^2}{4x+7}$

4) $\dfrac{x^2+1}{6} = \dfrac{5}{3}$

5) $\dfrac{5}{3x+4} = \dfrac{2}{3}$

6) $\dfrac{x}{x-4} = \dfrac{x+9}{5}$

7) $\dfrac{28}{3x-2} = 4$

8) $\dfrac{x+5}{x+2} = \dfrac{x+9}{x+5}$

7.5 Combine Algebraic Fractions

Recall from Sec. 1.7 that the way to add or subtract two fractions is to make a common denominator. For example, the least common denominator for $\frac{5}{6} - \frac{4}{9}$ is 18:

$$\frac{5}{6} - \frac{4}{9} = \frac{5}{6}\frac{3}{3} - \frac{4}{9}\frac{2}{2} = \frac{15}{18} - \frac{8}{18} = \frac{7}{18}$$

We multiplied $\frac{5}{6}$ by $\frac{3}{3}$ and multiplied $\frac{4}{9}$ by $\frac{2}{2}$ to make a common denominator of 18.

The same idea applies to add or subtract fractions that include variables. For example, for $\frac{1}{2x} + \frac{3}{x^2}$ we can make a common denominator of $2x^2$. Multiply $\frac{1}{2x}$ by $\frac{x}{x}$ and multiply $\frac{3}{x^2}$ by $\frac{2}{2}$:

$$\frac{1}{2x} + \frac{3}{x^2} = \frac{1}{2x}\frac{x}{x} + \frac{3}{x^2}\frac{2}{2} = \frac{x}{2x^2} + \frac{6}{2x^2} = \frac{x+6}{2x^2}$$

Example 1. $\frac{3}{x} - \frac{x}{2} = \frac{3}{x}\frac{2}{2} - \frac{x}{2}\frac{x}{x} = \frac{6}{2x} - \frac{x^2}{2x} = \frac{6-x^2}{2x} = \frac{-x^2+6}{2x}$

Note: We reordered the terms in the numerator because it is customary to express a polynomial in order of decreasing powers.

Example 2. $\frac{4}{x^2} + \frac{2}{3x} = \frac{4}{x^2}\frac{3}{3} + \frac{2}{3x}\frac{x}{x} = \frac{12}{3x^2} + \frac{2x}{3x^2} = \frac{12+2x}{3x^2} = \frac{2x+12}{3x^2}$

Example 3. $\frac{2}{x+3} + \frac{3}{x-2} = \frac{2}{x+3}\frac{x-2}{x-2} + \frac{3}{x-2}\frac{x+3}{x+3} = \frac{2(x-2)}{(x+3)(x-2)} + \frac{3(x+3)}{(x+3)(x-2)}$

$= \frac{2x-4}{x^2-2x+3x-6} + \frac{3x+9}{x^2-2x+3x-6} = \frac{2x-4+3x+9}{x^2+x-6} = \frac{5x+5}{x^2+x-6}$ Alternate answer: $\frac{5(x+1)}{(x+3)(x-2)}$

Example 4. $\frac{4}{2x-1} - \frac{2}{3x+1} = \frac{4}{2x-1}\frac{3x+1}{3x+1} - \frac{2}{3x+1}\frac{2x-1}{2x-1} = \frac{4(3x+1)}{(2x-1)(3x+1)} - \frac{2(2x-1)}{(2x-1)(3x+1)}$

$= \frac{12x+4}{6x^2+2x-3x-1} - \frac{4x-2}{6x^2+2x-3x-1} = \frac{12x+4-(4x-2)}{6x^2-x-1} = \frac{12x+4-4x-(-2)}{6x^2-x-1} = \frac{12x+4-4x+2}{6x^2-x-1} = \frac{8x+6}{6x^2-x-1}$

Alternate answer: $\frac{2(4x+3)}{(2x-1)(3x+1)}$

Exercise Set 7.5

Directions: Combine the fractions by making the least common denominator.

1) $\dfrac{3}{4x} - \dfrac{2}{3x} =$

2) $\dfrac{x}{6} + \dfrac{4}{x} =$

3) $\dfrac{2}{x} + \dfrac{1}{x+1} =$

4) $\dfrac{8}{x-3} - \dfrac{6}{x+3} =$

5) $\dfrac{5}{3x^2} - \dfrac{2}{5x^4} =$

6) $\dfrac{x}{2} + \dfrac{4}{3x+5} =$

7) $\dfrac{5}{x-8} + \dfrac{3}{x+5} =$

8) $\dfrac{6}{2x-1} - \dfrac{3}{x+7} =$

7.6 Multiply by the Least Common Denominator

If an equation with fractions has more than two terms, like $\frac{1}{2x} + \frac{5}{12} = \frac{2}{x}$, you can't cross multiply unless you first add or subtract fractions until there is just one term on each side, like $\frac{5}{12} = \frac{3}{2x}$. However, there is an alternative. Instead of adding or subtracting fractions, we may multiply the entire equation by a common denominator of all of the terms (on both sides).

- For example, in the equation $\frac{1}{2x} + \frac{5}{12} = \frac{2}{x}$, a common denominator of all three terms is $24x$ because $2x$, 12, and x all evenly divide into $24x$.
- Multiply by $24x$ on both sides: $\frac{24x}{2x} + \frac{5(24x)}{12} = \frac{2(24x)}{x}$. Simplify: $\frac{24x}{2x} + \frac{120x}{12} = \frac{48x}{x}$.
- If you do this correctly, there will no longer be any fractions: $12 + 10x = 48$.
- Subtract 12 from both sides: $10x = 36$.
- Divide by 10 on both sides: $x = \frac{36}{10}$. Reduce the answer: $x = \frac{18}{5}$.

Check the answer by plugging it into the original equation: $\frac{1}{2x} + \frac{5}{12} = \frac{1}{2\left(\frac{18}{5}\right)} + \frac{5}{12} = \frac{5}{36} + \frac{15}{36}$

$= \frac{20}{36} = \frac{5}{9}$ agrees with $\frac{2}{x} = \frac{2}{18/5} = \frac{2}{1}\frac{5}{18} = \frac{10}{18} = \frac{5}{9}$. To divide by a fraction, multiply by its reciprocal. The reciprocal of $\frac{18}{5}$ equals $\frac{5}{18}$. Note that $\frac{20}{36}$ reduces to $\frac{5}{9}$ by dividing 20 and 36 each by 4.

Example 1. $\frac{1}{x} + \frac{1}{6} = \frac{1}{2}$ Multiply by $6x$ on both sides: $\frac{6x}{x} + \frac{6x}{6} = \frac{6x}{2}$. Simplify: $6 + x = 3x$.
Subtract x from both sides: $6 = 2x$. Divide by 2 on both sides: $3 = x$.
Check the answer. Plug 3 in for x in the original equation: $\frac{1}{x} + \frac{1}{6} = \frac{1}{3} + \frac{1}{6} = \frac{2}{6} + \frac{1}{6} = \frac{3}{6} = \frac{1}{2}$.

Alternate solution: Instead of multiplying by the least common denominator of all three terms, we could subtract $\frac{1}{6}$ from both sides: $\frac{1}{x} = \frac{1}{2} - \frac{1}{6}$.
Make a common denominator: $\frac{1}{x} = \frac{3}{6} - \frac{1}{6}$. Simplify: $\frac{1}{x} = \frac{2}{6}$. Reduce: $\frac{1}{x} = \frac{1}{3}$.
Cross multiply: $3 = x$.

Example 2. $\frac{3}{x+2} - \frac{1}{x} = \frac{1}{2x}$ Multiply by $2x(x+2)$ on both sides: $\frac{3(2x)(x+2)}{x+2} - \frac{2x(x+2)}{x} = \frac{2x(x+2)}{2x}$.
Simplify: $3(2x) - 2(x+2) = x + 2$. Distribute: $6x - 2x - 2(2) = x + 2$.
Simplify: $6x - 2x - 4 = x + 2$. Subtract x from both sides: $6x - 2x - x - 4 = 2$.
Add 4 to both sides: $6x - 2x - x = 2 + 4$. Combine like terms: $3x = 6$.
Divide by 3 on both sides: $x = 2$.

Check the answer. Plug 2 in for x in the original equation: $\frac{3}{x+2} - \frac{1}{x} = \frac{3}{2+2} - \frac{1}{2} = \frac{3}{4} - \frac{1}{2} =$
$\frac{3}{4} - \frac{2}{4} = \frac{1}{4}$ agrees with $\frac{1}{2x} = \frac{1}{2(2)} = \frac{1}{4}$.

Exercise Set 7.6

Directions: Solve for the variable in each equation.

1) $\frac{3}{8} + \frac{7}{4x} = \frac{2}{3}$

2) $\frac{5}{4x} = \frac{3}{4} - \frac{5}{2x}$

3) $\frac{16}{3x^2} + \frac{5}{3} = 3$

4) $\frac{4}{x} + \frac{3}{x^2} = \frac{13}{3x}$

5) $\dfrac{5x}{6} = \dfrac{5x}{8} - \dfrac{5}{4}$

6) $\dfrac{4}{x-4} + \dfrac{7}{6} = \dfrac{5}{2}$

7) $\dfrac{1}{x+3} + \dfrac{1}{x-3} = \dfrac{18}{x^2-9}$

8) $\dfrac{3}{x+8} - \dfrac{2}{3x} = \dfrac{5}{6x}$

7.7 Quadratics from Fractions

As we'll see in the examples and problems of this section, equations with fractions can lead to quadratic equations. It may help to review Chapter 6 before proceeding.

Example 1. $\frac{x+3}{2} = \frac{5}{x}$ Cross multiply: $x(x+3) = 2(5)$. Distribute: $x^2 + 3x = 10$.

Put the equation in standard form (Sec. 6.4): $x^2 + 3x - 10 = 0$.

Either factor this equation or use the quadratic formula: $a = 1, b = 3, c = -10$.

$$x = \frac{-b \pm \sqrt{b^2 - 4ac}}{2a} = \frac{-3 \pm \sqrt{3^2 - 4(1)(-10)}}{2(1)}$$

$$= \frac{-3 \pm \sqrt{9 + 40}}{2} = \frac{-3 \pm \sqrt{49}}{2} = \frac{-3 \pm 7}{2}$$

$$x = \frac{-3 + 7}{2} = \frac{4}{2} = 2 \quad \text{or} \quad x = \frac{-3 - 7}{2} = \frac{-10}{2} = -5$$

Check: $\frac{x+3}{2} = \frac{2+3}{2} = \frac{5}{2}$ agrees with $\frac{5}{x} = \frac{5}{2}$ and $\frac{x+3}{2} = \frac{-5+3}{2} = -\frac{2}{2} = -1$ agrees with $\frac{5}{x} = \frac{5}{-5} = -1$.

Example 2. $\frac{4}{x} - \frac{6}{x^2} = \frac{1}{2}$ Multiply by $2x^2$ on both sides: $\frac{8x^2}{x} - \frac{12x^2}{x^2} = \frac{2x^2}{2}$.

Simplify: $8x - 12 = x^2$. Put the equation in standard form: $-x^2 + 8x - 12 = 0$.

Either factor this equation or use the quadratic formula: $a = -1, b = 8, c = -12$.

$$x = \frac{-b \pm \sqrt{b^2 - 4ac}}{2a} = \frac{-8 \pm \sqrt{8^2 - 4(-1)(-12)}}{2(-1)}$$

$$x = \frac{-8 \pm \sqrt{64 - 48}}{-2} = \frac{-8 \pm \sqrt{16}}{-2} = \frac{-8 \pm 4}{-2}$$

$$x = \frac{-8 + 4}{-2} = \frac{-4}{-2} = 2 \quad \text{or} \quad x = \frac{-8 - 4}{-2} = \frac{-12}{-2} = 6$$

Check: $\frac{4}{x} - \frac{6}{x^2} = \frac{4}{2} - \frac{6}{2^2} = \frac{4}{2} - \frac{6}{4} = \frac{4}{2} - \frac{3}{2} = \frac{1}{2}$ and $\frac{4}{x} - \frac{6}{x^2} = \frac{4}{6} - \frac{6}{6^2} = \frac{4}{6} - \frac{1}{6} = \frac{3}{6} = \frac{1}{2}$.

Exercise Set 7.7

Directions: Solve for the variable in each equation.

1) $\dfrac{x-16}{3} = -\dfrac{21}{x}$

2) $\dfrac{5}{x} = \dfrac{x}{x+10}$

3) $\dfrac{1}{4} - \dfrac{2}{x^2} = \dfrac{7}{4x}$

4) $\dfrac{x}{x-5} = \dfrac{x-12}{7x+1}$

5) $\frac{1}{2} + 3x = \frac{1}{2x}$

6) $\frac{24}{x^5} + \frac{1}{x^3} = \frac{11}{x^4}$

7) $\frac{x+6}{x+8} = \frac{5x}{x-6}$

8) $\frac{15-x^2}{4x^2} = \frac{1}{2x}$

7.8 Word Problems with Ratios

A **ratio** expresses a fixed relationship between two quantities in the form of a fraction. Two numbers separated by a colon like 3:4 represents a ratio. For example, the ratio 3:4 is equivalent to the fraction $\frac{3}{4}$. A **proportion** expresses that two ratios are equal. For example, if the ratio of boys to girls at a school is 3:4 and there are 150 boys, we can determine the number of girls by setting up a proportion:

$$\frac{3}{4} = \frac{150}{x}$$

Cross multiply: $3x = 4(150)$. Simplify: $3x = 600$. Divide by 3 on both sides: $x = 200$ girls. Check the answer by plugging $x = 200$ into the equation: $\frac{150}{x} = \frac{150}{200} = \frac{150/50}{200/50} = \frac{3}{4}$.

Beware that a ratio may be expressed part-to-part, part-to-whole, or whole-to-part. If a school has 150 boys and 200 girls, one part is 150, another part is 200, and the whole is $150 + 200 = 350$.

- The ratio of boys to girls is part-to-part. Since $\frac{150}{200} = \frac{150/50}{200/50} = \frac{3}{4}$, this ratio is 3:4.

- The ratio of boys to students is part-to-whole. Since $\frac{150}{350} = \frac{150/50}{350/50} = \frac{3}{7}$, this ratio is 3:7. The ratio of girls to students is also part-to-whole. Since $\frac{200}{350} = \frac{200/50}{350/50} = \frac{4}{7}$, this ratio is 4:7.

- The ratio of students to boys is whole-to-part. Since $\frac{350}{150} = \frac{350/50}{150/50} = \frac{7}{3}$, this ratio is 7:3. The ratio of students to girls is also whole-to-part. Since $\frac{350}{200} = \frac{350/50}{200/50} = \frac{7}{4}$, this ratio is 7:4.

The parts add up to the whole. In this example, 150 boys plus 200 girls equals 350 students.

The word problems in this section involve ratios. It may help to review Sec. 2.13 (which has tips for solving word problems), Sec. 1.4, and Sec. 3.12.

Example 1. The ratio of pens to pencils in a pouch is 2:3. If there are 14 pens in the pouch, how many pencils are in the pouch?

- x = the number of pencils in the pouch.
- The 2 and the 14 both correspond to pens. The 3 and x both correspond to pencils. The ratio 2:3 is pens to pencils: $\frac{2}{3} = \frac{14}{x}$. Pens are in the numerator.
- Cross multiply: $2x = 42$. Divide by 2 on both sides: $x = 21$ pencils.
- Check the answer: $\frac{14}{x} = \frac{14}{21} = \frac{14/7}{21/7} = \frac{2}{3}$. Since the ratio of pens to pencils is 2:3, it should make sense that there are more pencils (21) than pens (14).

Example 2. A computer store sells laptops and desktops in the ratio 3:5. If the store sells 120 computers, how many will be laptops?

- Note that laptops and desktops are parts, whereas computers are the whole.
- x = the number of laptops sold.
- The parts add up to the whole. The 3 and 5 correspond to laptops and desktops, which are parts. Add these together to get 8, corresponding to the whole. The ratio of laptops to computers is 3:8.
- The 3 and x both correspond to laptops. The 8 and 120 both correspond to computers. The ratio 3:8 is laptops to computers: $\frac{3}{8} = \frac{x}{120}$. Laptops are in the numerator.
- Cross multiply: $360 = 8x$. Divide by 8 on both sides: $45 = x$.
- Check the answer: $\frac{x}{120} = \frac{45}{120} = \frac{45/15}{120/15} = \frac{3}{8}$.
- The number of desktops sold is $120 - x = 120 - 45 = 75$. The ratio of laptops to desktops is $\frac{45}{75} = \frac{45/15}{75/15} = \frac{3}{5}$. It should make sense that fewer laptops (45) were sold than desktops (75).

Alternate solution:

- x = the number of laptops sold and $120 - x$ = the number of desktops sold.
- The ratio 3:5 is laptops to desktops: $\frac{3}{5} = \frac{x}{120-x}$.
- Cross multiply: $3(120 - x) = 5x$. Distribute: $360 - 3x = 5x$. Add $3x$ to both sides: $360 = 8x$. Divide by 8 on both sides: $45 = x$.

Exercise Set 7.8

Directions: Use algebra to solve each word problem.

1) The ratio of apples to oranges in a barrel is 9:5. If there are 45 apples in the barrel, how many oranges are in the barrel?

2) A movie theater sold children's tickets and adult tickets in the ratio 3:7. If the movie theater sold 150 tickets, how many tickets of each kind were sold?

3) A jar contains pennies and nickels in the ratio 11:4. If the value of the nickels is $3, what is the total value of the coins in the jar (if it only contains pennies and nickels)?

4) A box contains red, white, and blue buttons in the ratio 5:3:4. There are 480 buttons in the box. How many buttons of each color are there?

7.9 Word Problems with Rates

A **rate** is a fraction where the numerator and denominator are expressed in different units. For example, the rate of speed has units of distance in the numerator and units of time in the denominator, like $\frac{400 \text{ miles}}{7 \text{ hours}}$. Rates are typically expressed in decimal form. For example, $\frac{8 \text{ meters}}{5 \text{ seconds}}$ would typically be expressed as 1.6 m/s (since $\frac{8}{5} = \frac{8 \cdot 2}{5 \cdot 2} = \frac{16}{10} = 1.6$).

If an object travels with constant speed, the distance traveled and time are related by the **rate equation**:

$$r = \frac{d}{t}$$

- r represents the speed (which is a rate). In a word problem, the speed has a ratio of units, like m/s, km/hr, or mph (miles per hour).
- d represents the distance traveled. It is expressed in units of distance, like m, km, or miles.
- t represents time. It is expressed in units like seconds or hours.

The units need to be consistent. For example, if the speed is given in m/s, the distance needs to be expressed in meters and the time needs to be expressed in seconds. Note that 1 km = 1000 m, 1 m = 100 cm, 1 m = 1000 mm, 1 cm = 10 mm, 1 yard = 3 ft., 1 ft. = 12 in., 1 hr. = 60 min., 1 min. = 60 s, and 1 hr. = 3600 s.

Example 1. A girl rides a bicycle with a constant speed of 12 m/s. How far will she travel in 2 minutes?

- $r = 12$ m/s is the speed.
- $t = 2$ min. is the time. Since the speed is in m/s, we need to convert the time to minutes. Since 1 min. = 60 s, the time is $t = 2(60) = 120$ s.
- $r = \frac{d}{t}$ Plug $r = 12$ m/s and $t = 120$ s into the rate equation: $12 = \frac{d}{120}$.
- Multiply by 120 on both sides: 1440 m = d.

Exercise Set 7.9

Directions: Use the rate equation to solve each word problem.

1) A car travels with a constant speed of 48 mph for 30 minutes. How far does it travel?

2) A bug travels 75 inches in 5 seconds. What is the bug's speed?

3) A bird travels 24 kilometers with a speed of 16 km/hr. How much time does this trip take?

8 SYSTEMS OF EQUATIONS

8.1 Substitution

If a system of two equations has two unknowns, one way to solve for the variables is to use the method of **substitution**. This means to solve for one variable in terms of the other in one equation, and then plug this expression in place of the variable in the other equation. For example, consider the system of equations below.

$$2x + y = 11 \quad \text{(Eq. 1)}$$
$$5x - 2y = 5 \quad \text{(Eq. 2)}$$

We can isolate y in Eq. 1 by subtracting $2x$ from both sides:

$$y = 11 - 2x \quad \text{(Eq. 3)}$$

This allows us to replace y with the expression $11 - 2x$ in Eq. 2:

$$5x - 2(11 - 2x) = 5 \quad \text{(Eq. 4)}$$

Distribute like we did in Chapter 4:

$$5x - 2(11) - 2(-2x) = 5 \quad \text{(Eq. 5)}$$

Simplify:

$$5x - 22 + 4x = 5 \quad \text{(Eq. 6)}$$

Combine like terms:

$$9x - 22 = 5 \quad \text{(Eq. 7)}$$

Add 22 to both sides:

$$9x = 27 \quad \text{(Eq. 8)}$$

Divide by 9 on both sides:

$$x = \frac{27}{9} = \boxed{3}$$

Now that we know x, we can plug this into any equation that has both x and y to solve for y. Choose wisely. The simplest equation would be Eq. 3:

$$y = 11 - 2x = 11 - 2(3) = 11 - 6 = \boxed{5}$$

Check the answers by plugging them into the original equations.

- Eq. 1: $2x + y = 2(3) + 5 = 6 + 5 = 11$
- Eq. 2: $5x - 2y = 5(3) - 2(5) = 15 - 10 = 5$

Example 1. Solve for the unknowns in the system of equations below.

$$7x - 4y = 36 \quad \text{(Eq. 1)}$$
$$2x + 5y = 41 \quad \text{(Eq. 2)}$$

- Subtract $5y$ from both sides of Eq. 2: $2x = 41 - 5y$ (Eq. 3)
- Divide by 2 on both sides of Eq. 3: $x = \frac{41}{2} - \frac{5y}{2}$ (Eq. 4)
- Replace x with $\frac{41}{2} - \frac{5y}{2}$ in Eq. 1: $7\left(\frac{41}{2} - \frac{5y}{2}\right) - 4y = 36$ (Eq. 5)
- Distribute: $\frac{7(41)}{2} - \frac{7(5y)}{2} - 4y = 36$ (Eq. 6)
- Simplify: $\frac{287}{2} - \frac{35y}{2} - 4y = 36$ (Eq. 7)
- Multiply by 2 on both sides to eliminate the fractions:
  $287 - 35y - 8y = 72$ (Eq. 8)
- Combine like terms: $287 - 43y = 72$ (Eq. 9)
- Subtract 287 from both sides: $-43y = -215$ (Eq. 10)
- Divide by -43 on both sides: $y = \frac{-215}{-43} = \boxed{5}$
- Plug $y = 5$ into Eq. 4: $x = \frac{41}{2} - \frac{5y}{2} = \frac{41}{2} - \frac{5(5)}{2} = \frac{41}{2} - \frac{25}{2} = \frac{16}{2} = \boxed{8}$

Check the answers by plugging $x = 8$ and $y = 5$ into the original equations.

- Eq. 1: $7x - 4y = 7(8) - 4(5) = 56 - 20 = 36$
- Eq. 2: $2x + 5y = 2(8) + 5(5) = 16 + 25 = 41$

Exercise Set 8.1

Directions: Use the method of substitution to solve for each variable.

1)

$$7x + 3y = 13$$
$$9x + 8y = -4$$

2)

$$6x - y = 48$$
$$5x + 3y = 63$$

3)

$$8x + 5y = 11$$
$$5x - 3y = 13$$

4)

$6x - 5y = -2$

$7x - 9y = -34$

5)

$7x - 3y = 35$

$-6x + 5y = -47$

6)

$$5x - 7y = 20$$
$$2x + 3y = 95$$

7)

$$8x + 6y = 22$$
$$-6x + 9y = 6$$

Want more practice? Solve the problems of Sec.'s 8.4 and 8.6 on scratch paper.

8.2 Substitution with Three Unknowns

The method of substitution may also be applied to a system of three equations with three different variables. Following is an outline of the strategy:

- Choose one equation. Solve for one variable (such as z) in terms of the other two variables.
- Replace that variable with the expression from the first step in each of the other two equations.
- Now there will be two equations with two unknowns, which can be solved like the problems from Sec. 8.1.

Tip: Don't plug an expression back into an equation that you have already used. For example, if you solve for z in terms of x and y in the top equation and replace z with this expression in the other equations, and then you solve for y in terms of x in the middle equation, you need to replace y with this expression in the only equation where you haven't yet solved for a variable (which, in this example, is the bottom equation). If you plug an expression into an equation where you have already solved for a variable, the algebra will lead to something trivial like $0 = 0$ or like $2 = 2$ (if you carry out the algebra correctly; if you make a mistake, you'll get something worse, like $1 = 3$).

However, once you get a numerical answer for a variable, then you may plug this value into any equation, whether or not it has already been used.

Example 1. Solve for the unknowns in the system of equations below.

$$5x + 3y + 2z = 36 \quad \text{(Eq. 1)}$$
$$4x - 5y + 3z = 21 \quad \text{(Eq. 2)}$$
$$8x - 2y - 5z = 3 \quad \text{(Eq. 3)}$$

- Subtract $5x$ and $3y$ from both sides of Eq. 1: $2z = 36 - 5x - 3y$ (Eq. 4)
- Divide by 2 on both sides of Eq. 4: $z = 18 - \frac{5x}{2} - \frac{3y}{2}$ (Eq. 5)
- Replace z with $18 - \frac{5x}{2} - \frac{3y}{2}$ in Eq.'s 2 and 3 (the as-of-yet unused equations). We now have two equations with two variables, like the problems in Sec. 8.1.

$$4x - 5y + 3\left(18 - \frac{5x}{2} - \frac{3y}{2}\right) = 21 \quad \text{(Eq. 6)}$$

$$8x - 2y - 5\left(18 - \frac{5x}{2} - \frac{3y}{2}\right) = 3 \quad \text{(Eq. 7)}$$

- Distribute. Remember to distribute the minus sign of the -5.

$$4x - 5y + 54 - \frac{15x}{2} - \frac{9y}{2} = 21 \quad \text{(Eq. 8)}$$

$$8x - 2y - 90 + \frac{25x}{2} + \frac{15y}{2} = 3 \quad \text{(Eq. 9)}$$

- Multiply by 2 on both sides to eliminate the fractions:

$$8x - 10y + 108 - 15x - 9y = 42 \quad \text{(Eq. 10)}$$

$$16x - 4y - 180 + 25x + 15y = 6 \quad \text{(Eq. 11)}$$

- Combine like terms:

$$-7x - 19y = -66 \quad \text{(Eq. 12)}$$

$$41x + 11y = 186 \quad \text{(Eq. 13)}$$

- Subtract $41x$ from both sides of Eq. 13: $11y = 186 - 41x$ (Eq. 14)

- Divide by 11 on both sides: $y = \frac{186}{11} - \frac{41x}{11}$ (Eq. 15)

- Replace y with $\frac{186}{11} - \frac{41x}{11}$ in Eq. 12 (since that is the equation where a variable hasn't yet been isolated): $-7x - 19\left(\frac{186}{11} - \frac{41x}{11}\right) = -66$ (Eq. 16)

- Distribute: $-7x - \frac{3534}{11} + \frac{779x}{11} = -66$ (Eq. 17)

- Multiply by 11 on both sides: $-77x - 3534 + 779x = -726$ (Eq. 18)

- Combine like terms: $702x = 2808$ (Eq. 19)

- Divide by 702 on both sides: $x = \frac{2808}{702} = \boxed{4}$

- Plug $x = 4$ into Eq. 15: $y = \frac{186}{11} - \frac{41x}{11} = \frac{186}{11} - \frac{41(4)}{11} = \frac{186}{11} - \frac{164}{11} = \frac{22}{11} = \boxed{2}$

- Plug $x = 4$ and $y = 2$ into Eq. 5: $z = 18 - \frac{5x}{2} - \frac{3y}{2} = 18 - \frac{5(4)}{2} - \frac{3(2)}{2} = 18 - 10 - 3 = \boxed{5}$

Check the answers by plugging $x = 4$, $x = 2$, and $z = 5$ into the original equations.

- Eq. 1: $5x + 3y + 2z = 5(4) + 3(2) + 2(5) = 20 + 6 + 10 = 36$

- Eq. 2: $4x - 5y + 3z = 4(4) - 5(2) + 3(5) = 16 - 10 + 15 = 21$

- Eq. 3: $8x - 2y - 5z = 8(4) - 2(2) - 5(5) = 32 - 4 - 25 = 3$

Exercise Set 8.2

Directions: Use the method of substitution to solve for each variable.

1)

$$5x + 2z = 24$$
$$3x + 4y = 40$$
$$3y - 7z = 7$$

2)

$$5x + 4y + 3z = 90$$
$$10x - 3y - 6z = 5$$
$$9x - 4y - 5z = 2$$

3)

$$4x - 5z = 13$$
$$5x + 4y + 3z = 2$$
$$9x - 3y - 6z = -21$$

4)

$$4x + 8y - 9z = -2$$
$$-6x - 4y + 3z = 10$$
$$7x + 6y + 6z = -3$$

Want more practice? Solve the problems of Sec. 8.7 on scratch paper.

8.3 Eliminating Unknowns from Formulas

In applications of algebra, such as science and engineering, it is often convenient to use substitution to eliminate a variable from a set of formulas. For example, consider the formulas $PV = nRT$ and $d = \frac{nM}{V}$. Suppose that we know the numerical values for R, T, M, and d, and that we wish to calculate P, but we don't know n or V. We can solve for n in the second equation to get $\frac{Vd}{M} = n$, which allows us to replace n with $\frac{Vd}{M}$ in the first formula: $PV = \left(\frac{Vd}{M}\right) RT$, which is equivalent to $PV = \frac{VdRT}{M}$. If we divide by V on both sides, this symbol cancels out, and we get: $P = \frac{dRT}{M}$. We made a substitution that effectively eliminated the variables n and V, giving us a formula for P in terms of R, T, M, and d. We could now plug in the numerical values for R, T, M, and d in order to determine the numerical value for P (like we did in Sec. 1.13).

Example 1. Eliminate W from $P = 2L + 2W$ and $A = LW$. Solve for A.
- Subtract $2L$ from both sides of the left equation: $P - 2L = 2W$.
- Divide by 2 on both sides: $\frac{P}{2} - L = W$.
- Replace W with $\frac{P}{2} - L$ in the right equation: $A = L\left(\frac{P}{2} - L\right)$.
- Distribute: $A = \frac{LP}{2} - L^2$.

As directed, the final answer has A isolated and doesn't involve W.

Example 2. Eliminate V from $Q = CV$ and $U = \frac{1}{2}QV$. Solve for C.
- Divide by C on both sides of the left equation: $\frac{Q}{C} = V$.
- Replace V with $\frac{Q}{C}$ in the right equation: $U = \frac{1}{2}Q\left(\frac{Q}{C}\right) = \frac{Q^2}{2C}$.
- Multiply by C on both sides: $CU = \frac{Q^2}{2}$.
- Divide by U on both sides: $C = \frac{Q^2}{2U}$.

As directed, the final answer has C isolated and doesn't involve V.

Exercise Set 8.3

Directions: Solve for the indicated variable as directed.

1) Eliminate R from $D = 2R$ and $A = \pi R^2$. Solve for A.

2) Eliminate I from $V = IR$ and $P = IV$. Solve for R.

3) Eliminate L from $V = L^3$ and $S = 6L^2$. Solve for V.

4) Eliminate E from $E = mgh$ and $E = \frac{1}{2}kx^2$. Solve for x.

5) Eliminate t from $v = \frac{s}{t}$ and $\theta = \omega t$. Solve for s.

6) Eliminate p from $\frac{1}{p} + \frac{1}{q} = \frac{1}{f}$ and $q = 3p$. Solve for f.

7) Eliminate t from $v_f = v_i + at$ and $x = v_i t + \frac{1}{2}at^2$. Solve for v_f.

8) Eliminate v from $v = \frac{2\pi R}{T}$ and $G\frac{m}{R^2} = \frac{v^2}{R}$. Solve for T.

8.4 Simultaneous Equations

An alternative method for solving a system of equations is to apply a method known as **simultaneous equations**. This method is often more efficient than substitution. The main idea behind the method of simultaneous equations is to multiply each equation by the factor needed to make equal and opposite coefficients for the same variable. For example, consider the system of equations below.

$$2x + 3y = 16 \quad (\text{Eq. 1})$$
$$7x - 6y = 23 \quad (\text{Eq. 2})$$

If we multiply every term of the top equation by 2, the coefficients of y in the two equations will be 6 and -6.

$$4x + 6y = 32 \quad (\text{Eq. 3})$$

Now we can add Eq.'s 2 and 3 together. We set the sum of the left-hand sides equal to the sum of the right-hand sides.

$$7x - 6y + 4x + 6y = 23 + 32 \quad (\text{Eq. 4})$$

Combine like terms. Note that y cancels out because $-6y + 6y = 0$.

$$11x = 55 \quad (\text{Eq. 5})$$

Divide by 11 on both sides: $x = \frac{55}{11} = 5$. Plug $x = \boxed{5}$ into any of the equations to solve for y. We'll choose Eq. 1: $2x + 3y = 16 \rightarrow 2(5) + 3y = 16 \rightarrow 10 + 3y = 16 \rightarrow 3y = 6 \rightarrow y = \frac{6}{3} = \boxed{2}$. Check the answers by plugging them into the original equations.

- Eq. 1: $2x + 3y = 2(5) + 3(2) = 10 + 6 = 16$
- Eq. 2: $7x - 6y = 7(5) - 6(2) = 35 - 12 = 23$

Why is it okay to add two equations together? Consider the equations $a = b$ and $c = d$. If we add the left-hand sides and right-hand sides of these equations together, we get $a + c = b + d$. Since $a = b$ and $c = d$, this is equivalent to $b + d = b + d$. The idea is this: Since the two sides of each equation are equal, the sum of the two sides will also be equal.

When you multiply an equation, be careful to multiply every term on both sides by the same factor. Similarly, when you add two equations together, be careful to add *all* of the terms on both sides. It's a common mistake for a student to forget to multiply or add one term on one side of one equation. You can avoid this be paying attention closely when you perform these steps.

Example 1. Solve for the unknowns in the system of equations below.

$$5x + 8y = 62 \quad \text{(Eq. 1)}$$
$$3x - 2y = 10 \quad \text{(Eq. 2)}$$

- Multiply by 4 both sides of Eq. 2: $12x - 8y = 40$ (Eq. 3)
- Add Eq.'s 1 and 3 together. Set the sum of the left-hand sides equal to the sum of the right-hand sides: $5x + 8y + 12x - 8y = 62 + 40$ (Eq. 4)
- Combine like terms. Note that y cancels out: $17x = 102$ (Eq. 5)
- Divide by 17 on both sides: $x = \frac{102}{17} = \boxed{6}$ Plug $x = 6$ into Eq. 2: $3x - 2y = 10 \rightarrow$

$$3(6) - 2y = 10 \rightarrow 18 - 2y = 10 \rightarrow -2y = -8 \rightarrow y = \frac{-8}{-2} = \boxed{4}.$$

Check the answers by plugging $x = 6$ and $y = 4$ into the original equations.

- Eq. 1: $5x + 8y = 5(6) + 8(4) = 30 + 32 = 62$
- Eq. 2: $3x - 2y = 3(6) - 2(4) = 18 - 8 = 10$

Example 2. Solve for the unknowns in the system of equations below.

$$6x + 5y = 48 \quad \text{(Eq. 1)}$$
$$4x + 3y = 30 \quad \text{(Eq. 2)}$$

- Multiply both sides of Eq. 1 by 2 and both sides of Eq. 2 by -3. The minus sign is needed to make opposite coefficients.

$$12x + 10y = 96 \quad \text{(Eq. 3)}$$
$$-12x - 9y = -90 \quad \text{(Eq. 4)}$$

- Add Eq.'s 3 and 4 together: $12x + 10y - 12x - 9y = 96 - 90$ (Eq. 5)
- Combine like terms. Note that x cancels out: $y = \boxed{6}$ Plug $y = 6$ into Eq. 1:

$$6x + 5y = 48 \rightarrow 6x + 5(6) = 48 \rightarrow 6x + 30 = 48 \rightarrow 6x = 18 \rightarrow x = \frac{18}{6} = \boxed{3}$$

Check the answers by plugging $x = 3$ and $y = 6$ into the original equations.

- Eq. 1: $6x + 5y = 6(3) + 5(6) = 18 + 30 = 48$
- Eq. 2: $4x + 3y = 4(3) + 3(6) = 12 + 18 = 30$

Exercise Set 8.4

Directions: Use the method of simultaneous equations to solve for each variable.

1)

$$8x - 3y = 11$$
$$5x + 9y = 83$$

2)

$$4x + 5y = 57$$
$$-3x + 8y = 16$$

3)

$$7x + 5y = 85$$
$$4x + 6y = 58$$

4)

$$6x - 3y = 75$$
$$4x - 8y = 92$$

5)

$$8x - 5y = -13$$
$$-9x + 3y = 33$$

6)

$$-8x + 5y = 68$$
$$-3x + 4y = 51$$

7)

$$8x + 9y = 16$$
$$-4x + 3y = 2$$

8)

$$-10x - 12y = 3$$
$$-15x - 8y = -3$$

Want more practice? Solve the problems of Sec.'s 8.1 and 8.6 on scratch paper.

8.5 Determinants

A **matrix** has numbers arranged in an array. The **determinant** of a 2 × 2 matrix is a single number found by multiplying the two diagonals and subtracting as shown below. When we enclose an array in parentheses ( ), this refers to the matrix. If we write the letters "det" before the matrix, or if we enclose an array in straight lines | |, this means to find the determinant of the matrix.

$$\det \begin{pmatrix} a & b \\ c & d \end{pmatrix} = \begin{vmatrix} a & b \\ c & d \end{vmatrix} = ad - bc$$

For a 3 × 3 matrix, the determinant can be found by multiplying along six diagonals. The three diagonals going down to the left are subtracted from the three diagonals going up to the right. To help visualize all six diagonals, it helps to repeat the first two columns to the right.

$$\det \begin{pmatrix} a & b & c \\ d & e & f \\ g & h & i \end{pmatrix} = \begin{vmatrix} a & b & c \\ d & e & f \\ g & h & i \end{vmatrix} = \begin{matrix} a & b & c & a & b \\ d & e & f & d & e \\ g & h & i & g & h \end{matrix} = aei + bfg + cdh - ceg - afh - bdi$$

Example 1.

$$\det \begin{pmatrix} 2 & 3 \\ 4 & 5 \end{pmatrix} = \begin{vmatrix} 2 & 3 \\ 4 & 5 \end{vmatrix} = 2(5) - 3(4) = 10 - 12 = -2$$

Example 2.

$$\det \begin{pmatrix} 2 & 3 & 4 \\ 5 & 6 & 7 \\ 8 & 9 & 1 \end{pmatrix} = \begin{vmatrix} 2 & 3 & 4 \\ 5 & 6 & 7 \\ 8 & 9 & 1 \end{vmatrix} = \begin{matrix} 2 & 3 & 4 & 2 & 3 \\ 5 & 6 & 7 & 5 & 6 \\ 8 & 9 & 1 & 8 & 9 \end{matrix}$$

$$= 2(6)(1) + 3(7)(8) + 4(5)(9) - 4(6)(8) - 2(7)(9) - 3(5)(1)$$

$$= 12 + 168 + 180 - 192 - 126 - 15 = 360 - 333 = 27$$

Example 3.

$$\det \begin{pmatrix} 7 & 5 & 9 \\ 1 & 3 & 4 \\ 6 & 8 & 2 \end{pmatrix} = \begin{vmatrix} 7 & 5 & 9 \\ 1 & 3 & 4 \\ 6 & 8 & 2 \end{vmatrix} = \begin{matrix} 7 & 5 & 9 & 7 & 5 \\ 1 & 3 & 4 & 1 & 3 \\ 6 & 8 & 2 & 6 & 8 \end{matrix}$$

$$= 7(3)(2) + 5(4)(6) + 9(1)(8) - 9(3)(6) - 7(4)(8) - 5(1)(2)$$

$$= 42 + 120 + 72 - 162 - 224 - 10 = 234 - 396 = -162$$

Exercise Set 8.5

Directions: Evaluate each determinant.

1)

$$\begin{vmatrix} 8 & 5 \\ 7 & 6 \end{vmatrix} =$$

2)

$$\begin{vmatrix} -3 & 6 \\ 4 & 9 \end{vmatrix} =$$

3)

$$\begin{vmatrix} 4 & -5 \\ 9 & 8 \end{vmatrix} =$$

4)

$$\begin{vmatrix} 7 & -6 \\ -3 & 7 \end{vmatrix} =$$

5)

$$\begin{vmatrix} 8 & 12 \\ 6 & 9 \end{vmatrix} =$$

6)

$$\begin{vmatrix} -6 & -7 \\ 8 & -9 \end{vmatrix} =$$

7)

$$\begin{vmatrix} 4 & 3 & 8 \\ 9 & 5 & 1 \\ 2 & 7 & 6 \end{vmatrix} =$$

8)

$$\begin{vmatrix} 5 & -2 & 7 \\ 8 & 4 & 1 \\ -6 & 0 & 3 \end{vmatrix} =$$

9)

$$\begin{vmatrix} 2 & 2 & 2 \\ 2 & 2 & -2 \\ 2 & -2 & -2 \end{vmatrix} =$$

10)

$$\begin{vmatrix} 9 & 4 & 1 \\ 3 & 8 & -13 \\ -5 & 2 & -9 \end{vmatrix} =$$

8.6 Cramer's Rule

Cramer's rule is a method for solving a system of equations that uses determinants. To use Cramer's rule, first write the equations in the following form. This form has the variable terms on the left, in order, and the constant terms on the right. Note that the constants may be negative values.

$$a_1 x + b_1 y = c_1$$
$$a_2 x + b_2 y = c_2$$

For two equations with two unknowns, we need to make three determinants:

- The first determinant, which we will call D_c, uses the coefficients of x and y.

$$D_c = \begin{vmatrix} a_1 & b_1 \\ a_2 & b_2 \end{vmatrix}$$

- The second determinant, which we will call D_x, replaces the coefficients of x with the constant terms from the right side of the equation.

$$D_x = \begin{vmatrix} c_1 & b_1 \\ c_2 & b_2 \end{vmatrix}$$

- The third determinant, which we will call D_y, replaces the coefficients of y with the constant terms from the right side of the equation.

$$D_y = \begin{vmatrix} a_1 & c_1 \\ a_2 & c_2 \end{vmatrix}$$

The variables can be found by taking ratios of these determinants:

$$x = \frac{D_x}{D_c} \quad , \quad y = \frac{D_y}{D_c}$$

Example 1. Solve for the unknowns in the system of equations below.

$$3x + 2y = 16$$
$$10x - 3y = 5$$

$$D_c = \begin{vmatrix} a_1 & b_1 \\ a_2 & b_2 \end{vmatrix} = \begin{vmatrix} 3 & 2 \\ 10 & -3 \end{vmatrix} = 3(-3) - 2(10) = -9 - 20 = -29$$

$$D_x = \begin{vmatrix} c_1 & b_1 \\ c_2 & b_2 \end{vmatrix} = \begin{vmatrix} 16 & 2 \\ 5 & -3 \end{vmatrix} = 16(-3) - 2(5) = -48 - 10 = -58$$

$$D_y = \begin{vmatrix} a_1 & c_1 \\ a_2 & c_2 \end{vmatrix} = \begin{vmatrix} 3 & 16 \\ 10 & 5 \end{vmatrix} = 3(5) - 16(10) = 15 - 160 = -145$$

$$x = \frac{D_x}{D_c} = \frac{-58}{-29} = \boxed{2} \quad , \quad y = \frac{D_y}{D_c} = \frac{-145}{-29} = \boxed{5}$$

Check the answers by plugging $x = 2$ and $y = 5$ into the original equations.

- $3x + 2y = 3(2) + 2(5) = 6 + 10 = 16$
- $10x - 3y = 10(2) - 3(5) = 20 - 15 = 5$

Exercise Set 8.6

Directions: Use Cramer's rule to solve for each variable.

1)

$$4x - 6y = 2$$
$$3x - 5y = -1$$

2)

$$7x + 3y = 15$$
$$3x + 7y = -45$$

3)

$$5x + 9y = 7$$
$$6x - 2y = -30$$

4)

$8x + y = 70$

$-3x + 5y = 49$

5)

$6x - 8y = -1$

$5x + 2y = 10$

6)

$$4x - 6y = 68$$
$$-5x + 3y = -67$$

7)

$$-8x + 9y = -12$$
$$-2x - 3y = -17$$

Want more practice? Solve the problems of Sec.'s 8.1 and 8.4 on scratch paper.

8.7 Cramer's Rule with Three Unknowns

When Cramer's rule is applied to three equations with three variables, the determinants involve 3×3 matrices. First put the equations in the following form:

$$a_1 x + b_1 y + c_1 z = d_1$$
$$a_2 x + b_2 y + c_2 z = d_2$$
$$a_3 x + b_3 y + c_3 z = d_3$$

Note that the d's are the constant terms on the right, while the c's are the coefficients of z. The determinants are:

$$D_c = \begin{vmatrix} a_1 & b_1 & c_1 \\ a_2 & b_2 & c_2 \\ a_3 & b_3 & c_3 \end{vmatrix}, D_x = \begin{vmatrix} d_1 & b_1 & c_1 \\ d_2 & b_2 & c_2 \\ d_3 & b_3 & c_3 \end{vmatrix}, D_y = \begin{vmatrix} a_1 & d_1 & c_1 \\ a_2 & d_2 & c_2 \\ a_3 & d_3 & c_3 \end{vmatrix}, D_z = \begin{vmatrix} a_1 & b_1 & d_1 \\ a_2 & b_2 & d_2 \\ a_3 & b_3 & d_3 \end{vmatrix}$$

$$x = \frac{D_x}{D_c} \quad, \quad y = \frac{D_y}{D_c} \quad, \quad z = \frac{D_z}{D_c}$$

Example 1. Solve for the unknowns in the system of equations below.

$$5x + 2y + 3z = 21$$
$$3x + 4y - 6z = 16$$
$$2x - y + 5z = 5$$

$$D_c = \begin{vmatrix} a_1 & b_1 & c_1 \\ a_2 & b_2 & c_2 \\ a_3 & b_3 & c_3 \end{vmatrix} = \begin{vmatrix} 5 & 2 & 3 \\ 3 & 4 & -6 \\ 2 & -1 & 5 \end{vmatrix} = \begin{matrix} 5 & 2 & 3 & 5 & 2 \\ 3 & 4 & -6 & 3 & 4 \\ 2 & -1 & 5 & 2 & -1 \end{matrix}$$

$$D_c = 5(4)(5) + 2(-6)(2) + 3(3)(-1) - 3(4)(2) - 5(-6)(-1) - 2(3)(5)$$

$$D_c = 100 - 24 - 9 - 24 - 30 - 30 = 100 - 117 = -17$$

$$D_x = \begin{vmatrix} d_1 & b_1 & c_1 \\ d_2 & b_2 & c_2 \\ d_3 & b_3 & c_3 \end{vmatrix} = \begin{vmatrix} 21 & 2 & 3 \\ 16 & 4 & -6 \\ 5 & -1 & 5 \end{vmatrix} = \begin{matrix} 21 & 2 & 3 & 21 & 2 \\ 16 & 4 & -6 & 16 & 4 \\ 5 & -1 & 5 & 5 & -1 \end{matrix}$$

$$D_x = 21(4)(5) + 2(-6)(5) + 3(16)(-1) - 3(4)(5) - 21(-6)(-1) - 2(16)(5)$$

$$D_x = 420 - 60 - 48 - 60 - 126 - 160 = 420 - 454 = -34$$

$$D_y = \begin{vmatrix} a_1 & d_1 & c_1 \\ a_2 & d_2 & c_2 \\ a_3 & d_3 & c_3 \end{vmatrix} = \begin{vmatrix} 5 & 21 & 3 \\ 3 & 16 & -6 \\ 2 & 5 & 5 \end{vmatrix} = \begin{matrix} 5 & 21 & 3 & 5 & 21 \\ 3 & 16 & -6 & 3 & 16 \\ 2 & 5 & 5 & 2 & 5 \end{matrix}$$

$$D_y = 5(16)(5) + 21(-6)(2) + 3(3)(5) - 3(16)(2) - 5(-6)(5) - 21(3)(5)$$

$$D_y = 400 - 252 + 45 - 96 + 150 - 315 = 595 - 663 = -68$$

$$D_z = \begin{vmatrix} a_1 & b_1 & d_1 \\ a_2 & b_2 & d_2 \\ a_3 & b_3 & d_3 \end{vmatrix} = \begin{vmatrix} 5 & 2 & 21 \\ 3 & 4 & 16 \\ 2 & -1 & 5 \end{vmatrix} = \begin{matrix} 5 & 2 & 21 \\ 3 & 4 & 16 \\ 2 & -1 & 5 \end{matrix} \begin{matrix} 5 & 2 \\ 3 & 4 \\ 2 & -1 \end{matrix}$$

$$D_z = 5(4)(5) + 2(16)(2) + 21(3)(-1) - 21(4)(2) - 5(16)(-1) - 2(3)(5)$$

$$D_z = 100 + 64 - 63 - 168 + 80 - 30 = 244 - 261 = -17$$

$$x = \frac{D_x}{D_c} = \frac{-34}{-17} = \boxed{2} \quad , \quad y = \frac{D_y}{D_c} = \frac{-68}{-17} = \boxed{4} \quad , \quad z = \frac{D_z}{D_c} = \frac{-17}{-17} = \boxed{1}$$

Check the answers by plugging $x = 2$, $y = 4$, and $z = 1$ into the original equations.

- $5x + 2y + 3z = 5(2) + 2(4) + 3(1) = 10 + 8 + 3 = 21$
- $3x + 4y - 6z = 3(2) + 4(4) - 6(1) = 6 + 16 - 6 = 16$
- $2x - y + 5z = 2(2) - 4 + 5(1) = 4 - 4 + 5 = 5$

Exercise Set 8.7

Directions: Use Cramer's rule to solve for each variable.

1)

$$3x + 2y + 2z = 20$$
$$2x + 8y + 3z = 25$$
$$x + 5y + 2z = 15$$

2)

$$5x + 4y + 9z = 150$$
$$10x - 3z = 50$$
$$3x - 10y + 5z = 24$$

3)

$$3x + 4y + 3z = 15$$
$$4x - 3y - 5z = -10$$
$$-9x + 5y + 7z = 28$$

Want more practice? Solve the problems of Sec. 8.2 on scratch paper.

8.8 Special Solutions for Systems of Equations

Beware that not every system of equations has a unique set of solutions. For example:

- $x + y = 5$ and $x + y = 8$ have **no solution**. If you isolate y in the first equation, you get $y = 5 - x$. If you plug this into the second equation, you get $x + 5 - x = 8$. Since $x - x = 0$, x cancels out, leaving $5 = 8$, which isn't true. Another way to see that there is no solution to this problem is that $x + y$ can't equal 5 and also equal 8. The sum $x + y$ must be single-valued.

- The solution to $2x + 2y = 4$ and $3x + 3y = 6$ is **indeterminate**. This system doesn't provide enough information to solve for a unique solution. If we divide both sides of the first equation by 2, we get $x + y = 2$, and if we divide both sides of the second equation by 3, we also get $x + y = 2$, showing that these two equations are effectively the same. In order to obtain a unique solution to two equations with two unknowns, the two equations must be independent. These equations aren't independent.

- $x + y = y + x$ and $x + 1 - y = 1 + x - y$ are solved by **all real numbers**. If we subtract x and y from both sides of the first equation, we get $0 = 0$, and if we subtract x and add y to the second equation, we get $1 = 1$. In each case, x and y cancel out, and the resulting equation is true regardless of the values of x and y.

- The solution to $x + y + z = 1$ and $x + y - z = 7$ is **indeterminate**. We can solve for z, but there isn't enough information to determine x or y. From the first equation, $x + y = 1 - z$. Plug this into the second equation: $1 - z - z = 7 \rightarrow -2z = 6 \rightarrow z = -3$. Now the equations are $x + y - 3 = 1 \rightarrow x + y = 4$ and $x + y - (-3) = 7 \rightarrow x + y = 4$, which are effectively the same. Since we don't have two independent equations for x and y, the values of x and y can't be determined. It takes N independent equations to solve for N unknowns. This problem gives only 2 equations, but there are 3 unknowns.

- $x + y = 10$, $x - y = 6$, and $2x + y = 5$ has **no solution**. There isn't a pair of values for x and y that satisfies all three of these equations. The third equation contradicts the first two equations. From the second equation, $x = y + 6$. Plug this into the first equation to get $y + y + 6 = 10 \rightarrow 2y = 4 \rightarrow y = 2$. Plug this into $x = y + 6 = 2 + 6 = 8$. Plug these values into the third equation to see that $2x + y = 2(8) + 2 = 16 + 2 = 18$, which doesn't equal 5. This problem gives 3 independent equations, but there are only 2 unknowns.

Exercise Set 8.8

Directions: Determine whether each system of equations has a unique set of answers, has no solution, has an indeterminate solution, or is solved by all real numbers.

1)

$$2x - y = 8$$
$$-2x + y = -8$$

2)

$$3x + 2y = 6$$
$$6x + 4y = 7$$

3)

$$3x + y + 2z = 5$$
$$4x - y + 3z = 9$$

4)

$$2x + 3y = x + 3y + x$$
$$x + y + 3 = 3x + y + 3 - 2x$$

5)

$$5x + 2y = 20$$
$$8x - 3y = 1$$
$$-3x + 7y = 15$$

6)

$$4x - 3y = 1 - 3y + 4x$$
$$2x + 5y + 7 = x + 5y + x$$

8.9 Word Problems with Two Unknowns

The word problems in this section involve two unknowns. It may help to review Sec. 2.13 (which has tips for solving word problems) and Sec. 1.4. Use two variables in your solutions. For example, you might use x and y for two different unknowns.

Example 1. Julia is three times as old as her daughter, Linda. Julia is 24 years older than Linda. How old is each person?
- J = Julia's age and L = Linda's age.
- $J = 3L$ and $J = L + 24$.
- Replace J with $L + 24$ in the first equation: $L + 24 = 3L$.
- Subtract L from both sides: $24 = 2L$.
- Divide by 2 on both sides: $12 = L$ (Linda is 12 years old).
- Plug $L = 12$ into either of the original equations: $J = 3L = 3(12) = 36$ (Julia is 36 years old).

Check: Julia (36) is 3 times as old as Linda (12) since $3(12) = 36$ and Julia is 24 years older than Linda since $12 + 24 = 36$.

Example 2. Two numbers have a sum of 50 and a difference of 18. What are the numbers?
- x = the smaller number and y = the larger number.
- $x + y = 50$ and $y - x = 18$.
- Since the coefficients of x are already equal and opposite, these equations are ready to be added together: $x + y + y - x = 50 + 18$.
- Combine like terms. Note that x cancels out: $2y = 68$.
- Divide by 2 on both sides: $y = 34$. Plug $y = 34$ into either of the original equations: $x + 34 = 50$. Subtract 34 from both sides: $x = 50 - 34 = 16$.

Check: 16 and 34 have a sum of $16 + 34 = 50$ and a difference of $34 - 16 = 18$.

Exercise Set 8.9

Directions: Use algebra with two different variables to solve each word problem.

1) Two numbers have a sum of 21. When the smaller number is doubled, the result is 3 less than the larger number.

2) Brian's age plus twice Kevin's age equals 30 years. Twice Brian's age minus Kevin's age equals 15 years. How old are Brian and Kevin?

3) A student buys four pencils and nine erasers for $4.25. Another student buys eight pencils and three erasers for $4.75. How much does it cost for each pencil and for each eraser? (All of the pencils and erasers are identical. There is no sales tax.)

4) A rectangle has a perimeter of 35 cm and an area of 75 square cm. What are the length and width of the rectangle?

9 INEQUALITIES

9.1 Interpreting Simple Inequalities

The following symbols express common inequalities:

- The greater than symbol ($>$) means that the quantity on the left is larger than the quantity on the right. For example, $x > 3$ means that x is greater than 3 and $5 > y$ means that 5 is greater than y.

- The less than symbol ($<$) means that the quantity on the left is smaller than the quantity on the right. For example, $x < 7$ means that x is smaller than 7 and $2 < y$ means that 2 is smaller than y.

- The greater than or equal to symbol ($\geq$) means that the quantity on the left is either greater than the quantity on the right or equal to the quantity on the right. For example, $x \geq 4$ means that x is at least as large as 4 (it could equal 4, or it could be greater than 4) and $9 \geq y$ means that 9 is at least as large as y (it could be greater than y, or it could be equal to y).

- The less than or equal to symbol ($\leq$) means that the quantity on the left is either smaller than the quantity on the right or equal to the quantity on the right. For example, $x \leq 8$ means that x is no greater than 8 (it could equal 8, or it could be smaller than 8) and $1 \leq y$ means that 1 is no greater than y (it could be smaller than y, or it could be equal to y).

- The inequal sign ($\neq$) means that the quantities on each side aren't equal to one another. We don't know which side is greater, but we do know that they are not equal. For example, $x \neq 6$ means that x is different from 6.

The vertex of the $<$ or $>$ symbol always points to the smaller quantity. For example, $x < 2$ means that x is smaller than 2 and $y > 8$ means that 8 is smaller than y (which is equivalent to saying that y is greater than 8).

Any inequality that uses a greater than symbol (or a greater than or equal to symbol) can be expressed using a less than symbol in an equivalent way. For example, $x > 4$ is equivalent to $4 < x$. If we swap the left and right sides of an inequality, we must also reverse the direction of the inequality. For example, $y < 3$ is equivalent to $3 > y$.

Example 1. If $x > 5$, which of these values could x be? $3, 8, 25$

8 and 25 are greater than 5.

Example 2. If $x \leq 2$, which of these values could x be? $1, 2, 3$

1 and 2 are less than or equal to 2.

Exercise Set 9.1

Directions: Which of the listed numbers is consistent with the given inequality?

1) $x < 4$ List: $-5, -3, 0, 3, 4, 5$

2) $x > 3$ List: $-4.5, -2.5, 0, 2.5, 3, 4.5$

3) $x \leq 5$ List: $-6, -4, 0, 4, 5, 6$

4) $x \geq 2$ List: $-3, -1, 1, 2, 3$

5) $6 < x$ List: $-7, -5, 5, 6, 7$

6) $7 > x$ List: $-8.2, -6.4, 6.4, 7, 8.2$

7) $x < -1$ List: $-2, -1, -0.5, 0.5, 1, 2$

8) $x \geq -8$ List: $-9, -8, -7, 7, 8, 9$

9.2 Inequalities with Positive Coefficients

You can isolate an unknown in an inequality the same way that you can isolate an unknown in an equation (Chapter 2), except when you need to multiply or divide both sides of an inequality by a negative number. We'll focus on this exception in Sec. 9.3. In the current section, you can use the strategy from Chapter 2 to isolate the unknown (provided that you avoid multiplying or dividing both sides by a negative number).

Example 1. $3x - 4 < 5$

- Add 4 to both sides: $3x < 9$.
- Divide by 3 on both sides: $x < 3$.

One way to check the answer is to find a decimal value that barely satisfies the answer and plug that into the original inequality, using a calculator. In this case, 2.9 is a little less than 3. Plug 2.9 in for x in the original inequality: $3x - 4 = 3(2.9) - 4 = 4.7$ is a little less than 5.

Example 2. $8x > 6x + 10$

- Subtract $6x$ from both sides: $2x > 10$.
- Divide by 2 on both sides: $x > 5$.

Check the answer: 5.1 is a little more than 5. Plug 5.1 in for x in the original inequality: $8x = 8(5.1) = 40.8$ is a little more than $6x + 10 = 6(5.1) + 10 = 40.6$.

Exercise Set 9.2

Directions: Isolate the variable in each inequality.

1) $x - 8 > 7$

2) $3x \leq 27$

3) $6x - 9 < 15$

4) $\frac{x}{6} \leq -7$

5) $7x \geq 40 - x$

6) $9x + 8 > 32 + 6x$

7) $\frac{x}{4} + 9 < 16$

8) $25 - 2x \leq 3x - 25$

9) $7x + 2 > 3x - 2$

10) $5x + 6 < 9x + 4$

11) $8x - 3 \leq x - 7$

12) $5x + 6 \geq 3x + 6$

9.3 Inequalities with Negative Coefficients

If you multiply or divide both sides of an inequality by a negative number, the direction of the inequality reverses. For example, if we divide by -2 on both sides of $-2x > -4$, the $>$ sign changes into a $<$ sign and we get $x < \frac{-4}{-2}$, which simplifies to $x < 2$.

To see why, let's go back to the original inequality: $-2x > -4$. This time, we'll isolate the variable using a longer method. Add $2x$ to both sides: $0 > 2x - 4$. Add 4 to both sides: $4 > 2x$. Divide by 2 on both sides: $2 > x$. Note that $2 > x$ is equivalent to $x < 2$. We could achieve the same answer more efficiently by dividing by -2 on both sides of $-2x > -4$, reversing the direction of the inequality in the process to get $x < 2$.

You can also understand the logic behind this idea as follows. Consider the inequality $x > 5$, which states that x is greater than 5. If we multiply by -1 on both sides, we'll get $-x$ on the left and -5 on the right. Since $x > 5$, it follows that $-x$ will be more negative than -5. This means that $-x < -5$. For example, suppose that $x = 8$. In this case, we get $8 > 5$. When we multiply both sides by -1, we get $-8 < -5$ when we reverse the direction of the inequality. Indeed, -8 is less than -5 because -8 is more negative than -5.

Note that multiplying or dividing by a *positive* value doesn't change the direction of an inequality, but multiplying or dividing by a *negative* value does change the direction of an inequality. Also note that *adding* or *subtracting* a negative value doesn't change the direction of an inequality.

Example 1. $3 - 2x > -5$
- Subtract 3 from both sides: $-2x > -8$.
- Divide by -2 (which requires reversing the inequality) on both sides: $x < 4$.

Check the answer: 3.9 is a little less than 4. Plug 3.9 in for x in the original inequality: $3 - 2x = 3 - 2(3.9) = -4.8$ is a little more than -5 (because -4.8 is less negative than -5, since -4.8 is closer to zero than -5).

Exercise Set 9.3

Directions: Isolate the variable in each inequality.

1) $-x < 1$

2) $-2x > -16$

3) $3 - 6x \geq 21$

4) $-\frac{x}{7} < 8$

5) $-8x - 10 > -50$

6) $70 \leq 14 - 7x$

7) $-24 < -9x + 12$

8) $-52 - 4x > -100$

9.4 Reciprocals of Inequalities

If you reciprocate both sides of an inequality, the direction of the inequality reverses if both sides are positive or if both sides are negative (but not when only one side is negative). For example, compare $\frac{1}{2} > \frac{1}{4}$ (equivalent to $0.5 > 0.25$) with $2 < 4$. If both sides are negative, we get the similar inequalities $-\frac{1}{2} < -\frac{1}{4}$ and $-2 > -4$ (found by multiplying both sides of the previous inequalities by -1). Note that $-\frac{1}{2}$ is less than $-\frac{1}{4}$ because $-\frac{1}{2}$ is more negative (it is farther from zero). Similarly, -2 is greater than -4 because -2 is less negative (it is closer to zero).

However, if only one side is positive, the direction of the inequality doesn't reverse when reciprocating both sides. For example, compare $-\frac{1}{2} < \frac{1}{4}$ with $-2 < 4$. No matter what, the negative quantity will always be smaller than the positive quantity.

Now let's consider inequalities where a variable is involved.

- When we reciprocate both sides of $\frac{1}{x} > \frac{1}{4}$, we get $x < 4$. Both sides are positive in the original inequality, so the direction of the inequality reverses. You can verify this with numbers. For example, let x equal 2. Observe that $2 < 4$ and that $\frac{1}{2} > \frac{1}{4}$ (since $0.5 > 0.25$). Another way to see this is to work it out in steps: Multiply by 4 on both sides of $\frac{1}{x} > \frac{1}{4}$ to get $\frac{4}{x} > 1$, then multiply by x on both sides to get $4 > x$, which is equivalent to $x < 4$. Since x is positive, $0 < x < 4$.

- When we reciprocate both sides of $\frac{1}{x} < -\frac{1}{3}$, we get $x > -3$. Both sides are negative in the original inequality, so the direction of the inequality reverses. Even though you only "see" one minus sign in $\frac{1}{x} < -\frac{1}{3}$, both sides are clearly negative, meaning that x is negative; there is no way that x could be positive and satisfy $\frac{1}{x} < -\frac{1}{3}$. For example, let x equal -2. Observe that $-2 > -3$ (since -2 is less negative) and that $-\frac{1}{2} < -\frac{1}{3}$ (since $-\frac{1}{2}$ is more negative). Since x is negative, the answer to $\frac{1}{x} < -\frac{1}{3}$ is $0 > x > -3$ (equivalent to $-3 < x < 0$).

- The case $\frac{1}{x} > -\frac{1}{3}$ is interesting. Note how this differs from the previous bullet point. This time, x could be positive and it could also be negative. A positive value of x will definitely satisfy $\frac{1}{x} > -\frac{1}{3}$, but there are also some negative values of x that will satisfy $\frac{1}{x} > -\frac{1}{3}$. For example, $-\frac{1}{4} > -\frac{1}{3}$ (since $-\frac{1}{4}$ is less negative) and $\frac{1}{4} > -\frac{1}{3}$ (since a positive number is always greater than a negative number). In the case of $-\frac{1}{4} > -\frac{1}{3}$, since both sides are negative, we reverse the direction of the inequality when we reciprocate both sides: $-4 < -3$. In the case of $\frac{1}{4} > -\frac{1}{3}$, since only one side is negative, we don't reverse the inequality: $4 > -3$. This means that there are two possible solutions to $\frac{1}{x} > -\frac{1}{3}$, since x may be positive or negative. If x is negative, $\frac{1}{x} > -\frac{1}{3}$ becomes $x < -3$ (we reverse the direction of the inequality in this case, since both sides are negative), whereas another solution is for x to be positive: $x > 0$. (In the case where x is positive, we don't reverse the direction of the inequality when we reciprocate both sides because only one side is negative. This leads to $x > -3$. However, since x is positive in this case, we don't have to worry about x lying between 0 and -3: since x is positive, we know that $x > 0$.) The two ways that x can satisfy $\frac{1}{x} > -\frac{1}{3}$ are (i) for x to be positive and (ii) for x to be less than -3.

- The case $\frac{1}{x} < \frac{1}{4}$ is similarly interesting. Note how this differs from the first bullet point. This time, x can be negative or it can be positive. If x is positive, both sides of the inequality are positive, so we reverse the direction of the inequality when we reciprocate to get $x > 4$. If x is negative, only one side of the inequality is negative, so we don't reverse the inequality. In this case, we get $x < 4$, but since x is negative in this case, it is more precise to restrict this to $x < 0$. The two solutions to $\frac{1}{x} < \frac{1}{4}$ are $x > 4$ and $x < 0$.

It may help to review Sec. 7.1, which involved reciprocating both sides of equations.

Example 1. $\frac{1}{x} > \frac{3}{4}$ Both sides must be positive (x can't be negative). When we reciprocate both sides, the direction of the inequality reverses: $x < \frac{4}{3}$. Since x is positive, $0 < x < \frac{4}{3}$. Check the answer: 1.3 is a little less than $\frac{4}{3}$ (since $\frac{4}{3} \approx 1.333$), consistent with $x < \frac{4}{3}$, and $\frac{1}{x} = \frac{1}{1.3} \approx 0.769$ is a little more than $\frac{3}{4}$ (since $\frac{3}{4} = 0.75$), consistent with $\frac{1}{x} > \frac{3}{4}$.

Example 2. $\frac{1}{x} < -\frac{2}{3}$ Both sides must be negative (x can't be positive). When we reciprocate both sides, the direction of the inequality reverses: $x > -\frac{3}{2}$. Since x is negative, $0 > x > -\frac{3}{2}$, which is equivalent to $-\frac{3}{2} < x < 0$. Check the answer: -1.4 is a little more than $-\frac{3}{2}$ (since $\frac{3}{2} \approx 1.5$ and since -1.4 is less negative than -1.5), consistent with $x > -\frac{3}{2}$, and $\frac{1}{x} = \frac{1}{-1.4} \approx -0.714$ is a little less than $-\frac{2}{3}$ (since $\frac{2}{3} \approx 0.667$ and since -0.714 is more negative), consistent with $\frac{1}{x} < -\frac{2}{3}$.

Example 3. $\frac{1}{x} < \frac{2}{5}$ Any negative value of x will satisfy this, so one solution is $x < 0$. For the positive solutions, when we reciprocate both sides, the direction of the inequality reverses because both sides are positive: $x > \frac{5}{2}$. The two solutions are $x < 0$ or $x > \frac{5}{2}$. Check the answers: 2.6 is a little more than $\frac{5}{2}$ (since $\frac{5}{2} = 2.5$), consistent with $x > \frac{5}{2}$, and $\frac{1}{x} = \frac{1}{2.6} \approx 0.385$ is a little less than $\frac{2}{5}$ (since $\frac{2}{5} = 0.4$), consistent with $\frac{1}{x} < \frac{2}{5}$. Also, any negative value of x will satisfy $\frac{1}{x} < \frac{2}{5}$.

Example 4. $\frac{1}{x} > -\frac{1}{2}$ Any positive value of x will satisfy this, so one solution is $x > 0$. For the negative solutions, when we reciprocate both sides, the direction of the inequality reverses because both sides are negative: $x < -2$. The two solutions are $x > 0$ or $x < -2$.
Check the answers: -2.1 is a little less than -2 (since -2.1 is more negative), consistent with $x < -2$, and $\frac{1}{x} = \frac{1}{-2.1} \approx -0.476$ is a little more than $-\frac{1}{2}$ (since $\frac{1}{2} = 0.5$ and since -0.476 is less negative than -0.5), consistent with $\frac{1}{x} > -\frac{1}{2}$. Also, any positive value of x will satisfy $\frac{1}{x} > -\frac{1}{2}$.

Exercise Set 9.4

Directions: Isolate the variable in each inequality.

1) $\frac{1}{x} > \frac{3}{2}$

2) $\frac{1}{x} < -\frac{1}{3}$

3) $\frac{1}{x} \leq \frac{4}{5}$

4) $\frac{1}{x} > 4$

5) $-\frac{8}{3} < \frac{1}{x}$

6) $2 \geq \frac{1}{x}$

7) $\frac{1}{x} > -\frac{5}{6}$

8) $\frac{1}{x} < 0$

9) $-\frac{7}{2} \geq \frac{1}{x}$

10) $\frac{9}{4} < \frac{1}{x}$

11) $-\frac{1}{x} < \frac{1}{8}$

12) $-\frac{1}{x} \leq -\frac{3}{10}$

9.5 Cross Multiplying with Inequalities

Note that cross multiplying (Chapter 7) is equivalent to multiplying both sides of the equation by both denominators. If exactly one denominator is negative, this reverses the direction of the inequality. If both denominators are negative (or both are positive), the direction of the inequality remains unchanged. If either denominator involves a variable, you'll need to determine whether or not the denominator may be negative to correctly work out the signs (similar to what we did in Sec. 9.4).

Note: The above paragraph applies if you only multiply by the denominators (and not by any minus signs that may appear in front of the fraction or in the numerator).

Example 1. $\frac{2}{x} > \frac{3}{5}$ Both sides must be positive (x can't be negative). When we cross multiply, the inequality doesn't change: $10 > 3x \to \frac{10}{3} > x$. Since x is positive, $\frac{10}{3} > x$ > 0, which is equivalent to $0 < x < \frac{10}{3}$.

Check the answer: 3.2 is a little less than $\frac{10}{3}$ (since $\frac{10}{3} \approx 3.333$), consistent with $\frac{10}{3} > x$, and $\frac{2}{x} = \frac{2}{3.2} = 0.625$ is a little more than $\frac{3}{5}$ (since $\frac{3}{5} = 0.6$), consistent with $\frac{2}{x} > \frac{3}{5}$.

Example 2. $\frac{3}{x} > -\frac{1}{2}$ Any positive value of x will satisfy this, so one solution is $x > 0$. For negative values of x, when we cross multiply, the left denominator (which equals x) is negative (since x is negative for this case) while the right denominator (2) is positive, so the direction of the inequality changes in this case: $6 < -x$. When we multiply both sides by -1, the direction of the inequality reverses again (as in Sec. 9.3): $-6 > x$. The two solutions are $x > 0$ or $-6 > x$.

Check the answers: -6.1 is a little less than -6 (since -6.1 is more negative), consistent with $-6 > x$, and $\frac{3}{x} = \frac{3}{-6.1} \approx -0.492$ is a little more than $-\frac{1}{2}$ (since $\frac{1}{2} = 0.5$ and since -0.492 is less negative than -0.5), consistent with $\frac{3}{x} > -\frac{1}{2}$. Also, any positive value of x will satisfy $\frac{3}{x} > -\frac{1}{2}$.

Exercise Set 9.5

Directions: Isolate the variable in each inequality.

1) $\dfrac{4}{x} < -\dfrac{2}{3}$

2) $\dfrac{3}{x} > 2$

3) $\dfrac{6}{x} \geq -\dfrac{4}{3}$

4) $\dfrac{5}{x} < \dfrac{1}{5}$

5) $-\dfrac{4}{5} \leq -\dfrac{2}{x}$

6) $\dfrac{4}{x} > \dfrac{x}{9}$

7) $\dfrac{3}{x} < -\dfrac{6}{x^2}$

8) $\dfrac{4}{x-2} < \dfrac{5}{x-3}$

9.6 Word Problems with Inequalities

The word problems in this section involve inequalities. It may help to review Sec. 2.13 (which has tips for solving word problems) and Sec. 1.4.

Example 1. One number is twice another number. The sum of the numbers is greater than 150.

- x = the smaller number and $2x$ = the larger number.
- $x + 2x > 150 \rightarrow 3x > 150 \rightarrow x > \frac{150}{3} \rightarrow x > 50$ and $2x > 100$.
- One number is greater than 50. The other number is greater than 100.

Example 2. A small carton contains at least 60 beads and a large carton contains at least 150 beads. A person purchases two large cartons and three small cartons. How many beads will the person have?

- x = the number of beads in a small carton, y = the number of beads in a large carton, and z = the total number of beads.
- $x \geq 60, y \geq 150$, and $z = 2y + 3x$.
- Substitute the inequalities for x and y into the equation for z:
- $z \geq 2(150) + 3(60) \rightarrow z \geq 300 + 180 \rightarrow z \geq 480$
- The person will have at least 480 beads.

Example 3. A car gets 40 mpg (miles per gallon) on the highway and 30 mpg in the city. How far will the car travel with 15 gallons of gas?

- x = the minimum distance traveled, y = the maximum distance traveled, and z = the actual distance traveled.
- $x = 30(15) = 450$ miles and $y = 40(15) = 600$ miles
- $x \leq z \leq y \rightarrow 450 \leq z \leq 600$
- The car will travel at least 450 miles and as far as 600 miles.

Exercise Set 9.6

Directions: Use algebra and inequalities to solve each word problem.

1) The sum of two positive numbers is at least 20. One number is four times the other number. Find inequalities for each number.

2) A woman has 12 square yards of fabric. If she makes a rectangle with a length of more than 4 yards, what is the maximum width that the rectangle can have?

3) A toy car travels with a speed greater than 4 m/s and less than 5 m/s. How far will the toy car travel in one minute?

4) Jenny's mother is at least 30 years old. Jenny's father is no more than 40 years old. Jenny is at least half as old as her mother, but Jenny isn't half as old as her father. Jenny's father is older than Jenny's mother. Express Jenny's age with an inequality.

ANSWER KEY

Chapter 1 Getting Ready

Exercise Set 1.2

1) $(x - 1)^2 = 16$ is an equation because it has an equal $(=)$ sign.

2) $\frac{3}{x^2} - \frac{4}{x} + \frac{1}{8}$ is an expression because it doesn't have an equal $(=)$ sign.

3) There are four terms: $x^3, 8x^2, 3x$, and 6. Note: We'll accept $-3x$ instead of $3x$ since $x^3 + 8x^2 + (-3x) + 6$ is equivalent to $x^3 + 8x^2 - 3x + 6$. Adding a negative number is equivalent to subtracting a positive number.

4) There are three terms: $9, x$, and 4. Note: We'll accept $-x$ instead of x since $9 + (-x)$ is equivalent to $9 - x$.

5) There are three terms: $5xy^2, 7y^3$, and 3. Note: We'll accept $-7y^3$ instead of $7y^3$ since $5xy^2 + (-7y^3) + 3$ is equivalent to $5xy^2 - 7y^3 + 3$.

6) The variable is x. The coefficient is 3. The coefficient multiplies the variable.

7) The variables are x and y. The coefficients are 5 and 2.

Exercise Set 1.3

1) $9 \cdot 7 = 63$

2) $\frac{48}{8} = 6$

3) $27/3 = 9$

4) $7(8) = 56$

5) $(4)(2)(3) = 24$

6) $\frac{12}{1} = 12$

7) $\frac{32}{4} = 8$

8) $5(6)(3) = 90$

9) $1xx = x^2$

10) $\frac{x^4}{1} = x^4$

11) $5x^2y$ means 5 times x squared times y

12) $\frac{4x^3}{7}$ means the quantity 4 times x cubed divided by 7

13) $\frac{x/2}{y/5}$ means the quantity x divided by 2 is divided by the quantity y divided by 5

Exercise Set 1.4

1) $\frac{x}{2}$ represents the length of each piece of the log.

2) $t - 4$ represents the time that the blue ball has been rolling. (The blue ball has been rolling for less time than the red ball because the blue ball started rolling after the red ball started.)

3) $2y - 50$ represents Pat's paycheck.

4) $\sqrt{x}$ represents the square root of the number.

5) $x(x + 2) = 80$

Exercise Set 1.5

1) $2 + 8^2/4 - 3 \cdot 2^2 = 2 + 64/4 - 3 \cdot 4 = 2 + 16 - 12 = 18 - 12 = 6$

2) $3(12 - 2 \cdot 4) - 2(3^2 - 2^2) = 3(12 - 8) - 2(9 - 4) = 3(4) - 2(5) = 12 - 10 = 2$

3) $6 + 4(20 - 5 \times 3) = 6 + 4(20 - 15) = 6 + 4(5) = 6 + 20 = 26$

4) $(3^2 - 5)^3(8 - 2 \cdot 3)^2 = (9 - 5)^3(8 - 6)^2 = (4)^3(2)^2 = (64)(4) = 256$

5) $6\sqrt{8 \cdot 3 + 5^2} - 3\sqrt{6 \cdot 4 - 5 \cdot 3 + 4^2} = 6\sqrt{8 \cdot 3 + 25} - 3\sqrt{6 \cdot 4 - 5 \cdot 3 + 16}$

$= 6\sqrt{24 + 25} - 3\sqrt{24 - 15 + 16} = 6\sqrt{49} - 3\sqrt{9 + 16} = 6\sqrt{49} - 3\sqrt{25}$

$= 6 \cdot 7 - 3 \cdot 5 = 42 - 15 = 27$

6) $4(5)(2)^3 - 7(3)(2)^2 = 4(5)(8) - 7(3)(4) = 20(8) - 21(4) = 160 - 84 = 76$

7) $(9 \cdot 8 - 7 \cdot 6)/3 + 6 = (72 - 42)/3 + 6 = 30/3 + 6 = 10 + 6 = 16$

8) $\frac{64}{4} + \frac{36}{6} = 16 + 6 = 22$

9) $\frac{8 \cdot 6 \cdot 4}{4 \cdot 3 \cdot 2} = \frac{48 \cdot 4}{12 \cdot 2} = \frac{192}{24} = 8$ If you apply the commutative property of multiplication (see Sec. 1.6), you can solve this problem as $\frac{8 \cdot 6 \cdot 4}{4 \cdot 3 \cdot 2} = \frac{8}{4} \cdot \frac{6}{3} \cdot \frac{4}{2} = 2 \cdot 2 \cdot 2 = 8$.

10) $\frac{12 + 6 \cdot 3}{2 \cdot 8 - 6} = \frac{12 + 18}{16 - 6} = \frac{30}{10} = 3$

11) $\sqrt{(5 + 2 \cdot 4)^2 - (9 - 2 \cdot 2)^2} = \sqrt{(5 + 8)^2 - (9 - 4)^2} = \sqrt{(13)^2 - (5)^2} = \sqrt{169 - 25} = \sqrt{144} = 12$ (Technically, negative 12 is also a solution, but in algebra it is conventional to take the positive root unless a $-$ sign or a $\pm$ sign appears before the square root. Otherwise, the solution to Exercise 5 would have been more interesting.)

12) $\left(\frac{4\cdot7+8}{3^2}\right)^3 = \left(\frac{4\cdot7+8}{9}\right)^3 = \left(\frac{28+8}{9}\right)^3 = \left(\frac{36}{9}\right)^3 = (4)^3 = 64$

13) $(3^3 - 7\cdot3)^{2\cdot5-2^3} = (27 - 7\cdot3)^{2\cdot5-8} = (27 - 21)^{10-8} = (6)^2 = 36$

14) $\sqrt{\frac{6\cdot7-5\cdot2}{2\cdot7-3\cdot2}} = \sqrt{\frac{42-10}{14-6}} = \sqrt{\frac{32}{8}} = \sqrt{4} = 2$ (See the note with the solution to Exercise 11

about negative 2 as a possible solution.)

Exercise Set 1.6

1) $xy + xz = x(y + z)$ is the distributive property. We applied the reflexive property to swap the two sides of the equation from $x(y + z) = xy + xz$ to $xy + xz = x(y + z)$.

2) $0 + x = x$ is the identity property of addition. We applied the commutative property of addition to change the order from $x + 0$ to $0 + x$.

3) $(x + y) + z = x + (y + z) = y + (x + z)$ is the associative property. The difference is that this equation includes $y + (x + z)$. If we swap the rules of x and y, $(x + y) + z = x + (y + z)$ becomes $(y + x) + z = y + (x + z)$. The commutative property of addition tells us that $x + y = y + x$, from which it follows that $(x + y) + z$ equals both $x + (y + z)$ and $y + (x + z)$. (We applied the transitive property in this last step.)

4) $x - y = -y + x$ expresses the commutative property of addition and the inverse property of addition. Start by writing the commutative property of addition as $x + z = z + x$. Then let $z = -y$ to get $x - y = -y + x$. Alternatively, start with $x + y = y + x$ and subtract $2y$ from each side of the equation to get $x + y - y - y = y - y - y + x$, for which the inverse property of addition tells us that $y + (-y) = y - y = 0$. Another way to describe $x - y = -y + x$ is to state that subtraction is anti-commutative. That is, $x - y + (y - x) = 0$ or $x - y = -(y - x)$.

Exercise Set 1.7

1) $\frac{8}{20} = \frac{8/4}{20/4} = \frac{2}{5}$

2) $\frac{16}{14} = \frac{16/2}{14/2} = \frac{8}{7}$

3) $\frac{28}{42} = \frac{28/14}{42/14} = \frac{2}{3}$

4) $\frac{27}{36} = \frac{27/9}{36/9} = \frac{3}{4}$

5) $\frac{64}{48} = \frac{64/16}{48/16} = \frac{4}{3}$

6) $\frac{56}{40} = \frac{56/8}{40/8} = \frac{7}{5}$

7) $\frac{4}{5} + \frac{7}{3} = \frac{4\cdot3}{5\cdot3} + \frac{7\cdot5}{3\cdot5} = \frac{12}{15} + \frac{35}{15} = \frac{47}{15}$

8) $\frac{5}{6} - \frac{2}{9} = \frac{5\cdot3}{6\cdot3} - \frac{2\cdot2}{9\cdot2} = \frac{15}{18} - \frac{4}{18} = \frac{11}{18}$

9) $\frac{8}{15} + \frac{9}{20} = \frac{8\cdot4}{15\cdot4} + \frac{9\cdot3}{20\cdot3} = \frac{32}{60} + \frac{27}{60} = \frac{59}{60}$

10) $\frac{7}{4} - \frac{5}{6} = \frac{7\cdot3}{4\cdot3} - \frac{5\cdot2}{6\cdot2} = \frac{21}{12} - \frac{10}{12} = \frac{11}{12}$

11) $\frac{9}{8}$

12) 3

13) $\frac{1}{6}$

14) $\frac{11}{7}$

15) $\frac{9}{5}\frac{7}{2} = \frac{9\cdot7}{5\cdot2} = \frac{63}{10}$

16) $\frac{8}{9}\frac{5}{4} = \frac{8\cdot5}{9\cdot4} = \frac{40}{36} = \frac{40/4}{36/4} = \frac{10}{9}$

17) $\frac{2}{9}\frac{3}{8} = \frac{2\cdot3}{9\cdot8} = \frac{6}{72} = \frac{6/6}{72/6} = \frac{1}{12}$

18) $\frac{2}{3}6 = \frac{2}{3}\frac{6}{1} = \frac{2\cdot6}{3\cdot1} = \frac{12}{3} = 4$

19) $\frac{7/5}{9/2} = \frac{7}{5} \div \frac{9}{2} = \frac{7}{5}\frac{2}{9} = \frac{7\cdot2}{5\cdot9} = \frac{14}{45}$

20) $\frac{4/3}{2/5} = \frac{4}{3} \div \frac{2}{5} = \frac{4}{3}\frac{5}{2} = \frac{4\cdot5}{3\cdot2} = \frac{20}{6} = \frac{20/2}{6/2} = \frac{10}{3}$

21) $\frac{12/5}{4} = \frac{12}{5} \div 4 = \frac{12}{5} \div \frac{4}{1} = \frac{12}{5}\frac{1}{4} = \frac{12\cdot1}{5\cdot4} = \frac{12}{20} = \frac{12/4}{20/4} = \frac{3}{5}$

22) $\frac{9}{2/3} = 9 \div \frac{2}{3} = \frac{9}{1} \div \frac{2}{3} = \frac{9}{1}\frac{3}{2} = \frac{9\cdot3}{1\cdot2} = \frac{27}{2}$

Exercise Set 1.8

1) $3x - 2 = 16$ (multiply by 10)
2) $5200 - 24x = 2390$ (multiply by 1000)
3) $80x^2 + 23x = 400$ (multiply by 100)

3) $0.65 + 0.57 = 1.22$ 4) $2.8 + 0.73 = 3.53$
5) $7.2 - 3.45 = 3.75$ 6) $8 - 1.76 = 6.24$
7) $(0.8)(0.9) = 0.72$ 8) $7(0.125) = 0.875$

9) $\frac{2.4}{9.6} = \frac{2.4(10)}{9.6(10)} = \frac{24}{96} = \frac{24/24}{96/24} = \frac{1}{4}$ (which is equivalent to 0.25)

10) $\frac{0.45}{2.5} = \frac{0.45(100)}{2.5(100)} = \frac{45}{250} = \frac{45/5}{250/5} = \frac{9}{50}$ (which is equivalent to 0.18)

Exercise Set 1.9

1) $45\% = \frac{45\%}{100\%} = 0.45$ and $45\% = \frac{45/5}{100/5} = \frac{9}{20}$

2) $120\% = \frac{120\%}{100\%} = 1.2$ and $120\% = \frac{120}{100} = \frac{120/20}{100/20} = \frac{6}{5}$

3) $2.5\% = \frac{2.5\%}{100\%} = 0.025$ and $2.5\% = \frac{2.5}{100} = \frac{2.5(10)}{100(10)} = \frac{25}{1000} = \frac{25/25}{1000/25} = \frac{1}{40}$

4) $350\% = \frac{350\%}{100\%} = 3.5$ and $350\% = \frac{350}{100} = \frac{350/50}{100/50} = \frac{7}{2}$

5) $87.5\% = \frac{87.5\%}{100\%} = 0.875$ and $87.5\% = \frac{87.5}{100} = \frac{87.5(10)}{100(10)} = \frac{875}{1000} = \frac{875/125}{1000/125} = \frac{7}{8}$

6) $76\% = \frac{76\%}{100\%} = 0.76$ and $76\% = \frac{76}{100} = \frac{76/4}{100/4} = \frac{19}{25}$

7) $16.8\% = \frac{16.8\%}{100\%} = 0.168$ and $16.8\% = \frac{16.8}{100} = \frac{16.8(10)}{100(10)} = \frac{168}{1000} = \frac{168/8}{10,000/8} = \frac{21}{125}$

8) $6.25\% = \frac{6.25\%}{100\%} = 0.0625$ and $6.25\% = \frac{6.25}{100} = \frac{6.25(100)}{100(100)} = \frac{625}{10,000} = \frac{625/625}{10,000/625} = \frac{1}{16}$

Exercise Set 1.10

1) $18 + (-6) = 18 - 6 = 12$ 2) $24 - (-8) = 24 + 8 = 32$
3) $-32 + 18 = -14$ 4) $-17 - 15 = -32$
5) $-(-11) = 11$ 6) $9 + (-16) = -7$
7) $-25 + 40 = 15$ 8) $0 + (-5) = -5$

9) $-12 - (-4) = -12 + 4 = -8$

10) $15 - (-25) = 15 + 25 = 40$

11) $48 - (-24) = 48 + 24 = 72$

12) $6 + (-6) = 6 - 6 = 0$

13) $-9 - (-9) = -9 + 9 = 0$

14) $-8 + (-8) = -16$

15) $-16 - 12 = -28$

16) $0 - 14 = -14$

17) $8(-8) = -64$

18) $-6(-9) = 54$

19) $-7(-5) = 35$

20) $9(-7) = -63$

21) $\frac{56}{-7} = -8$

22) $\frac{-42}{6} = -7$

23) $\frac{-81}{-9} = 9$

24) $-\frac{49}{-7} = 7$

25) $-2(-3)(-4) = 6(-4) = -24$

26) $-4(5)(-6) = -20(-6) = 120$

27) $\frac{9(-8)}{-6} = \frac{-72}{-6} = 12$

28) $\frac{-8(-7)}{-4} = \frac{56}{-4} = -14$

29) $\frac{(-8)(-6)}{(-2)(-3)} = \frac{48}{6} = 8$

30) $\frac{(6)(-6)}{(-2)(-9)} = \frac{-36}{18} = -2$

Exercise Set 1.11

1) $8^2 = 8(8) = 64$

2) $4^3 = 4(4)(4) = 16(4) = 64$

3) $5^4 = 5(5)(5)(5) = 25(25) = 625$

4) $2^5 = 2(2)(2)(2)(2) = 4(4)(2) = 16(2) = 32$

5) $(-4)^4 = -4(-4)(-4)(-4) = 16(16) = 256$

6) $(-3)^5 = -3(-3)(-3)(-3)(-3) = 9(9)(-3) = 81(-3) = -243$

7) $9^0 = 1$

8) $8^1 = 8$

9) $\sqrt{49} = 7$ since $7^2 = 7(7) = 49$

10) $\sqrt{100} = 10$ since $10^2 = 10(10) = 100$

Note: Only the positive roots are given here.

11) $\sqrt[3]{125} = 5$ since $5^3 = 5(5)(5) = 125$

12) $\sqrt[3]{512} = 8$ since $8^3 = 8(8)(8) = 64(8) = 512$

13) $\sqrt[3]{-64} = -4$ since $(-4)^3 = -4(-4)(-4) = 16(-4) = -64$

14) $\sqrt[4]{16} = 2$ since $2^4 = 2(2)(2)(2) = 4(4) = 16$

15) $64^{1/2} = 8$ since $\sqrt{64} = 8$ and $8^2 = 64$

16) $1000^{1/3} = 10$ since $\sqrt[3]{1000} = 10$ and $10^3 = 1000$

17) $\left(\frac{4}{7}\right)^{-1} = \frac{7}{4}$

18) $7^{-1} = \frac{1}{7}$

19) $27^{2/3} = \left(27^{1/3}\right)^2 = 3^2 = 3(3) = 9$

20) $36^{3/2} = \left(36^{1/2}\right)^3 = 6^3 = 6(6)(6) = 36(6) = 216$

21) $16^{3/4} = \left(16^{1/4}\right)^3 = 2^3 = 2(2)(2) = 4(2) = 8$

Note: $16^{1/4} = 2$ since $16^{1/4} = \sqrt[4]{16}$ and $2^4 = 2(2)(2)(2) = 4(4) = 16$

22) $5^{-2} = \left(\frac{1}{5}\right)^2 = \frac{1}{5^2} = \frac{1}{5(5)} = \frac{1}{25}$

23) $4^{-3} = \left(\frac{1}{4}\right)^3 = \frac{1}{4^3} = \frac{1}{4(4)(4)} = \frac{1}{16(4)} = \frac{1}{64}$

24) $\left(\frac{1}{2}\right)^{-4} = \left(\frac{2}{1}\right)^4 = \frac{2^4}{1^4} = \frac{2(2)(2)(2)}{1} = \frac{4(4)}{1} = \frac{16}{1} = 16$

25) $25^{-1/2} = \left(\frac{1}{25}\right)^{1/2} = \frac{1}{25^{1/2}} = \frac{1}{\sqrt{25}} = \frac{1}{5}$

26) $\left(\frac{1}{36}\right)^{-1/2} = \left(\frac{36}{1}\right)^{1/2} = \frac{36^{1/2}}{1^{1/2}} = \frac{\sqrt{36}}{\sqrt{1}} = \frac{6}{1} = 6$

27) $\left(\frac{4}{9}\right)^{-3/2} = \left(\frac{9}{4}\right)^{3/2} = \frac{9^{3/2}}{4^{3/2}} = \frac{\left(9^{1/2}\right)^3}{\left(4^{1/2}\right)^3} = \frac{3^3}{2^3} = \frac{3(3)(3)}{2(2)(2)} = \frac{27}{8}$

28) $\left(\frac{27}{64}\right)^{-2/3} = \left(\frac{64}{27}\right)^{2/3} = \frac{64^{2/3}}{27^{2/3}} = \frac{\left(64^{1/3}\right)^2}{\left(27^{1/3}\right)^2} = \frac{4^2}{3^2} = \frac{4(4)}{3(3)} = \frac{16}{9}$

Exercise Set 1.12

1) 0.5 is real and rational

2) 16,300 is a real rational integer

3) -20 is a real rational integer

4) $\frac{5}{2}$ is real and rational

5) $\sqrt{25}$ is a real rational integer since $\sqrt{25} = 5$

6) $\sqrt{19}$ is real and irrational

7) 7.2 is positive, nonnegative, and nonzero

8) 0.01 is positive, nonnegative, and nonzero

9) -1.8 is negative and nonzero

10) zero is nonnegative

Exercise Set 1.13

1) $v = \dfrac{d}{t} = \dfrac{80}{16} = 5$

2) $F = \dfrac{9}{5}C + 32 = \dfrac{9}{5}(20) + 32 = \dfrac{180}{5} + 32 = 36 + 32 = 68$

3) $a = \dfrac{v^2}{R} = \dfrac{6^2}{4} = \dfrac{36}{4} = 9$

4) $A = \dfrac{(a+b)h}{2} = \dfrac{(4+12)(5)}{2} = \dfrac{16(5)}{2} = \dfrac{80}{2} = 40$

5) $P = \dfrac{V^2}{R} = \dfrac{12^2}{18} = \dfrac{144}{18} = 8$

6) $U = \dfrac{1}{2}k(x-c)^2 = \dfrac{1}{2}(8)(32-12)^2 = \dfrac{1}{2}(8)(20)^2 = \dfrac{1}{2}(8)(400) = 4(400) = 1600$

7) $E = \dfrac{|A-B|}{A}100\% = \dfrac{|5-5.5|}{5}100\% = \dfrac{|-0.5|}{5}100\% = \dfrac{0.5}{5}(100\%) = \dfrac{50}{5}\% = 10\%$

Note: $|-0.5| = 0.5$ is the absolute value of -0.5. The absolute value is positive (Sec. 1.10).

8) $c = \sqrt{a^2 + b^2} = \sqrt{12^2 + 5^2} = \sqrt{144 + 25} = \sqrt{169} = 13$

9) $T = c\sqrt{\dfrac{L}{g}} = 6\sqrt{\dfrac{90}{10}} = 6\sqrt{9} = 6(3) = 18$

10) $x = vt + \dfrac{1}{2}at^2 = 40(5) + \dfrac{1}{2}(8)(5)^2 = 200 + 4(25) = 200 + 100 = 300$

11) $F = k\dfrac{q^2}{R^2} = 18\dfrac{4^2}{6^2} = \dfrac{18(16)}{36} = \dfrac{288}{36} = 8$

12) $z = \left(\dfrac{1}{x} + \dfrac{1}{y}\right)^{-1} = \left(\dfrac{1}{4} + \dfrac{1}{12}\right)^{-1} = \left(\dfrac{1\cdot3}{4\cdot3} + \dfrac{1}{12}\right)^{-1} = \left(\dfrac{3}{12} + \dfrac{1}{12}\right)^{-1} = \left(\dfrac{4}{12}\right)^{-1} = \dfrac{12}{4} = 3$

Note: It may help to review Sec. 1.11 regarding reciprocals.

2 Solving Linear Equations

Exercise Set 2.1

1) $x = 12 - 5 = \boxed{7}$ Check: $7 + 5 = 12$

2) $x = 8 + 6 = \boxed{14}$ Check: $14 - 6 = 8$

3) $x = \frac{36}{4} = \boxed{9}$ Check: $4(9) = 36$

4) $x = 9(7) = \boxed{63}$ Check: $\frac{63}{7} = 9$

5) $x = 5 + 6 = \boxed{11}$ Check: $11 - 6 = 5$

6) $x = \frac{42}{7} = \boxed{6}$ Check: $7(6) = 42$

7) $x = 9(5) = \boxed{45}$ Check: $\frac{45}{9} = 5$

8) $x = 16 - 8 = \boxed{8}$ Check: $8 + 8 = 16$

Note: $8 + x = 16$ is equivalent to $x + 8 = 16$ due to the commutative property of addition (Sec. 1.6).

9) $x = \frac{72}{8} = \boxed{9}$ Check: $8(9) = 72$

10) $9 + 7 = \boxed{16} = x$ Check: $16 - 7 = 9$

Note: $9 = x - 7$ is equivalent to $x - 7 = 9$ due to the reflexive property (Sec. 1.6).

11) $13 - 4 = \boxed{9} = x$ Check: $9 + 4 = 13$

12) $x = 5(4) = \boxed{20}$ Check: $\frac{20}{5} = 4$

13) $x = 7 + 7 = \boxed{14}$ Check: $14 - 7 = 7$

14) $x = 1 - 1 = \boxed{0}$ Check: $0 + 1 = 1$

15) $6(8) = \boxed{48} = x$ Check: $\frac{48}{8} = 6$

16) $\frac{30}{5} = \boxed{6} = x$ Check: $5(6) = 30$

17) $32 - 18 = \boxed{14} = x$ Check: $14 + 18 = 32$

18) $x = 40 + 25 = \boxed{65}$ Check: $65 - 25 = 40$

19) $x = \frac{0}{3} = \boxed{0}$ Check: $3(0) = 0$

20) $x = 2(9) = \boxed{18}$ Check: $\frac{18}{2} = 9$

Exercise Set 2.2

1) $9x^2 + 8x^2 = 17x^2$

2) $8x + 4x + x = 13x$

3) $3x + 6 + 2x + 4 + x = 6x + 10$

4) $5x^2 + 9 - 2x^2 - 3 = 3x^2 + 6$

5) $6x - 4 - 5x - 2 = x - 6$

6) $7x^2 + 8x - 5x^2 + 7x = 2x^2 + 15x$

7) $8x + 6 - 4x + 4 + 3x + 1 = 7x + 11$

8) $5x^2 - 8 + 4x^2 + 7 - 3x^2 + 5 = 6x^2 + 4$

9) $x^3 + 2x^2 + x^3 - x^2 = 2x^3 + x^2$

10) $x^2 + x - x^2 + x = 2x$

11) $6x^2 + 5x + 4 + 3x^2 + 2x + 1 = 9x^2 + 7x + 5$

12) $4x^3 + 3x^2 - 2x + 2x^3 - 2x^2 - x^3 - x + x^2 - x = 5x^3 + 2x^2 - 4x$

13) $8x^4 + 9x^2 + 6 + 7x^4 - 2x^2 - 4 + 5x^4 - 2x^2 - 4 = 20x^4 + 5x^2 - 2$

14) $9x^2 - x + 8x^2 - 2x + 7x^2 - 3x + 6x^2 - 4x = 30x^2 - 10x$

15) $x + 2 + 3x + 4 + 5x + 6 + 7x + 8 + 9x + 10 = 25x + 30$

Exercise Set 2.3

1) $4x + 9 = 21$ yes, both

2) $3x = 15$ yes, both

3) $x + 12 = 22$ yes, both

4) $3x = 24$ yes, both

5) $63 = 9x$ yes, both

6) $27 = 3x$ yes, both

7) $10 = 2x$ yes, both

8) $x = 2$ yes, both

9) $28 = 4x$ yes, both

Exercise Set 2.4

1) $5x + 10 = 45$ Subtract 10 from both sides: $5x = 35$

Divide by 5 on both sides: $x = \boxed{7}$

Check: $5x + 10 = 5(7) + 10 = 35 + 10 = 45$

2) $18 = 6x - 12$ Add 12 to both sides: $30 = 6x$

Divide by 6 on both sides: $\boxed{5} = x$ (equivalent to $x = 5$)

Check: $6x - 12 = 6(5) - 12 = 30 - 12 = 18$

3) $6x = 14 + 5x$ Subtract $5x$ from both sides: $x = \boxed{14}$

Check: $6x = 6(14) = 84$ agrees with $14 + 5x = 14 + 5(14) = 14 + 70 = 84$

4) $2x + 28 = 9x$ Subtract $2x$ from both sides: $28 = 7x$

Divide by 7 on both sides: $\boxed{4} = x$ (equivalent to $x = 4$)

Check: $2x + 28 = 2(4) + 28 = 8 + 28 = 36$ agrees with $9x = 9(4) = 36$

5) $18 + 6x = 90 - 3x$ Add $3x$ to both sides: $18 + 9x = 90$

Subtract 18 from both sides: $9x = 72$

Divide by 9 on both sides: $x = \boxed{8}$

Check: $18 + 6x = 18 + 6(8) = 18 + 48 = 66$ agrees with

$90 - 3x = 90 - 3(8) = 90 - 24 = 66$

Exercise Set 2.5

1) $8x - 18 + 18 = 30 + 18$ simplifies to $8x = 48$

$\frac{8x}{8} = \frac{48}{8}$ simplifies to $x = \boxed{6}$

Check: $8x - 18 = 8(6) = 48 - 18 = 30$

2) $7x + 15 - 15 = 50 - 15$ simplifies to $7x = 35$

$\frac{7x}{7} = \frac{35}{7}$ simplifies to $x = \boxed{5}$

Check: $7x + 15 = 7(5) + 15 = 35 + 15 = 50$

3) $3x + 2x = 20 - 2x + 2x$ simplifies to $5x = 20$

$\frac{5x}{5} = \frac{20}{5}$ simplifies to $x = \boxed{4}$

Check: $3x = 3(4) = 12$ agrees with $20 - 2x = 20 - 2(4) = 20 - 8 = 12$

4) $24 + 4x - 24 = 72 - 24$ simplifies to $4x = 48$

$\frac{4x}{4} = \frac{48}{4}$ simplifies to $x = \boxed{12}$

Check: $24 + 4x = 24 + 4(12) = 24 + 48 = 72$

5) $5x + 4x = 63$ simplifies to $9x = 63$

$\frac{9x}{9} = \frac{63}{9}$ simplifies to $x = \boxed{7}$

Check: $5x + 4x = 5(7) + 4(7) = 35 + 28 = 63$

6) $9x - x = x + 64 - x$ simplifies to $8x = 64$

$\frac{8x}{8} = \frac{64}{8}$ simplifies to $x = \boxed{8}$

Check: $9x = 9(8) = 72$ agrees with $x + 64 = 8 + 64 = 72$

7) $\frac{x}{6} + 12 - 12 = 19 - 12$ simplifies to $\frac{x}{6} = 7$

$\frac{6x}{6} = 7(6)$ simplifies to $x = \boxed{42}$ (in this step, we multiplied both sides by 6)

Check: $\frac{x}{6} + 12 = \frac{42}{6} + 12 = 7 + 12 = 19$

8) $32 - 5x + 5x = 3x + 5x$ simplifies to $32 = 8x$

$\frac{32}{8} = \frac{8x}{8}$ simplifies to $\boxed{4} = x$ (equivalent to $x = 4$)

Check: $32 - 5x = 32 - 5(4) = 32 - 20 = 12$ agrees with $3(4) = 12$

9) $40 + 9 = 7x - 9 + 9$ simplifies to $49 = 7x$

$\frac{49}{7} = \frac{7x}{7}$ simplifies to $\boxed{7} = x$ (equivalent to $x = 7$)

Check: $7x - 9 = 7(7) - 9 = 49 - 9 = 40$

10) $18 - 9 = \frac{x}{3} + 9 - 9$ simplifies to $9 = \frac{x}{3}$

$9(3) = \frac{3x}{3}$ simplifies to $\boxed{27} = x$ (in this step, we multiplied both sides by 3)

Check: $\frac{x}{3} + 9 = \frac{27}{3} + 9 = 9 + 9 = 18$

11) $15x - 6x = 54 + 6x - 6x$ simplifies to $9x = 54$

$\frac{9x}{9} = \frac{54}{9}$ simplifies to $x = \boxed{6}$

Check: $15x = 15(6) = 90$ agrees with $54 + 6x = 54 + 6(6) = 54 + 36 = 90$

12) $12 + 9 = 3x - 9 + 9$ simplifies to $21 = 3x$

$\frac{21}{3} = \frac{3x}{3}$ simplifies to $\boxed{7} = x$ (equivalent to $x = 7$)

Check: $3x - 9 = 3(7) - 9 = 21 - 9 = 12$

13) $11 + \frac{x}{5} - 11 = 20 - 11$ simplifies to $\frac{x}{5} = 9$

$\frac{5x}{5} = 9(5)$ simplifies to $x = \boxed{45}$ (in this step, we multiplied both sides by 5)

Check: $11 + \frac{x}{5} = 11 + \frac{45}{5} = 11 + 9 = 20$

14) $20 - x + x = 4x + x$ simplifies to $20 = 5x$

$\frac{20}{5} = \frac{5x}{5}$ simplifies to $\boxed{4} = x$ (equivalent to $x = 4$)

Check: $20 - x = 20 - 4 = 16$ agrees with $4x = 4(4) = 16$

15) $9 + 6x + 4 = 7x - 4 + 4$ simplifies to $13 + 6x = 7x$

$13 + 6x - 6x = 7x - 6x$ simplifies to $\boxed{13} = x$ (equivalent to $x = 13$)

Check: $9 + 6x = 9 + 6(13) = 9 + 78 = 87$ agrees with

$7x - 4 = 7(13) - 4 = 91 - 4 = 87$

16) $\frac{x}{4} - 5 + 5 = 5 + 5$ simplifies to $\frac{x}{4} = 10$

$\frac{4x}{4} = 10(4)$ simplifies to $x = \boxed{40}$ (in this step, we multiplied both sides by 4)

Check: $\frac{x}{4} - 5 = \frac{40}{4} - 5 = 10 - 5 = 5$

17) $6x - 30 + 30 = 0 + 30$ simplifies to $6x = 30$

$\frac{6x}{6} = \frac{30}{6}$ simplifies to $x = \boxed{5}$

Check: $6x - 30 = 6(5) - 30 = 30 - 30 = 0$

18) $56 = 8x + x - 2x$ simplifies to $56 = 7x$ (combine like terms)

$\frac{56}{7} = \frac{7x}{7}$ simplifies to $\boxed{8} = x$ (equivalent to $x = 8$)

Check: $8x + x - 2x = 8(8) + 8 - 2(8) = 64 + 8 - 16 = 72 - 16 = 56$

19) $x + \frac{x}{2} - x = 8 + x - x$ simplifies to $\frac{x}{2} = 8$ (we subtracted x on both sides)

$\frac{2x}{2} = 8(2)$ simplifies to $x = \boxed{16}$ (in this step, we multiplied both sides by 2)

Check: $x + \frac{x}{2} = 16 + \frac{16}{2} = 16 + 8 = 24$ agrees with $8 + x = 8 + 16 = 24$

20) $8 + 4x + 7 = 5x - 7 + 7$ simplifies to $15 + 4x = 5x$

$15 + 4x - 4x = 5x - 4x$ simplifies to $\boxed{15} = x$ (equivalent to $x = 15$)

Check: $8 + 4x = 8 + 4(15) = 8 + 60 = 68$ agrees with

$5x - 7 = 5(15) - 7 = 75 - 7 = 68$

Exercise Set 2.6

1) Subtract 18 from both sides to get $8x = 24 + 2x$

Subtract $2x$ from both sides to get $6x = 24$

Divide by 6 on both sides to get $x = \boxed{4}$

Check: $18 + 8x = 18 + 8(4) = 18 + 32 = 50$ agrees with

$42 + 2x = 42 + 2(4) = 42 + 8 = 50$

2) Subtract 13 from both sides to get $6x = 27 - 3x$

Add $3x$ to both sides to get $9x = 27$

Divide by 9 on both sides to get $x = \boxed{3}$

Check: $6x + 13 = 6(3) + 13 = 18 + 13 = 31$ agrees with

$40 - 3x = 40 - 3(3) = 40 - 9 = 31$

3) Add 36 to both sides to get $7x = 4x + 24$

Subtract $4x$ from both sides to get $3x = 24$

Divide by 3 on both sides to get $x = \boxed{8}$

Check: $7x - 36 = 7(8) - 36 = 56 - 36 = 20$ agrees with

$4x - 12 = 4(8) - 12 = 32 - 12 = 20$

4) Subtract 8 from both sides to get $48 - x = 5x$

Add x to both sides to get $48 = 6x$

Divide by 6 on both sides to get $\boxed{8} = x$ (equivalent to $x = 8$)

Check: $56 - x = 56 - 8 = 48$ agrees with $5x + 8 = 5(8) + 8 = 40 + 8 = 48$

5) Subtract 18 from both sides to get $72 - 4x = 4x$

Add $4x$ to both sides to get $72 = 8x$

Divide by 8 on both sides to get $\boxed{9} = x$ (equivalent to $x = 9$)

Check: $90 - 4x = 90 - 4(9) = 90 - 36 = 54$ agrees with

$18 + 4x = 18 + 4(9) = 18 + 36 = 54$

6) Subtract 28 from both sides to get $15x = 42 + 8x$

Subtract $8x$ from both sides to get $7x = 42$

Divide by 7 on both sides to get $x = \boxed{6}$

Check: $15x + 28 = 15(6) + 28 = 90 + 28 = 118$ agrees with

$70 + 8x = 70 + 8(6) = 70 + 48 = 118$

7) Add 3 to both sides to get $6x + 6 = 12x$

Subtract $6x$ from both sides to get $6 = 6x$

Divide by 6 on both sides to get $\boxed{1} = x$ (equivalent to $x = 1$)

Check: $6x + 3 = 6(1) + 3 = 6 + 3 = 9$ agrees with $12x - 3 = 12(1) - 3 = 12 - 3 = 9$

8) Subtract 12 from both sides to get $8x - 3x = 36 - 4x$

Combine like terms on the left side to get $5x = 36 - 4x$

Add $4x$ to both sides to get $9x = 36$

Divide by 9 on both sides to get $x = \boxed{4}$

Check: $8x + 12 - 3x = 8(4) + 12 - 3(4) = 32 + 12 - 12 = 32$ agrees with

$48 - 4x = 48 - 4(4) = 48 - 16 = 32$

9) Subtract 100 from both sides to get $4x = 250 - 6x$

Add $6x$ to both sides to get $10x = 250$

Divide by 10 on both sides to get $x = \boxed{25}$

Check: $100 + 4x = 100 + 4(25) = 100 + 100 = 200$ agrees with

$350 - 6x = 350 - 6(25) = 350 - 150 = 200$

10) Add 4 to both sides to get $x + x = 12 - x$

Combine like terms on the left side to get $2x = 12 - x$

Add x to both sides to get $3x = 12$

Divide by 3 on both sides to get $x = \boxed{4}$

Check: $x - 4 + x = 4 - 4 + 4 = 0 + 4 = 4$ agrees with $8 - x = 8 - 4 = 4$

11) Add 8 to both sides to get $22 + 7x = 3x + 50$

Subtract 22 from both sides to get $7x = 3x + 28$

Subtract $3x$ from both sides to get $4x = 28$

Divide by 4 on both sides to get $x = \boxed{7}$

Check: $22 + 7x - 8 = 22 + 7(7) - 8 = 22 + 49 - 8 = 71 - 8 = 63$ agrees with

$3x + 42 = 3(7) + 42 = 21 + 42 = 63$

12) Add 19 to both sides to get $8x - 3x + 24 = 8x$

Combine like terms on the left side to get $5x + 24 = 8x$

Subtract $3x$ from both sides to get $24 = 3x$

Divide by 3 on both sides to get $\boxed{8} = x$ (equivalent to $x = 8$)

Check: $8x - 3x + 5 = 8(8) - 3(8) + 5 = 64 - 24 + 5 = 40 + 5 = 45$ agrees with

$8x - 19 = 8(8) - 19 = 64 - 19 = 45$

13) Add 16 to both sides to get $6x + 6x = 56 + 4x$

Combine like terms on the left side to get $12x = 56 + 4x$

Subtract $4x$ from both sides to get $8x = 56$

Divide by 8 on both sides to get $x = \boxed{7}$

Check: $6x - 16 + 6x = 6(7) - 16 + 6(7) = 42 - 16 + 42 = 26 + 42 = 68$ agrees

with $40 + 4x = 40 + 4(7) = 40 + 28 = 68$

14) Subtract 4 from both sides to get $7 - 2x = 7 + 3x$

Subtract 7 from both sides to get $-2x = 3x$

Add $2x$ to both sides to get $0 = 5x$

Divide by 5 on both sides to get $\boxed{0} = x$ (equivalent to $x = 0$)

Check: $7 - 2x + 4 = 7 - 2(0) + 4 = 7 - 0 + 4 = 11$ agrees with

$11 + 3x = 11 + 3(0) = 11 + 0 = 11$

15) Add 22 to both sides to get $44 - 8x = 9x - 6x$

Combine like terms on the right side to get $44 - 8x = 3x$

Add $8x$ to both sides to get $44 = 11x$

Divide by 11 on both sides to get $\boxed{4} = x$ (equivalent to $x = 4$)

Check: $22 - 8x = 22 - 8(4) = 22 - 32 = -10$ agrees with

$9x - 22 - 6x = 9(4) - 22 - 6(4) = 36 - 22 - 24 = 14 - 24 = -10$

16) Combine like terms on each side to get $5x + 18 = 90 - x$

Subtract 18 from both sides to get $5x = 72 - x$

Add x to both sides to get $6x = 72$

Divide by 6 on both sides to get $x = \boxed{12}$

Check: $7x + 18 - 2x = 7(12) + 18 - 2(12) = 84 + 18 - 24 = 102 - 24 = 78$

agrees with $56 - x + 34 = 56 - 12 + 34 = 44 + 34 = 78$

Exercise Set 2.7

1) Add 23 to both sides to get $-7x = 63$

Divide both sides by -7 to get $x = \boxed{-9}$

Check: $-7x - 23 = -7(-9) - 23 = 63 - 23 = 40$

2) Subtract 8 from both sides to get $42 = -6x$

Divide both sides by -6 to get $\boxed{-7} = x$ (equivalent to $x = -7$)

Check: $8 - 6x = 8 - 6(-7) = 8 + 42 = 50$

3) Subtract 20 from both sides to get $5x = -40$

Divide both sides by 5 to get $x = \boxed{-8}$

Check: $5x + 20 = 5(-8) + 20 = -40 + 20 = -20$

4) Add 16 to both sides to get $-24 = -4x$

Divide both sides by -4 to get $6 = x$ (equivalent to $x = 6$)

Check: $-4x - 16 = -4(6) - 16 = -24 - 16 = -40$

5) Subtract 1 from both sides to get $4 - 3x = -2x$

Add $3x$ to both sides to get $\boxed{4} = x$ (equivalent to $x = 4$)

Check: $5 - 3x = 5 - 3(4) = 5 - 12 = -7$ agrees with

$-2x + 1 = -2(4) + 1 = -8 + 1 = -7$

6) Subtract 36 from both sides to get $3x + 64 = -5x$

Subtract $3x$ from both sides to get $64 = -8x$

Divide both sides by -8 to get $\boxed{-8} = x$ (equivalent to $x = -8$)

Check: $3x + 100 = 3(-8) + 100 = -24 + 100 = 76$ agrees with

$-5x + 36 = -5(-8) + 36 = 40 + 36 = 76$

7) Subtract 2 from both sides to get $6 + x = -x$

Subtract x from both sides to get $6 = -2x$

Divide both sides by -2 to get $\boxed{-3} = x$ (equivalent to $x = -3$)

Check: $8 + x = 8 + (-3) = 5$ agrees with $2 - x = 2 - (-3) = 2 + 3 = 5$

8) Add 20 to both sides to get $-7x - 25 = -2x$

Add $7x$ to both sides to get $-25 = 5x$

Divide both sides by 5 to get $\boxed{-5} = x$ (equivalent to $x = -5$)

Check: $-7x - 45 = -7(-5) - 45 = 35 - 45 = -10$ agrees with

$-20 - 2x = -20 - 2(-5) = -20 + 10 = -10$

9) Subtract 5 from both sides to get $-\frac{x}{6} = 7$

Multiply both sides by -6 to get $x = \boxed{-42}$

Check: $5 - \frac{x}{6} = 5 - \left(\frac{-42}{6}\right) = 5 - (-7) = 5 + 7 = 12$

10) Combine like terms on the right side to get $2x = -45 - 7x$

Add $7x$ to both sides to get $9x = -45$

Divide both sides by 9 to get $x = \boxed{-5}$

Check: $2x = 2(-5) = -10$ agrees with

$-45 - 3x - 4x = -45 - 3(-5) - 4(-5) = -45 + 15 + 20 = -45 + 35 = -10$

11) Combine like terms on the right side to get $-21 = -3x$

Divide both sides by -3 to get $7 = x$ (equivalent to $x = 7$)

Check: $-5x + 2x = -5(7) + 2(7) = -35 + 14 = -21$

12) Subtract 9 from both sides to get $-5 = -\frac{x}{7}$

Multiply both sides by -7 to get $\boxed{35} = x$ (equivalent to $x = 35$)

Check: $9 - \frac{x}{7} = 9 - \frac{35}{7} = 9 - 5 = 4$

13) Add $4x$ to both sides to get $-111 + 30 = 9x$

Combine like terms on the left side to get $-81 = 9x$

Divide both sides by 9 to get $\boxed{-9} = x$ (equivalent to $x = -9$)

Check: $-111 - 4x + 30 = -111 - 4(-9) + 30 = -111 + 36 + 30 = -45$ agrees

with $5x = 5(-9) = -45$

14) Add 72 to both sides to get $-17x + 48 = -9x$

Add $17x$ to both sides to get $48 = 8x$

Divide both sides by 8 to get $\boxed{6} = x$ (equivalent to $x = 6$)

Check: $-17x - 24 = -17(6) - 24 = -102 - 24 = -126$ agrees with

$-9x - 72 = -9(6) - 72 = -54 - 72 = -126$

15) Add 18 to both sides to get $-3x = 24 + 3x$

Subtract $3x$ from both sides to get $-6x = 24$

Divide both sides by -6 to get $x = \boxed{-4}$

Check: $-18 - 3x = -18 - 3(-4) = -18 + 12 = -6$ agrees with

$6 + 3x = 6 + 3(-4) = 6 - 12 = -6$

16) Subtract 10 from both sides to get $-7x = -12x - 40$

Add $12x$ to both sides to get $5x = -40$

Divide both sides by 5 to get $x = \boxed{-8}$

Check: $10 - 7x = 10 - 7(-8) = 10 + 56 = 66$ agrees with

$-12x - 30 = -12(-8) - 30 = 96 - 30 = 66$

Exercise Set 2.8

1) Multiply both sides by -1 to get $x = \boxed{11}$

Check: $-x = -11$

2) Subtract 5 from both sides to get $3 = -x$

Multiply both sides by -1 to get $\boxed{-3} = x$ (equivalent to $x = -3$)

Check: $5 - x = 5 - (-3) = 5 + 3 = 8$

3) Subtract 9 from both sides to get $8 = -x$

Multiply both sides by -1 to get $\boxed{-8} = x$ (equivalent to $x = -8$)

Check: $-x + 9 = -(-8) + 9 = 8 + 9 = 17$

4) Add 4 to both sides to get $-x = -2$

Multiply both sides by -1 to get $x = \boxed{2}$

Check: $-4 - x = -4 - 2 = -6$

5) Subtract 1 from both sides to get $-x = 9$

Multiply both sides by -1 to get $x = \boxed{-9}$

Check: $-x + 1 = -(-9) + 1 = 9 + 1 = 10$

6) Subtract 2 from both sides to get $7 - 6x = -7x$

Add $6x$ to both sides to get $7 = -x$

Multiply both sides by -1 to get $\boxed{-7} = x$ (equivalent to $x = -7$)

Check: $9 - 6x = 9 - 6(-7) = 9 + 42 = 51$ agrees with

$2 - 7x = 2 - 7(-7) = 2 + 49 = 51$

Exercise Set 2.9

1) Subtract 8 from both sides to get $3x = 4$

Divide both sides by 3 to get $x = \boxed{\dfrac{4}{3}}$

Check: $3\left(\dfrac{4}{3}\right) + 8 = 4 + 8 = 12$

2) Add 8 to both sides to get $17 = 5x$

Divide both sides by 5 to get $\boxed{\dfrac{17}{5}} = x$

Check: $5\left(\dfrac{17}{5}\right) - 8 = 17 - 8 = 9$

3) Subtract 7 from both sides to get $2x = -1$

Divide both sides by 2 to get $x = \boxed{-\dfrac{1}{2}}$

Check: $2x + 7 = 2\left(-\dfrac{1}{2}\right) + 7 = -1 + 7 = 6$

4) Subtract 6 from both sides to get $-8x = 14$

Divide both sides by -8 to get $x = -\dfrac{14}{8}$

Divide 14 and 8 each by 2 to reduce the answer to $x = \boxed{-\dfrac{7}{4}}$

Check: $6 - 8x = 6 - 8\left(-\dfrac{7}{4}\right) = 6 + \dfrac{56}{4} = 6 + 14 = 20$

5) Add 11 to both sides to get $9x = 4x + 8$

Subtract $4x$ from both sides to get $5x = 8$

Divide both sides by 5 to get $x = \boxed{\dfrac{8}{5}}$

Check: $9x - 11 = 9\left(\dfrac{8}{5}\right) - 11 = \dfrac{72}{5} - 11 = \dfrac{72}{5} - \dfrac{55}{5} = \dfrac{17}{5}$ agrees with

$4x - 3 = 4\left(\dfrac{8}{5}\right) - 3 = \dfrac{32}{5} - 3 = \dfrac{32}{5} - \dfrac{15}{5} = \dfrac{17}{5}$

6) Subtract 8 from both sides to get $2x + 10 = 5x$

Subtract $2x$ from both sides to get $10 = 3x$

Divide both sides by 3 to get $\boxed{\dfrac{10}{3}} = x$

Check: $2x + 18 = 2\left(\dfrac{10}{3}\right) + 18 = \dfrac{20}{3} + 18 = \dfrac{20}{3} + \dfrac{54}{3} = \dfrac{74}{3}$ agrees with

$5x + 8 = 5\left(\dfrac{10}{3}\right) + 8 = \dfrac{50}{3} + 8 = \dfrac{50}{3} + \dfrac{24}{3} = \dfrac{74}{3}$

7) Add 12 to both sides to get $2x - 4 = 8x$

Subtract $2x$ from both sides to get $-4 = 6x$

Divide both sides by 6 to get $-\dfrac{4}{6} = x$

Divide 4 and 6 each by 2 to reduce the answer to $\boxed{-\dfrac{2}{3}} = x$

Check: $2x - 16 = 2\left(-\dfrac{2}{3}\right) - 16 = -\dfrac{4}{3} - 16 = -\dfrac{4}{3} - \dfrac{48}{3} = -\dfrac{52}{3}$ agrees with

$8x - 12 = 8\left(-\dfrac{2}{3}\right) - 12 = -\dfrac{16}{3} - 12 = -\dfrac{16}{3} - \dfrac{36}{3} = -\dfrac{52}{3}$

8) Add 9 to both sides to get $5x = 20 - 3x$

Add $3x$ to both sides to get $8x = 20$

Divide both sides by 8 to get $x = \frac{20}{8}$

Divide 20 and 8 each by 4 to reduce the answer to $x = \boxed{\frac{5}{2}}$

Check: $5x - 9 = 5\left(\frac{5}{2}\right) - 9 = \frac{25}{2} - 9 = \frac{25}{2} - \frac{18}{2} = \frac{7}{2}$ agrees with

$11 - 3x = 11 - 3\left(\frac{5}{2}\right) = 11 - \frac{15}{2} = \frac{22}{2} - \frac{15}{2} = \frac{7}{2}$

Exercise Set 2.10

1) Subtract $\frac{1}{12}$ from both sides to get $3x = \frac{3}{4} - \frac{1}{12}$

Since $\frac{3}{4} - \frac{1}{12} = \frac{9}{12} - \frac{1}{12} = \frac{8}{12} = \frac{2}{3}$, this simplifies to $3x = \frac{2}{3}$

Divide by 3 on both sides to get $x = \boxed{\frac{2}{9}}$

Check: $3x + \frac{1}{12} = 3\left(\frac{2}{9}\right) + \frac{1}{12} = \frac{6}{9} + \frac{1}{12} = \frac{2}{3} + \frac{1}{12} = \frac{8}{12} + \frac{1}{12} = \frac{9}{12} = \frac{3}{4}$

2) Add 5 to both sides to get $\frac{2x}{3} = 12$

Multiply by 3 on both sides to get $2x = 36$

Divide by 2 on both sides to get $x = \boxed{18}$

Check: $\frac{2x}{3} - 5 = \frac{2(18)}{3} - 5 = \frac{36}{3} - 5 = 12 - 5 = 7$

3) Subtract $\frac{1}{6}$ from both sides to get $\frac{3}{4} - \frac{1}{6} = \frac{5x}{3}$

Since $\frac{3}{4} - \frac{1}{6} = \frac{9}{12} - \frac{2}{12} = \frac{7}{12}$, this simplifies to $\frac{7}{12} = \frac{5x}{3}$

Multiply by 3 on both sides to get $\frac{21}{12} = 5x$

Divide by 5 on both sides to get $\frac{21}{60} = x$

Divide 21 and 60 each by 3 to reduce the answer to $\boxed{\frac{7}{20}} = x$

Check: $\frac{1}{6} + \frac{5x}{3} = \frac{1}{6} + \frac{5}{3}\frac{7}{20} = \frac{1}{6} + \frac{35}{60} = \frac{10}{60} + \frac{35}{60} = \frac{45}{60} = \frac{3}{4}$

4) Subtract $\frac{5}{6}$ from both sides to get $-\frac{9x}{2} = \frac{8}{3} - \frac{5}{6}$

Since $\frac{8}{3} - \frac{5}{6} = \frac{16}{6} - \frac{5}{6} = \frac{11}{6}$, this simplifies to $-\frac{9x}{2} = \frac{11}{6}$

Multiply by -2 on both sides to get $9x = -\frac{22}{6}$

Divide by 9 on both sides to get $x = -\frac{22}{54}$

Divide 22 and 54 each by 2 to reduce the answer to $x = \boxed{-\frac{11}{27}}$

Check: $\frac{5}{6} - \frac{9x}{2} = \frac{5}{6} - \frac{9}{2}\left(-\frac{11}{27}\right) = \frac{5}{6} + \frac{99}{54} = \frac{45}{54} + \frac{99}{54} = \frac{144}{54} = \frac{8}{3}$

Note: We divided 144 and 54 each by 18 in the last step in order to reduce the answer.

5) Add $\frac{1}{3}$ to both sides to get $8x = 2x + \frac{5}{2} + \frac{1}{3}$

Since $\frac{5}{2} + \frac{1}{3} = \frac{15}{6} + \frac{2}{6} = \frac{17}{6}$, this simplifies to $8x = 2x + \frac{17}{6}$

Subtract $2x$ from both sides to get $6x = \frac{17}{6}$

Divide by 6 on both sides to get $x = \boxed{\frac{17}{36}}$

Check: $8x - \frac{1}{3} = 8\left(\frac{17}{36}\right) - \frac{1}{3} = \frac{136}{36} - \frac{1}{3} = \frac{136}{36} - \frac{12}{36} = \frac{124}{36} = \frac{31}{9}$ agrees with

$2x + \frac{5}{2} = 2\left(\frac{17}{36}\right) + \frac{5}{2} = \frac{34}{36} + \frac{5}{2} = \frac{34}{36} + \frac{90}{36} = \frac{124}{36} = \frac{31}{9}$

6) Subtract 4 from both sides to get $\frac{x}{5} = \frac{x}{8} + 6$

Subtract $\frac{x}{8}$ from both sides to get $\frac{x}{5} - \frac{x}{8} = 6$

Since $\frac{x}{5} - \frac{x}{8} = \frac{8x}{40} - \frac{5x}{40} = \frac{3x}{40}$, this simplifies to $\frac{3x}{40} = 6$

Multiply by 40 on both sides to get $3x = 240$

Divide by 3 on both sides to get $x = \boxed{80}$

Check: $\frac{x}{5} + 4 = \frac{80}{5} + 4 = 16 + 4 = 20$ agrees with

$\frac{x}{8} + 10 = \frac{80}{8} + 10 = 10 + 10 = 20$

7) Subtract $\frac{7}{6}$ from both sides to get $3x = \frac{3x}{2} - \frac{9}{4} - \frac{7}{6}$

Since $-\frac{9}{4} - \frac{7}{6} = -\frac{27}{12} - \frac{14}{12} = -\frac{41}{12}$, this simplifies to $3x = \frac{3x}{2} - \frac{41}{12}$

Subtract $\frac{3x}{2}$ from both sides to get $3x - \frac{3x}{2} = -\frac{41}{12}$

Since $3x - \frac{3x}{2} = \frac{6x}{2} - \frac{3x}{2} = \frac{3x}{2}$, this simplifies to $\frac{3x}{2} = -\frac{41}{12}$

Multiply by 2 on both sides to get $3x = -\frac{82}{12}$

Divide by 3 on both sides to get $x = -\frac{82}{36}$

Divide 82 and 36 each by 2 to reduce the answer to $x = \boxed{-\frac{41}{18}}$

Check: $3x + \frac{7}{6} = 3\left(-\frac{41}{18}\right) + \frac{7}{6} = -\frac{123}{18} + \frac{7}{6} = -\frac{123}{18} + \frac{21}{18} = -\frac{102}{18} = -\frac{17}{3}$ agrees with

$\frac{3x}{2} - \frac{9}{4} = \frac{3}{2}\left(-\frac{41}{18}\right) - \frac{9}{4} = -\frac{123}{36} - \frac{9}{4} = -\frac{123}{36} - \frac{81}{36} = -\frac{204}{36} = -\frac{17}{3}$

8) Add $\frac{7}{12}$ to both sides to get $\frac{5}{4} + \frac{7}{12} - \frac{3x}{16} = \frac{5x}{8}$

Since $\frac{5}{4} + \frac{7}{12} = \frac{15}{12} + \frac{7}{12} = \frac{22}{12} = \frac{11}{6}$, this simplifies to $\frac{11}{6} - \frac{3x}{16} = \frac{5x}{8}$

Add $\frac{3x}{16}$ to both sides to get $\frac{11}{6} = \frac{5x}{8} + \frac{3x}{16}$

Since $\frac{5x}{8} + \frac{3x}{16} = \frac{10x}{16} + \frac{3x}{16} = \frac{13x}{16}$, this simplifies to $\frac{11}{6} = \frac{13x}{16}$

Multiply by 16 on both sides to get $\frac{176}{6} = 13x$

Divide by 13 on both sides to get $\frac{176}{78} = x$

Divide 178 and 78 each by 2 to reduce the answer to $\boxed{\frac{88}{39}} = x$

Check: $\frac{5}{4} - \frac{3x}{16} = \frac{5}{4} - \frac{3}{16}\left(\frac{88}{39}\right) = \frac{5}{4} - \frac{264}{624} = \frac{780}{624} - \frac{264}{624} = \frac{516}{624} = \frac{43}{52}$ agrees with

$\frac{5x}{8} - \frac{7}{12} = \frac{5}{8}\left(\frac{88}{39}\right) - \frac{7}{12} = \frac{440}{312} - \frac{7}{12} = \frac{440}{312} - \frac{182}{312} = \frac{258}{312} = \frac{43}{52}$

Exercise Set 2.11

1) Add $2x$ to both sides to get $6x = 6x$

Divide by 6 on both sides to get $x = x$

The solution is all real numbers.

2) Subtract x from both sides to get $5 = 7$

Since 5 can never equal 7, there is no solution.

3) Combine like terms on the left side to get $2 = 0$

Since 2 can never equal 0, there is no solution.

4) Combine like terms on the left side to get $4x = 4x$

Divide by 4 on both sides to get $x = x$

The solution is all real numbers.

5) Combine like terms on the left side to get $7 + x = x + 7$

Subtract 7 from both sides to get $x = x$

The solution is all real numbers.

6) Multiply by 12 on both sides to get $\frac{24x}{3} = 8x$

Simplify the left side to get $8x = 8x$

Divide by 8 on both sides to get $x = x$

The solution is all real numbers.

Exercise Set 2.12

1) Divide both sides by I to get $\frac{V}{I} = R$

2) Multiply both sides by V to get $dV = m$

3) First multiply both sides by t to get $vt = d$

Then divide both sides by v to get $t = \frac{d}{v}$

4) Add 273 to both sides to get $C + 273 = K$

5) First add θ to both sides to get $\varphi + \theta = 90$

Then subtract φ from both sides to get $\theta = 90 - \varphi$

6) Divide both sides by nR to get $\frac{PV}{nR} = T$

7) First subtract 32 from both sides to get $F - 32 = \frac{9}{5}C$

Then multiply both sides by 5 to get $5F - 160 = 9C$

Note that both terms are multiplied by 5

Finally, divide both sides by 9 to get $\frac{5F}{9} - \frac{160}{9} = C$, which is equivalent to $\frac{5F-160}{9} = C$

It would also be correct to write $\frac{5}{9}(F - 32) = C$

8) First subtract b from both sides to get $y - b = mx$

Then divide both sides by m to get $\frac{y}{m} - \frac{b}{m} = x$

It would also be correct to write $\frac{y-b}{m} = x$

9) First multiply both sides by 2 to get $2A = b(h_1 + h_2)$

Then divide both sides by $h_1 + h_2$ to get $\frac{2A}{h_1+h_2} = b$

Note: In this case, it would be **incorrect** to separate this into $\frac{2A}{h_1}$ plus $\frac{2A}{h_2}$. We'll learn how to separate expressions properly in Chapters 4-5.

10) First subtract $3x$ from both sides to get $2y = 6z - 3x$

Then divide both sides by 2 to get $y = 3z - \frac{3x}{2}$

It would also be correct to write $y = \frac{6z-3x}{2}$ or to write $y = \frac{3}{2}(2z - x)$. We'll learn how to separate expressions properly in Chapters 4-5.

Exercise Set 2.13

1) $x =$ the amount of money (in dollars) that Alice has left over

$75 - 28 - 19 = x$ simplifies to $\boxed{28}$ (meaning that Alice has \$28 left over)

Check: $28 + 28 + 19 = 75$

2) $x =$ the smallest of the three consecutive numbers

$x + 2 =$ the middle number

$x + 4 =$ the largest of the three consecutive numbers

$x + x + 2 + x + 4 = 63$ simplifies to $3x + 6 = 63$

Subtract 6 from both sides to get $3x = 57$

Divide by 3 on both sides to get $x = 19$ (so the numbers are $\boxed{19}$, $\boxed{21}$, and $\boxed{23}$)

Check: $19 + 21 + 23 = 63$

3) $x =$ Dave's age

$3x =$ Susan's age

$x + 3x = 64$ simplifies to $4x = 64$

Divide by 4 on both sides to get $x = \boxed{16}$ (Dave's age)

Susan is $3x = 3(16) = \boxed{48}$ (Susan's age)

Check: Susan (48) is 3 times as old as Dave (16) and their ages add up to $16 + 48 = 64$

4) x = the number of beads that Melissa had originally

$\frac{x}{3}$ = the number of beads that Melissa gave away

$x - \frac{x}{3} = 24$ To subtract the fraction, make a common denominator: $\frac{3x}{3} - \frac{x}{3} = 24$

This simplifies to $\frac{2x}{3} = 24$

Multiply by 3 on both sides to get $2x = 72$

Divide by 2 on both sides to get $x = \boxed{36}$

Check: One-third of 36 equals 12, and $36 - 12 = 24$

5) x = the smaller number

$x + 25$ = the larger number (so that the difference between these numbers is 25)

$x + x + 25 = 47$ simplifies to $2x + 25 = 47$

Subtract 25 from both sides to get $2x = 22$

Divide by 2 on both sides to get $x = \boxed{11}$ (the smaller number)

The larger number is $x + 25 = 11 + 25 = \boxed{36}$ (the larger number)

Check: $36 - 11 = 25$ and $36 + 11 = 47$

6) x = the cost of one apple (in dollars)

$4x$ = the cost of 4 applies (in dollars)

$10 - 4x = 7$ Subtract 10 from both sides to get $-4x = -3$

Divide by -4 on both sides to get $x = \boxed{\frac{3}{4}}$ (three-quarters of a dollar, which is equivalent

to $\boxed{\$0.75}$ or $\boxed{75}$ cents)

Check: $4(\$0.75) = \$3$ and $10 - 3 = 7$

7) x = the smaller number

$8x$ = the larger number

$8x - x = 77$ simplifies to $7x = 77$

Divide by 7 on both sides to get $x = 11$ (so the numbers are $\boxed{11}$ and $\boxed{88}$)

Check: 88 is 8 times 11 and the difference is $88 - 11 = 77$

8) x = the cost of one child's ticket

$2x$ = the cost of one adult ticket

$2(2x) = 4x$ = the cost of two adult tickets

$4x + 5x = 63$ simplifies to $9x = 63$

Divide by 9 on both sides to get $x = 7$ (a child's ticket is $\boxed{\$7}$ and an adult ticket is $\boxed{\$14}$)

Check: 2 adult tickets plus 5 children's tickets costs $2(14) + 5(7) = 28 + 35 = 63$

3 Exponents

Exercise Set 3.1

1) $x^3x^2 = x^{3+2} = x^5$

2) $x^7x^3 = x^{7+3} = x^{10}$

3) $x^8x^5 = x^{8+5} = x^{13}$

4) $x^6x^6 = x^{6+6} = x^{12}$

5) $x^4x = x^4x^1 = x^{4+1} = x^5$

6) $xx = x^1x^1 = x^{1+1} = x^2$

7) $x^9x^6 = x^{9+6} = x^{15}$

8) $x^4x^4x^4 = x^{4+4+4} = x^{12}$

9) $x^7x^5x^2 = x^{7+5+2} = x^{14}$

10) $x^8x^7x = x^8x^7x^1 = x^{8+7+1} = x^{16}$

11) $xxxx = x^1x^1x^1x^1 = x^{1+1+1+1} = x^4$

12) $x^9x^8x^7x^6 = x^{9+8+7+6} = x^{30}$

Exercise Set 3.2

1) $\dfrac{x^6}{x^2} = x^{6-2} = x^4$

2) $\dfrac{x^8}{x^5} = x^{8-5} = x^3$

3) $\dfrac{x^{12}}{x^7} = x^{12-7} = x^5$

4) $\dfrac{x^5}{x} = \dfrac{x^5}{x^1} = x^{5-1} = x^4$

5) $\dfrac{x^4x^3}{x^2} = \dfrac{x^{4+3}}{x^2} = \dfrac{x^7}{x^2} = x^{7-2} = x^5$

6) $\dfrac{x^9}{x^6x^2} = \dfrac{x^9}{x^{6+2}} = \dfrac{x^9}{x^8} = x^{9-8} = x^1 = x$

7) $\dfrac{x^7x}{x^6} = \dfrac{x^7x^1}{x^6} = \dfrac{x^{7+1}}{x^6} = \dfrac{x^8}{x^6} = x^{8-6} = x^2$

8) $\dfrac{x^{14}}{x^6x} = \dfrac{x^{14}}{x^6x^1} = \dfrac{x^{14}}{x^{6+1}} = \dfrac{x^{14}}{x^7} = x^{14-7} = x^7$

9) $\dfrac{x^9x^8}{x^6x^4} = \dfrac{x^{9+8}}{x^{6+4}} = \dfrac{x^{17}}{x^{10}} = x^{17-10} = x^7$

10) $\dfrac{x^9x}{x^3x^3} = \dfrac{x^9x^1}{x^3x^3} = \dfrac{x^{9+1}}{x^{3+3}} = \dfrac{x^{10}}{x^6} = x^{10-6} = x^4$

Exercise Set 3.3

1) $1^0 = 1$

2) $\left(\dfrac{1}{0}\right)^2$ is undefined since $\dfrac{1}{0}$ is undefined

3) $(-1)^0 = 1$

4) $0^{6(2)-4(3)} = 0^{12-12} = 0^0$ is indeterminate

Exercise Set 3.4

1) $x^9x^{-5} = x^{9+(-5)} = x^4$

2) $x^{-8}x^4 = x^{-8+4} = x^{-4} = \dfrac{1}{x^4}$

3) $x^7x^{-7} = x^{7+(-7)} = x^0 = 1$

4) $x^{-5}x^{-5} = x^{-5+(-5)} = x^{-10} = \dfrac{1}{x^{10}}$

5) $xx^{-4} = x^1x^{-4} = x^{1+(-4)} = x^{-3} = \dfrac{1}{x^3}$

6) $x^{-8}x^{-6} = x^{-8+(-6)} = x^{-14} = \dfrac{1}{x^{14}}$

7) $x^{12}x^{-4}x^{-3} = x^{12+(-4)+(-3)} = x^{12+(-7)} = x^5$

8) $x^7x^6x^{-5} = x^{7+6+(-5)} = x^{13+(-5)} = x^8$

9) $x^{-5}x^{-3}x^{-2} = x^{-5+(-3)+(-2)} = x^{-5+(-5)} = x^{-10} = \frac{1}{x^{10}}$

10) $x^8x^{-7}x = x^8x^{-7}x^1 = x^{8+(-7)+1} = x^{9+(-7)} = x^2$

11) $\frac{x^2}{x^{-4}} = x^{2-(-4)} = x^{2+4} = x^6$ 12) $\frac{x^{-9}}{x^7} = x^{-9-7} = x^{-16} = \frac{1}{x^{16}}$

13) $\frac{x^{-6}}{x^{-6}} = x^{-6-(-6)} = x^{-6+6} = x^0 = 1$ 14) $\frac{x^8}{x^{-8}} = x^{8-(-8)} = x^{8+8} = x^{16}$

15) $\frac{x^{-3}}{x^{-4}} = x^{-3-(-4)} = x^{-3+4} = x^1 = x$ 16) $\frac{x^{-2}}{x} = \frac{x^{-2}}{x^1} = x^{-2-1} = x^{-3} = \frac{1}{x^3}$

17) $\frac{x^8x^{-3}}{x^{-7}} = \frac{x^{8+(-3)}}{x^{-7}} = \frac{x^5}{x^{-7}} = x^{5-(-7)} = x^{5+7} = x^{12}$

18) $\frac{x^{-6}}{x^7x^{-5}} = \frac{x^{-6}}{x^{7+(-5)}} = \frac{x^{-6}}{x^2} = x^{-6-2} = x^{-8} = \frac{1}{x^8}$

19) $\frac{x^{-9}x}{x^{-8}x^{-5}} = \frac{x^{-9}x^1}{x^{-8}x^{-5}} = \frac{x^{-9+1}}{x^{-8+(-5)}} = \frac{x^{-8}}{x^{-13}} = x^{-8-(-13)} = x^{-8+13} = x^5$

20) $\frac{x^{-7}x^{-4}}{x^{-9}x^{-3}} = \frac{x^{-7+(-4)}}{x^{-9+(-3)}} = \frac{x^{-11}}{x^{-12}} = x^{-11-(-12)} = x^{-11+12} = x^1 = x$

Exercise Set 3.5

1) $(x^6)^4 = x^{6(4)} = x^{24}$ 2) $(x^3)^3 = x^{3(3)} = x^9$

3) $(x^7)^{-3} = x^{7(-3)} = x^{-21} = \frac{1}{x^{21}}$ 4) $(x^{-6})^{-8} = x^{-6(-8)} = x^{48}$

5) $(x^{-9})^6 = x^{-9(6)} = x^{-54} = \frac{1}{x^{54}}$ 6) $(x^7)^0 = x^{7(0)} = x^0 = 1$

7) $(x^{-5})^{-5} = x^{-5(-5)} = x^{25}$ 8) $(x^8)^2 = x^{8(2)} = x^{16}$

9) $(x^{-7})^1 = x^{-7(1)} = x^{-7} = \frac{1}{x^7}$ 10) $(x^{-2})^8 = x^{-2(8)} = x^{-16} = \frac{1}{x^{16}}$

11) $(x^{-3})^{-8} = x^{-3(-8)} = x^{24}$ 12) $(x^4)^{-1} = x^{4(-1)} = x^{-4} = \frac{1}{x^4}$

13) $(x^7)^6 = x^{7(6)} = x^{42}$ 14) $(x)^{-8} = (x^1)^{-8} = x^{1(-8)} = x^{-8} = \frac{1}{x^8}$

15) $(x^0)^{-2} = x^{0(-2)} = x^0 = 1$ 16) $(x^{-9})^{-9} = x^{-9(-9)} = x^{81}$

Exercise Set 3.6

1) $x^{1/2} = \sqrt{x}$, $49^{1/2} = \sqrt{49} = 7$ because $7^2 = 49$

2) $x^{1/3} = \sqrt[3]{x}$, $64^{1/3} = \sqrt[3]{64} = 4$ because $4^3 = 64$

3) $x^{-1/2} = \frac{1}{x^{1/2}} = \frac{1}{\sqrt{x}}$, $81^{-1/2} = \frac{1}{81^{1/2}} = \frac{1}{\sqrt{81}} = \frac{1}{9}$ because $9^2 = 81$

4) $x^{1/5} = \sqrt[5]{x}$, $32^{1/5} = \sqrt[5]{32} = 2$ because $2^5 = 32$

5) $x^{-1/4} = \frac{1}{x^{1/4}} = \frac{1}{\sqrt[4]{x}}$, $625^{-1/4} = \frac{1}{625^{1/4}} = \frac{1}{\sqrt[4]{625}} = \frac{1}{5}$ because $5^4 = 625$

Note: As explained in the text, in this context we are only giving the positive answers.

Exercise Set 3.7

1) $x^{4/5}x^{2/3} = x^{4/5+2/3} = x^{22/15}$ because $\frac{4}{5} + \frac{2}{3} = \frac{12}{15} + \frac{10}{15} = \frac{22}{15}$

2) $x^{5/6}x^{-3/8} = x^{5/6+(-3/8)} = x^{11/24}$ because $\frac{5}{6} - \frac{3}{8} = \frac{20}{24} - \frac{9}{24} = \frac{11}{24}$

3) $x^{-5/6}x^{1/2} = x^{-5/6+1/2} = x^{-1/3} = \frac{1}{x^{1/3}}$ because $-\frac{5}{6} + \frac{1}{2} = -\frac{5}{6} + \frac{3}{6} = -\frac{2}{6} = -\frac{1}{3}$

4) $x^{-5/2}x^{-4/3} = x^{-5/2-4/3} = x^{-23/6} = \frac{1}{x^{23/6}}$ because $-\frac{5}{2} - \frac{4}{3} = -\frac{15}{6} - \frac{8}{6} = -\frac{23}{6}$

5) $x^{3/4}x^{2/3}x = x^{3/4}x^{2/3}x^1 = x^{3/4+2/3+1} = x^{29/12}$ because $\frac{3}{4} + \frac{2}{3} + 1 = \frac{9}{12} + \frac{8}{12} + \frac{12}{12} = \frac{29}{12}$

6) $x^{3/8}x^{-1/4}x^{-1/2} = x^{3/8-1/4-1/2} = x^{-3/8} = \frac{1}{x^{3/8}}$ because $\frac{3}{8} - \frac{1}{4} - \frac{1}{2} = \frac{3}{8} - \frac{2}{8} - \frac{4}{8} = \frac{3-6}{8} = -$

7) $\frac{x^{5/6}}{x^{1/3}} = x^{5/6-1/3} = x^{1/2}$ because $\frac{5}{6} - \frac{1}{3} = \frac{5}{6} - \frac{2}{6} = \frac{3}{6} = \frac{1}{2}$

8) $\frac{x^{5/3}}{x^{-3/4}} = x^{5/3-(-3/4)} = x^{5/3+3/4} = x^{29/12}$ because $\frac{5}{3} + \frac{3}{4} = \frac{20}{12} + \frac{9}{12} = \frac{29}{12}$

9) $\frac{x^{-7/2}}{x^{9/4}} = x^{-7/2-9/4} = x^{-23/4} = \frac{1}{x^{23/4}}$ because $-\frac{7}{2} - \frac{9}{4} = -\frac{14}{4} - \frac{9}{4} = -\frac{23}{4}$

10) $\frac{x^{3/4}x^{3/4}}{x^{3/8}} = x^{3/4+3/4-3/8} = x^{9/8}$ because $\frac{3}{4} + \frac{3}{4} - \frac{3}{8} = \frac{6}{8} + \frac{6}{8} - \frac{3}{8} = \frac{12-3}{8} = \frac{9}{8}$

11) $\frac{x^{-4/5}}{x^{2/5}x^{1/2}} = x^{-4/5-2/5-1/2} = x^{-17/10} = \frac{1}{x^{17/10}}$ because $-\frac{4}{5} - \frac{2}{5} - \frac{1}{2} = -\frac{8}{10} - \frac{4}{10} - \frac{5}{10} = -\frac{1}{1}$

12) $\frac{x^{2/3}x^{1/2}}{x^{3/4}x^{-1/3}} = \frac{x^{2/3+1/2}}{x^{3/4-1/3}} = \frac{x^{7/6}}{x^{5/12}} = x^{3/4}$ because $\frac{2}{3} + \frac{1}{2} = \frac{4}{6} + \frac{3}{6} = \frac{7}{6}$,

$\frac{3}{4} - \frac{1}{3} = \frac{9}{12} - \frac{4}{12} = \frac{5}{12}$, and $\frac{7}{6} - \frac{5}{12} = \frac{14}{12} - \frac{5}{12} = \frac{9}{12} = \frac{3}{4}$

13) $\left(x^{3/4}\right)^6 = x^{(3/4)6} = x^{9/2}$ because $\left(\frac{3}{4}\right)6 = \frac{3(6)}{4} = \frac{18}{4} = \frac{9}{2}$

14) $\left(x^{5/6}\right)^{-3/2} = x^{(5/6)(-3/2)} = x^{-5/4} = \frac{1}{x^{5/4}}$ because $\frac{5}{6}\left(-\frac{3}{2}\right) = \frac{5(-3)}{6(2)} = -\frac{15}{12} = -\frac{5}{4}$

15) $\left(x^{-4/9}\right)^{3/2} = x^{(-4/9)(3/2)} = x^{-2/3} = \frac{1}{x^{2/3}}$ because $\left(-\frac{4}{9}\right)\frac{3}{2} = -\frac{4(3)}{9(2)} = -\frac{12}{18} = -\frac{2}{3}$

16) $(x^{-3})^{-5/6} = x^{-3(-5/6)} = x^{5/2}$ because $-3\left(-\frac{5}{6}\right) = \frac{3}{1}\left(\frac{5}{6}\right) = \frac{(3)(5)}{6} = \frac{15}{6} = \frac{5}{2}$

Exercise Set 3.8

1) $(9x^3)^2 = 9^2 x^{3(2)} = 81x^6$

2) $(-5x^8)^3 = (-5)^3 x^{8(3)} = -125x^{24}$

3) $(-4x^7)^4 = (-4)^4 x^{7(4)} = 256x^{28}$

4) $(2x)^{-5} = \left(\dfrac{1}{2x}\right)^5 = \dfrac{1}{2^5 x^5} = \dfrac{1}{32x^5}$

5) $(6x^7)^0 = 6^0 x^{7(0)} = 6^0 x^0 = 1(1) = 1$

6) $(8x^{-4})^{-1} = \left(\dfrac{1}{8x^{-4}}\right)^1 = \dfrac{1}{8x^{-4}} = \dfrac{x^4}{8}$ because $x^{-4} = \dfrac{1}{x^4}$ such that $x^{-4}x^4 = 1$ and $x^4 = \dfrac{1}{x^{-4}}$

7) $(64x^5)^{1/2} = 64^{1/2} x^{5/2} = \sqrt{64}(x^{5/2}) = 8x^{5/2}$

8) $\left(27x^{3/4}\right)^{2/3} = 27^{2/3} x^{(3/4)(2/3)} = \left(\sqrt[3]{27}\right)^2 x^{1/2} = 3^2 x^{1/2} = 9x^{1/2}$

since $\dfrac{3}{4}\dfrac{2}{3} = \dfrac{3(2)}{4(3)} = \dfrac{6}{12} = \dfrac{1}{2}$

9) $\left(\dfrac{x^4}{5}\right)^3 = \dfrac{x^{4(3)}}{5^3} = \dfrac{x^{12}}{125}$

10) $\left(\dfrac{3}{x^2}\right)^4 = \dfrac{3^4}{x^{2(4)}} = \dfrac{81}{x^8}$

11) $\left(-\dfrac{x^9}{4}\right)^5 = \dfrac{(-1)^5 x^{9(5)}}{4^5} = -\dfrac{x^{45}}{1024}$

12) $\left(-\dfrac{x^7}{6}\right)^2 = \dfrac{(-1)^2 x^{7(2)}}{6^2} = \dfrac{(1)x^{14}}{36} = \dfrac{x^{14}}{36}$

13) $\left(\dfrac{x^6}{5}\right)^{-2} = \left(\dfrac{5}{x^6}\right)^2 = \dfrac{5^2}{x^{6(2)}} = \dfrac{25}{x^{12}}$

Note: A base raised to a negative exponent is equivalent to raising the reciprocal of the base to the absolute value of the exponent, like Example 3. We did this in Sec. 1.11 using numbers. It may help to review Sec. 1.11 and Sec. 3.4.

14) $\left(\dfrac{x^8}{7}\right)^{-1} = \left(\dfrac{7}{x^8}\right)^1 = \dfrac{7}{x^8}$

15) $\left(\dfrac{x^8}{256}\right)^{3/4} = \dfrac{x^{8(3/4)}}{256^{3/4}} = \dfrac{x^6}{\left(\sqrt[4]{256}\right)^3} = \dfrac{x^6}{4^3} = \dfrac{x^6}{64}$ because $8\left(\dfrac{3}{4}\right) = \dfrac{8}{1}\dfrac{3}{4} = \dfrac{8(3)}{1(4)} = \dfrac{24}{4} = 6$

Note: It may help to review Sec. 1.11 and Sec. 3.7 regarding fractional exponents.

16) $\left(\dfrac{x^{10}}{32}\right)^{4/5} = \dfrac{x^{10(4/5)}}{32^{4/5}} = \dfrac{x^8}{\left(\sqrt[5]{32}\right)^4} = \dfrac{x^8}{2^4} = \dfrac{x^8}{16}$ because $10\left(\dfrac{4}{5}\right) = \dfrac{10}{1}\dfrac{4}{5} = \dfrac{10(4)}{1(5)} = \dfrac{40}{5} = 8$

Exercise Set 3.9

1) $518 = 5.18 \times 10^2$ (moved left 2 places)
2) $0.094 = 9.4 \times 10^{-2}$ (moved right 2 places)
3) $0.00675 = 6.75 \times 10^{-3}$ (moved right 3 places)
4) $639{,}415 = 6.39415 \times 10^5$ (moved left 5 places)
5) $1694 = 1.694 \times 10^3$ (moved left 3 places)
6) $0.00001487 = 1.487 \times 10^{-5}$ (moved right 5 places)
7) $0.62 = 6.2 \times 10^{-1}$ (moved right 1 place)
8) $81{,}300 = 8.13 \times 10^4$ (moved left 4 places)
9) $9{,}583{,}000 = 9.583 \times 10^6$ (moved left 6 places)
10) $0.000256 = 2.56 \times 10^{-4}$ (moved right 4 places)

Exercise Set 3.10

1) Subtract 18 from both sides to get $x^5 = 32$

Take the fifth root of both sides to get $x = 32^{1/5}$; the answer is $x = \boxed{2}$

Check: $2^5 + 18 = 32 + 18 = 50$

2) Add 9 to both sides to get $81 = x^4$

Take the fourth root of both sides to get $81^{1/4} = x$; the answers are $\boxed{\pm 3} = x$

Check: $x^4 - 9 = (\pm 3)^4 - 9 = 81 - 9 = 72$

3) Divide by 5 on both sides to get $x^{1/2} = 4$

Square both sides to get $x = 4^2$; the answer is $x = \boxed{16}$

Check: $5x^{1/2} = 5(16)^{1/2} = 5(4) = 20$

4) Subtract 1 from both sides to get $6 = 2x^{1/3}$

Divide by 2 on both sides to get $3 = x^{1/3}$

Cube both sides to get $3^3 = x$; the answer is $\boxed{27} = x$

Check: $2x^{1/3} + 1 = 2(27)^{1/3} + 1 = 2(3) + 1 = 6 + 1 = 7$

5) Multiply by 5 on both sides to get $x^3 = -125$

Cube root both sides to get $x = (-125)^{1/3}$; the answer is $x = \boxed{-5}$

Check: $\dfrac{x^3}{5} = \dfrac{(-5)^3}{5} = \dfrac{-125}{5} = -25$

6) Subtract 4 from both sides to get $6x^2 = 96$

Divide by 6 on both sides to get $x^2 = 16$

Square root both sides to get $x = 16^{1/2}$; the answers are $x = \boxed{\pm 4}$

Check: $4 + 6x^2 = 4 + 6(\pm 4)^2 = 4 + 6(16) = 4 + 96 = 100$

7) Add 1 to both sides to get $3x^4 = 3$

Divide by 3 on both sides to get $x^4 = 1$

Take the fourth root of both sides to get $x = 1^{1/4}$; the answers are $x = \boxed{\pm 1}$

Check: $3x^4 - 1 = 3(\pm 1)^4 - 1 = 3(1) - 1 = 3 - 1 = 2$

8) Subtract 36 from both sides to get $864 = -4x^3$

Divide by -4 on both sides to get $-216 = x^3$

Cube root both sides to get $(-216)^{1/3} = x$; the answer is $\boxed{-6} = x$

Check: $-4x^3 + 36 = -4(-6)^3 + 36 = -4(-216) + 36 = 864 + 36 = 900$

9) Multiply by 4 on both sides to get $x^{1/2} = 12$

Square both sides to get $x = 12^2$; the answer is $x = \boxed{144}$

Check: $\dfrac{x^{1/2}}{4} = \dfrac{144^{1/2}}{4} = \dfrac{12}{4} = 3$

10) Subtract x^3 from both sides to get $4x^3 = 500$

Divide by 4 on both sides to get $x^3 = 125$

Cube root both sides to get $x = 125^{1/3}$; the answer is $x = \boxed{5}$

Check: $5x^3 = 5(5)^3 = 5(125) = 625$ agrees with

$x^3 + 500 = 5^3 + 500 = 125 + 500 = 625$

11) Add 48 to both sides to get $5x^2 = 98 + 3x^2$

Subtract $3x^2$ from both sides to get $2x^2 = 98$

Divide by 2 on both sides to get $x^2 = 49$

Square root both sides to get $x = 49^{1/2}$; the answers are $x = \boxed{\pm 7}$

Check: $5x^2 - 48 = 5(\pm 7)^2 - 48 = 5(49) - 48 = 245 - 48 = 197$ agrees with

$50 + 3x^2 = 50 + 3(\pm 7)^2 = 50 + 3(49) = 50 + 147 = 197$

12) Multiply by 4 on both sides to get $x^{-2} = 36$

Raise both sides to the power of $-\dfrac{1}{2}$ to get $(x^{-2})^{-1/2} = 36^{-1/2}$

Since $-2\left(-\dfrac{1}{2}\right) = 1$ and $x^1 = 1$, this simplifies to $x = 36^{-1/2}$

Simplify: $x = 36^{-1/2} = \dfrac{1}{36^{1/2}} = \dfrac{1}{\pm 6} = \boxed{\pm \dfrac{1}{6}}$

Check: $\frac{x^{-2}}{4} = \frac{1}{4}\left(\pm\frac{1}{6}\right)^{-2} = \frac{1}{4}(\pm 6)^2 = \frac{1}{4}(36) = 9$

Recall that a negative power equates to raising the reciprocal of the base to the absolute value of the power (recall Sec. 1.11 and Sec. 3.4). The reciprocal of $\frac{1}{6}$ equals 6.

13) Divide by 2 on both sides to get $x^{3/4} = 27$

Raise both sides to the power of $\frac{4}{3}$ to get $\left(x^{3/4}\right)^{4/3} = 27^{4/3}$

Since $\frac{3}{4}\left(\frac{4}{3}\right) = \frac{12}{12} = 1$ and $x^1 = x$, this simplifies to $x = 27^{4/3}$

Simplify: $x = 27^{4/3} = \left(\sqrt[3]{27}\right)^4 = 3^4 = \boxed{81}$

Check: $2x^{3/4} = 2(81)^{3/4} = 2\left(\sqrt[4]{81}\right)^3 = 2(3)^3 = 2(27) = 54$

14) Divide by 3 on both sides to get $x^{-2/3} = 16$

Raise both sides to the power of $-\frac{3}{2}$ to get $\left(x^{-2/3}\right)^{-3/2} = 16^{-3/2}$

Since $-\frac{2}{3}\left(-\frac{3}{2}\right) = \frac{6}{6} = 1$ and $x^1 = x$, this simplifies to $x = 16^{-3/2}$

Simplify: $x = 16^{-3/2} = \frac{1}{16^{3/2}} = \frac{1}{\left(\sqrt{16}\right)^3} = \frac{1}{(\pm 4)^3} = \boxed{\pm\frac{1}{64}}$

Check: $3x^{-2/3} = 3\left(\pm\frac{1}{64}\right)^{-2/3} = 3(\pm 64)^{2/3} = 3\left(\pm\sqrt[3]{64}\right)^2 = 3(\pm 4)^2 = 3(16) = 48$

Recall that a negative power equates to raising the reciprocal of the base to the absolute value of the power (recall Sec. 1.11 and Sec. 3.4). The reciprocal of $\frac{1}{64}$ equals 64.

Exercise Set 3.11

1) Apply the rule $\frac{x^m}{x^n} = x^{m-n}$ to get $x^{5-5} = 1$ which simplifies to $x^0 = 1$

The answer is indeterminate because any nonzero value raised to the power of zero equals one.

2) Combine like terms to get $6x^2 + 8 = 6x^2$

Subtract $6x^2$ from both sides to get $8 = 0$

Since 8 will never equal 0, there is no solution.

3) Add 4 to both sides to get $x^2 = 4$

Square root both sides to get $x = 4^{1/2}$

The two answers are $x = \pm 2$ since $(\pm 2)^2 = 4$

4) Apply the order of operations (Sec. 1.5):

$$x = \left[\frac{8(2) - 4^2}{8(2) + 4^2}\right]^{2(3)-6} = \left(\frac{16 - 16}{16 + 16}\right)^{6-6} = \left(\frac{0}{32}\right)^0 = 0^0$$

The answer is indeterminate because 0^0 is indeterminate (Sec. 3.3).

5) Subtract 4 from both sides to get $x^2 = -4$

Square root both sides to get $x = \sqrt{-4}$

This has no real solution (just an imaginary solution).

6) Raise both sides to the power of -1 to get $(x^{-1})^{-1} = 0^{-1}$ (following the strategy of Sec. 3.10). Note that $(x^{-1})^{-1} = x^{(-1)(-1)} = x^1 = x$. The equation becomes $x = 0^{-1}$.

Since 0^{-1} is equivalent to $\frac{1}{0}$, the answer is undefined (Sec. 3.3).

Exercise Set 3.12

1) $x =$ the area of the carpet in square feet

The square is 12 feet wide: $x = 12^2 = 144$ square feet

2) $x =$ how far Steve's token will be from the finish line after he moves it

$x = 35 - 3^3 = 35 - 27 = 8$ steps away

3) $x =$ the number

$x^{1/2} =$ the square root of the number

According to the problem: $5x^{1/2} = 80$

Divide both sides by 5 to get $x^{1/2} = 16$

Square both sides to get $x = 16^2 = 256$

Check: $5x^{1/2} = 5(256)^{1/2} = 5(16) = 80$

4) $x =$ the number

According to the problem: $x^{2/3} = 6^2 - 20$

Simplify the right side to get $x^{2/3} = 16$ (since $6^2 - 20 = 36 - 20 = 16$)

Raise both sides of the equation to the power of $\frac{3}{2}$.

We get $x = 16^{3/2}$ because $x^{(2/3)(3/2)} = x^1 = x$ since $\frac{2}{3}\left(\frac{3}{2}\right) = \frac{6}{6} = 1$

Simplify: $x = 16^{3/2} = \left(\sqrt{16}\right)^3 = (4)^3 = 64$ (the problem specified "positive")

Check: $x^{2/3} = 64^{2/3} = \left(\sqrt[3]{64}\right)^2 = 4^2 = 16$ agrees with $6^2 - 20 = 36 - 20 = 16$

4 The Distributive Property

Exercise Set 4.1

1) $7(9x - 8) = 7(9x) + 7(-8) = 63x - 56$

2) $x(x - 1) = x(x) + x(-1) = x^2 - x$

3) $6x(4x + 7) = 6x(4x) + 6x(7) = 24x^2 + 42x$

4) $5x^2(4x^2 - 8x) = 5x^2(4x^2) + 5x^2(-8x) = 20x^4 - 40x^3$

5) $8x^3(6x^5 + 4x^3) = 8x^3(6x^5) + 8x^3(4x^3) = 48x^8 + 32x^6$

6) $9x(x^2 - 6x + 5) = 9x(x^2) + 9x(-6x) + 9x(5) = 9x^3 - 54x^2 + 45x$

7) $4x^2(2x^4 + 3x^2 - 6) = 4x^2(2x^4) + 4x^2(3x^2) + 4x^2(-6) = 8x^6 + 12x^4 - 24x^2$

8) $3x^5(-8x^6 - 7x^4 + x^2) = 3x^5(-8x^6) + 3x^5(-7x^4) + 3x^5(x^2) = -24x^{11} - 21x^9 + 3x^7$

9) $x(x^3 - x^2 + x - 1) = x(x^3) + x(-x^2) + x(x) + x(-1) = x^4 - x^3 + x^2 - x$

10) $2x^{11}(3x^{12} - 8x^7 + 6x^2) = 2x^{11}(3x^{12}) + 2x^{11}(-8x^7) + 2x^{11}(6x^2)$
$= 6x^{23} - 16x^{18} + 12x^{13}$

11) $5x^{-2}(3x^2 + 7x) = 5x^{-2}(3x^2) + 5x^{-2}(7x) = 15 + 35x^{-1}$ (Note that $x^{-2+2} = x^0 = 1$)

12) $9x^4(6x^{-2} + 4x^{-8}) = 9x^4(6x^{-2}) + 9x^4(4x^{-8}) = 54x^2 + 36x^{-4}$

13) $2x^{-3}(4x^{-2} - 6x^{-4}) = 2x^{-3}(4x^{-2}) + 2x^{-3}(-6x^{-4}) = 8x^{-5} - 12x^{-7}$

14) $8x^{-4}(3x^6 + 2x^4 - 4x^2) = 8x^{-4}(3x^6) + 8x^{-4}(2x^4) + 8x^{-4}(-4x^2)$
$= 24x^2 + 16 - 32x^{-2}$

15) $3x^{2/3}\left(2x^{3/4} - x^{1/3}\right) = 3x^{2/3}\left(2x^{3/4}\right) - 3x^{2/3}\left(-x^{1/3}\right) = 6x^{17/12} - 3x$

Notes: $\frac{2}{3} + \frac{3}{4} = \frac{8}{12} + \frac{9}{12} = \frac{17}{12}$ and $\frac{2}{3} + \frac{1}{3} = \frac{3}{3} = 1$ (and $x^1 = x$)

Exercise Set 4.2

1) $-6(4x - 8) = -6(4x) - 6(-8) = -24x + 48$

2) $-x(x^2 - 2x) = -x(x^2) - x(-2x) = -x^3 + 2x^2$

3) $-(-x - 5) = -(-x) - (-5) = x + 5$ Note: $-(-x - 5)$ is equivalent to $-1(-x - 5)$

4) $-7x(-4x^2 + 9) = -7x(-4x^2) - 7x(9) = 28x^3 - 63x$

5) $-x(-3x^2 - 2x + 4) = -x(-3x^2) - x(-2x) - x(4) = 3x^3 + 2x^2 - 4x$

6) $-8x^2(7x^5 + 5x^3 - 3x) = -8x^2(7x^5) - 8x^2(5x^3) - 8x^2(-3x) = -56x^7 - 40x^5 + 24x^3$

Exercise Set 4.3

1) $(x + 7)(x + 4) = x(x) + x(4) + 7(x) + 7(4) = x^2 + 4x + 7x + 28 = x^2 + 11x + 28$

2) $(x + 8)(x - 6) = x(x) + x(-6) + 8(x) + 8(-6) = x^2 - 6x + 8x - 48 = x^2 + 2x - 48$

3) $(x - 1)(x + 5) = x(x) + x(5) - 1(x) - 1(5) = x^2 + 5x - x - 5 = x^2 + 4x - 5$

4) $(-x + 9)(x + 3) = -x(x) - x(3) + 9(x) + 9(3) = -x^2 - 3x + 9x + 27 = -x^2 + 6x + 27$

5) $(x - 4)(x - 6) = x(x) + x(-6) - 4(x) - 4(-6) = x^2 - 6x - 4x + 24 = x^2 - 10x + 24$

6) $(-x - 2)(x + 8) = -x(x) - x(8) - 2(x) - 2(8)$
$= -x^2 - 8x - 2x - 16 = -x^2 - 10x - 16$

7) $(x - 5)(-x - 7) = x(-x) + x(-7) - 5(-x) - 5(-7)$
$= -x^2 - 7x + 5x + 35 = -x^2 - 2x + 35$

8) $(9 - x)(6 - x) = 9(6) + 9(-x) - x(6) - x(-x) = 54 - 9x - 6x + x^2 = x^2 - 15x + 54$

9) $(6x - 4)(3x + 7) = 6x(3x) + 6x(7) - 4(3x) - 4(7)$
$= 18x^2 + 42x - 12x - 28 = 18x^2 + 30x - 28$

10) $(5x + 2)(3x + 4) = 5x(3x) + 5x(4) + 2(3x) + 2(4)$
$= 15x^2 + 20x + 6x + 8 = 15x^2 + 26x + 8$

11) $(2x^2 + 6)(x^2 - 5) = 2x^2(x^2) + 2x^2(-5) + 6(x^2) + 6(-5)$
$= 2x^4 - 10x^2 + 6x^2 - 30 = 2x^4 - 4x^2 - 30$

12) $(3x^2 - 4x)(2x^2 + 6) = 3x^2(2x^2) + 3x^2(6) - 4x(2x^2) - 4x(6)$
$= 6x^4 + 18x^2 - 8x^3 - 24x = 6x^4 - 8x^3 + 18x^2 - 24x$

13) $(5x^6 + 7x^4)(8x^3 - 6x) = 5x^6(8x^3) + 5x^6(-6x) + 7x^4(8x^3) + 7x^4(-6x)$
$= 40x^9 - 30x^7 + 56x^7 - 42x^5 = 40x^9 + 26x^7 - 42x^5$

14) $(5x + 3x^{-1})(4x^{-3} - 2x^{-4}) = 5x(4x^{-3}) + 5x(-2x^{-4}) + 3x^{-1}(4x^{-3}) + 3x^{-1}(-2x^{-4})$
$= 20x^{-2} - 10x^{-3} + 12x^{-4} - 6x^{-5}$

15) $(x^{3/4} - 2x^{1/2})(3x^{1/2} + 5x^{1/4}) = x^{3/4}(3x^{1/2}) + x^{3/4}(5x^{1/4}) - 2x^{1/2}(3x^{1/2}) - 2x^{1/2}(5x^{1/4})$
$= 3x^{5/4} + 5x^1 - 6x^1 - 10x^{3/4} = 3x^{5/4} - x - 10x^{3/4}$

Notes: $\frac{3}{4} + \frac{1}{2} = \frac{3}{4} + \frac{2}{4} = \frac{5}{4}, \frac{3}{4} + \frac{1}{4} = \frac{4}{4} = 1, \frac{1}{2} + \frac{1}{2} = \frac{2}{2} = 1, \frac{1}{2} + \frac{1}{4} = \frac{2}{4} + \frac{1}{4} = \frac{3}{4}$, and $x^1 = x$

Exercise Set 4.4

1) $(x + 8)^2 = x^2 + 2(x)(8) + 8^2 = x^2 + 16x + 64$ since $u = x$ and $v = 8$

2) $(x - 7)(x - 7) = x^2 + 2(x)(-7) + (-7)^2 = x^2 - 14x + 49$ since $u = x$ and $v = -7$

3) $(4x + 9)^2 = (4x)^2 + 2(4x)(9) + 9^2 = 16x^2 + 72x + 81$ since $u = 4x$ and $v = 9$

4) $(6x - 5)(6x - 5) = (6x)^2 + 2(6x)(-5) + (-5)^2 = 36x^2 - 60x + 25$ since $u = 6x$ and $v = -5$

5) $(-x + 6)^2 = (-x)^2 + 2(-x)(6) + 6^2 = x^2 - 12x + 36$ since $u = -x$ and $v = 6$

6) $(-x - 4)^2 = (-x)^2 + 2(-x)(-4) + (-4)^2 = x^2 + 8x + 16$ since $u = -x$ and $v = -4$

7) $(1 - x)^2 = 1^2 + 2(1)(-x) + (-x)^2 = 1 - 2x + x^2 = x^2 - 2x + 1$ since $u = 1$ and $v = -x$

8) $(5x^2 + 8)^2 = (5x)^2 + 2(5x^2)(8) + 8^2 = 25x^4 + 80x^2 + 64$ since $u = 5x$ and $v = 8$

9) $(6x^2 - 3x)^2 = (6x)^2 + 2(6x^2)(-3x) + (-3x)^2 = 36x^4 - 36x^3 + 9x^2$ since $u = 6x^2$ and $v = -3x$

10) $(9x^5 + 4x^3)(9x^5 + 4x^3) = (9x^5)^2 + 2(9x^5)(4x^3) + (4x^3)^2 = 81x^{10} + 72x^8 + 16x^6$ since $u = 9x^5$ and $v = 4x^3$

Exercise Set 4.5

1) $(x + 5)(x - 5) = x^2 - 5^2 = x^2 - 25$ since $u = x$ and $v = 5$

2) $(x - 4)(x + 4) = x^2 - (-4)^2 = x^2 - 16$ since $u = x$ and $v = -4$
Notes: The minus sign in $(-4)^2$ gets squared: $(-4)^2 = (-4)(-4) = 16$ such that $-(-4)^2 = -16$ Also, note that $(x - 4)(x + 4)$ is the same as $(x + 4)(x - 4)$

3) $(x + 1)(x - 1) = x^2 - 1^2 = x^2 - 1$ since $u = x$ and $v = 1$

4) $(-x + 3)(-x - 3) = (-x)^2 - 3^2 = x^2 - 9$ since $u = -x$ and $v = 3$

5) $(6 + x)(6 - x) = 6^2 - x^2 = 36 - x^2$ since $u = 6$ and $v = x$

6) $(2x + 7)(2x - 7) = (2x)^2 - 7^2 = 4x^2 - 49$ since $u = 2x$ and $v = 7$

7) $(4x - 9)(4x + 9) = (4x)^2 - (-9)^2 = 16x^2 - 81$ since $u = 4x$ and $v = -9$
See the notes to the solution to Problem 2.

8) $(x^2 + x)(x^2 - x) = (x^2)^2 - x^2 = x^4 - x^2$ since $u = x^2$ and $v = x$

9) $(3x^5 + 4x^3)(3x^5 - 4x^3) = (3x^5)^2 - (4x^3)^2 = 9x^{10} - 16x^6$ since $u = 3x^5$ and $v = 4x^3$

10) $\left(x^{3/2} + x^{-1}\right)\left(x^{3/2} - x^{-1}\right) = \left(x^{3/2}\right)^2 - (x^{-1})^2 = x^3 - x^{-2} = x^3 - \frac{1}{x^2}$
since $u = x^{3/2}$ and $v = x^{-1}$ Notes: $\left(x^{3/2}\right)^2 = x^3$ since $\frac{3}{2} + \frac{3}{2} = \frac{6}{2} = 3$ Also, $x^{-2} = \frac{1}{x^2}$

Exercise Set 4.6

1) $(x + 4)(x^2 - 6x + 7) = x(x^2) + x(-6x) + x(7) + 4(x^2) + 4(-6x) + 4(7)$
$= x^3 - 6x^2 + 7x + 4x^2 - 24x + 28 = x^3 - 2x^2 - 17x + 28$

2) $(2x - 1)(3x^2 + 4x - 5) = 2x(3x^2) + 2x(4x) + 2x(-5) - 1(3x^2) - 1(4x) - 1(-5)$
$= 6x^3 + 8x^2 - 10x - 3x^2 - 4x + 5 = 6x^3 + 5x^2 - 14x + 5$

3) $(-x + 2)(x^2 - 8x - 9) = -x(x^2) - x(-8x) - x(-9) + 2(x^2) + 2(-8x) + 2(-9)$
$= -x^3 + 8x^2 + 9x + 2x^2 - 16x - 18 = -x^3 + 10x^2 - 7x - 18$

4) $(4x - 6)(x^2 + 7x - 3) = 4x(x^2) + 4x(7x) + 4x(-3) - 6(x^2) - 6(7x) - 6(-3)$
$= 4x^3 + 28x^2 - 12x - 6x^2 - 42x + 18 = 4x^3 + 22x^2 - 54x + 18$

5) $(3x + 8)(2x^2 - 5x + 9) = 3x(2x^2) + 3x(-5x) + 3x(9) + 8(2x^2) + 8(-5x) + 8(9)$
$= 6x^3 - 15x^2 + 27x + 16x^2 - 40x + 72 = 6x^3 + x^2 - 13x + 72$

6) $(x^2 - 4x - 7)(x^2 + 3x - 7) = x^2(x^2) + x^2(3x) + x^2(-7) - 4x(x^2) - 4x(3x) - 4x(-7)$
$-7(x^2) - 7(3x) - 7(-7) = x^4 + 3x^3 - 7x^2 - 4x^3 - 12x^2 + 28x - 7x^2 - 21x + 49$
$= x^4 - x^3 - 26x^2 + 7x + 49$

7) $(2x^2 - 3x + 6)(3x^2 - 4x - 5) = 2x^2(3x^2) + 2x^2(-4x) + 2x^2(-5) - 3x(3x^2) - 3x(-4x)$
$-3x(-5) + 6(3x^2) + 6(-4x) + 6(-5) = 6x^4 - 8x^3 - 10x^2 - 9x^3 + 12x^2 + 15x + 18x^2 - 24x - 30$
$= 6x^4 - 17x^3 + 20x^2 - 9x - 30$

Exercise Set 4.7

1) $(x - 5)^3 = (x - 5)(x - 5)(x - 5) = [x(x) + x(-5) - 5(x) - 5(-5)](x - 5)$
$= (x^2 - 5x - 5x + 25)(x - 5) = (x^2 - 10x + 25)(x - 5)$
$= x^2(x) + x^2(-5) - 10x(x) - 10x(-5) + 25(x) + 25(-5)$
$= x^3 - 5x^2 - 10x^2 + 50x + 25x - 125 = x^3 - 15x^2 + 75x - 125$

2) $(6x + 3)^3 = (6x + 3)(6x + 3)(6x + 3) = [6x(6x) + 6x(3) + 3(6x) + 3(3)](6x + 3)$
$= (36x^2 + 18x + 18x + 9)(6x + 3) = (36x^2 + 36x + 9)(6x + 3)$
$= 36x^2(6x) + 36x^2(3) + 36x(6x) + 36x(3) + 9(6x) + 9(3)$
$= 216x^3 + 108x^2 + 216x^2 + 108x + 54x + 27 = 216x^3 + 324x^2 + 162x + 27$

3) $(x - 2)^4 = (x - 2)(x - 2)(x - 2)(x - 2)$
$= [x(x) + x(-2) - 2(x) - 2(-2)](x - 2)(x - 2)$
$= (x^2 - 2x - 2x + 4)(x - 2)(x - 2) = (x^2 - 4x + 4)(x - 2)(x - 2)$

$$= [x^2(x) + x^2(-2) - 4x(x) - 4x(-2) + 4(x) + 4(-2)](x - 2)$$

$$= (x^3 - 2x^2 - 4x^2 + 8x + 4x - 8)(x - 2) = (x^3 - 6x^2 + 12x - 8)(x - 2)$$

$$= x^3(x) + x^3(-2) - 6x^2(x) - 6x^2(-2) + 12x(x) + 12x(-2) - 8(x) - 8(-2)$$

$$= x^4 - 2x^3 - 6x^3 + 12x^2 + 12x^2 - 24x - 8x + 16 = x^4 - 8x^3 + 24x^2 - 32x + 16$$

4) $(x^2 + 5x - 2)^3 = (x^2 + 5x - 2)(x^2 + 5x - 2)(x^2 + 5x - 2)$

$$= [x^2(x^2) + x^2(5x) + x^2(-2) + 5x(x^2) + 5x(5x) + 5x(-2) - 2(x^2) - 2(5x) - 2(-2)](x^2 + 5x - 2)$$

$$= (x^4 + 5x^3 - 2x^2 + 5x^3 + 25x^2 - 10x - 2x^2 - 10x + 4)(x^2 + 5x - 2)$$

$$= (x^4 + 10x^3 + 21x^2 - 20x + 4)(x^2 + 5x - 2)$$

$$= x^4(x^2) + x^4(5x) + x^4(-2) + 10x^3(x^2) + 10x^3(5x) + 10x^3(-2) + 21x^2(x^2) + 21x^2(5x)$$
$$\quad + 21x^2(-2) - 20x(x^2) - 20x(5x) - 20x(-2) + 4(x^2) + 4(5x) + 4(-2)$$

$$= x^6 + 5x^5 - 2x^4 + 10x^5 + 50x^4 - 20x^3 + 21x^4 + 105x^3 - 42x^2 - 20x^3 - 100x^2 + 40x$$
$$\quad + 4x^2 + 20x - 8 = x^6 + 15x^5 + 69x^4 + 65x^3 - 138x^2 + 60x - 8$$

Exercise Set 4.8

1) $(x + 9)^4 = u^4 + 4u^3v + 6u^2v^2 + 4uv^3 + v^4$

$$= x^4 + 4x^39 + 6x^29^2 + 4x9^3 + 9^4$$

$$= x^4 + 4x^39 + 6x^281 + 4x729 + 6561$$

$$= x^4 + 36x^3 + 486x^2 + 2916x + 6561 \text{ since } u = x \text{ and } v = 9$$

2) $(x - 3)^5 = u^5 + 5u^4v + 10u^3v^2 + 10u^2v^3 + 5uv^4 + v^5$

$$= x^5 + 5x^4(-3) + 10x^3(-3)^2 + 10x^2(-3)^3 + 5x(-3)^4 + (-3)^5$$

$$= x^5 + 5x^4(-3) + 10x^3(9) + 10x^2(-27) + 5x(81) + (-243)$$

$$= x^5 - 15x^4 + 90x^3 - 270x^2 + 405x - 243 \text{ since } u = x \text{ and } v = -3$$

3) $(2 + x)^6 = u^6 + 6u^5v + 15u^4v^2 + 20u^3v^3 + 15u^2v^4 + 6uv^5 + v^6$

$$= 2^6 + 6(2^5)x + 15(2^4)x^2 + 20(2^3)x^3 + 15(2^2)x^4 + 6(2)x^5 + x^6$$

$$= 64 + 6(32)x + 15(16)x^2 + 20(8)x^3 + 15(4)x^4 + 12x^5 + x^6$$

$$= x^6 + 12x^5 + 60x^4 + 160x^3 + 240x^2 + 192x + 64 \text{ since } u = 2 \text{ and } v = x$$

4) $(-x + 4)^7 = u^7 + 7u^6v + 21u^5v^2 + 35u^4v^3 + 35u^3v^4 + 21u^2v^5 + 7uv^6 + v^7$

$$= (-x)^7 + 7(-x)^64 + 21(-x)^54^2 + 35(-x)^44^3 + 35(-x)^34^4 + 21(-x)^24^5 + 7(-x)4^6 + 4^7$$

$$= -x^7 + 7x^6(4) - 21x^5(16) + 35x^4(64) - 35x^3(256) + 21x^2(1024) - 7x(4096) + 16{,}384$$

$$= -x^7 + 28x^6 - 336x^5 + 2240x^4 - 8960x^3 + 21{,}504x^2 - 28{,}672x + 16{,}384$$

since $u = -x$ and $v = 4$

5) $(3x + 2)^4 = u^4 + 4u^3v + 6u^2v^2 + 4uv^3 + v^4$

$= (3x)^4 + 4(3x)^3 2 + 6(3x)^2 2^2 + 4(3x)2^3 + 2^4$

$= 81x^4 + 4(27x^3)2 + 6(9x^2)4 + 4(3x)8 + 16$

$= 81x^4 + 216x^3 + 216x^2 + 96x + 16$ since $u = 3x$ and $v = 2$

6) $(2x - 4)^3 = u^3 + 3u^2v + 3uv^2 + v^3$

$= (2x)^3 + 3(2x)^2(-4) + 3(2x)(-4)^2 + (-4)^3$

$= 8x^3 + 3(4x^2)(-4) + 3(2x)(16) - 64$

$= 8x^3 - 48x^2 + 96x - 64$ since $u = 2x$ and $v = -4$

7) $(x + 10)^9 = u^9 + 9u^8v + 36u^7v^2 + 84u^6v^3 + 126u^5v^4 + 126u^4v^5 + 84u^3v^6 + 36u^2v^7 + 9uv^8 + v^9$

$= x^9 + 9x^8 10 + 36x^7 10^2 + 84x^6 10^3 + 126x^5 10^4 + 126x^4 10^5 + 84x^3 10^6 + 36x^2 10^7 + 9x10^8 + 10^9$

$= x^9 + 90x^8 + 3600x^7 + 84{,}000x^6 + 1{,}260{,}000x^5 + 12{,}600{,}000x^4 + 84{,}000{,}000x^3$

$\quad + 360{,}000{,}000x^2 + 900{,}000{,}000x + 1{,}000{,}000{,}000$ since $u = x$ and $v = 10$

One way to determine the coefficients for $(u + v)^9$ is to add together coefficients from $(u + v)^8$. For example, $8 + 28 = 36$, $28 + 56 = 84$, and $56 + 70 = 126$.

Exercise Set 4.9

1) $\frac{3}{4}(8x + 12) = \frac{3}{4}8x + \frac{3}{4}12 = \frac{24x}{4} + \frac{36}{4} = 6x + 9$

2) $24x\left(\frac{x}{6} - \frac{2}{3}\right) = 24x\frac{x}{6} + 24x\left(-\frac{2}{3}\right) = \frac{24x^2}{6} - \frac{48x}{3} = 4x^2 - 16x$

3) $\frac{x}{2}(2x^2 + 4x - 14) = \frac{x}{2}2x^2 + \frac{x}{2}4x + \frac{x}{2}(-14) = \frac{2x^3}{2} + \frac{4x^2}{2} - \frac{14x}{2} = x^3 + 2x^2 - 7x$

4) $\frac{2}{3}\left(\frac{x}{5} + \frac{9}{8}\right) = \frac{2}{3}\frac{x}{5} + \frac{2}{3}\frac{9}{8} = \frac{2x}{15} + \frac{18}{24} = \frac{2x}{15} + \frac{3}{4}$

5) $\frac{5}{x}(4x^3 - 3x^2 + 2x) = \frac{5}{x}4x^3 + \frac{5}{x}(-3x^2) + \frac{5}{x}2x = \frac{20x^3}{x} - \frac{15x^2}{x} + \frac{10x}{x} = 20x^2 - 15x + 10$

6) $\frac{3x}{8}\left(\frac{4x}{5} - \frac{2}{7}\right) = \frac{3x}{8}\frac{4x}{5} + \frac{3x}{8}\left(-\frac{2}{7}\right) = \frac{12x^2}{40} - \frac{6x}{56} = \frac{3x^2}{10} - \frac{3x}{28}$

Exercise Set 4.10

1) $9\frac{7x-9}{5} = \frac{9}{1}\frac{7x-9}{5} = \frac{9(7x-9)}{1(5)} = \frac{9(7x)-9(9)}{5} = \frac{63x-81}{5}$

2) $6x\frac{8x-1}{7} = \frac{6x}{1}\frac{8x-1}{7} = \frac{6x(8x-1)}{1(7)} = \frac{6x(8x)-6x(1)}{7} = \frac{48x^2-6x}{7}$

3) $\dfrac{1}{5}\dfrac{x^2}{3x^2+4} = \dfrac{1(x^2)}{5(3x^2+4)} = \dfrac{x^2}{5(3x^2)+5(4)} = \dfrac{x^2}{15x^2+20}$

4) $\dfrac{x}{3}\dfrac{4x+8}{9} = \dfrac{x(4x+8)}{3(9)} = \dfrac{x(4x)+x(8)}{27} = \dfrac{4x^2+8x}{27}$

5) $\dfrac{x^3}{5}\dfrac{2x^2-3x+6}{7x-4} = \dfrac{x^3(2x^2-3x+6)}{5(7x-4)} = \dfrac{x^3(2x^2)+x^3(-3x)+x^3(6)}{5(7x)-5(4)} = \dfrac{2x^5-3x^4+6x^3}{35x-20}$

6) $\dfrac{x+3}{x-4}\dfrac{x+3}{x+4} = \dfrac{(x+3)(x+3)}{(x-4)(x+4)} = \dfrac{x(x)+x(3)+3(x)+3(3)}{x(x)+x(4)-4(x)-4(4)} = \dfrac{x^2+3x+3x+9}{x^2+4x-4x-16} = \dfrac{x^2+6x+9}{x^2-16}$

7) $\dfrac{5x^2-2}{6x^2+3}\dfrac{3x+4}{2x^2-3} = \dfrac{(5x^2-2)(3x+4)}{(6x^2+3)(2x^2-3)} = \dfrac{5x^2(3x)+5x^2(4)-2(3x)-2(4)}{6x^2(2x^2)+6x^2(-3)+3(2x^2)+3(-3)}$

$= \dfrac{15x^3+20x^2-6x-8}{12x^4-18x^2+6x^2-9} = \dfrac{15x^3+20x^2-6x-8}{12x^4-12x^2-9}$

Exercise Set 4.11

1) $25(7x-8)^2 = \left[25^{1/2}(7x-8)\right]^2 = [5(7x-8)]^2 = (35x-40)^2$

2) $64(6x+5)^3 = \left[64^{1/3}(6x+5)\right]^3 = [4(6x+5)]^3 = (24x+20)^3$

3) $9(5x-9)^{1/2} = [9^2(5x-9)]^{1/2} = [81(5x-9)]^{1/2} = (405x-729)^{1/2}$

4) $x^{12}(6x^3-9x)^2 = \left[(x^{12})^{1/2}(6x^3-9x)\right]^2 = [x^6(6x^3-9x)]^2 = (6x^9-9x^7)^2$

5) $81x^8(x^2-3x+4)^4 = \left[(81x^8)^{1/4}(x^2-3x+4)\right]^4$

$= [3x^2(x^2-3x+4)]^4 = (3x^4-9x^3+12x^2)^4$

6) $36x^2\left(\dfrac{x}{3}-\dfrac{2}{x}\right)^2 = \left[(36x^2)^{1/2}\left(\dfrac{x}{3}-\dfrac{2}{x}\right)\right]^2 = \left[6x\left(\dfrac{x}{3}-\dfrac{2}{x}\right)\right]^2 = (2x^2-12)^2$

7) $x^6\sqrt{8x^2-4} = x^6(8x^2-4)^{1/2} = [(x^6)^2(8x^2-4)]^{1/2}$

$= [x^{12}(8x^2-4)]^{1/2} = (8x^{14}-4x^{12})^{1/2} = \sqrt{8x^{14}-4x^{12}}$

8) $6x(2x+8)^{-1} = \left[(6x)^{-1/1}(2x+8)\right]^{-1} = [(6x)^{-1}(2x+8)]^{-1} = \left[\dfrac{1}{6x}(2x+8)\right]^{-1}$

$= \left(\dfrac{2x}{6x}+\dfrac{8}{6x}\right)^{-1} = \left(\dfrac{1}{3}+\dfrac{4}{3x}\right)^{-1}$ Note: The answer is equivalent to $\left(\dfrac{x}{3x}+\dfrac{4}{3x}\right)^{-1} = \left(\dfrac{x+4}{3x}\right)^{-1} = \dfrac{3}{x+}$

Note: A reciprocal swaps the numerator and denominator of a fraction. Sec.'s 1.7, 3.4, and 3.8 show that $\left(\dfrac{x}{y}\right)^{-1} = \dfrac{y}{x}$.

9) $4x^4(3x-5)^{-2} = \left[(4x^4)^{-1/2}(3x-5)\right]^{-2}$

$= \left[\dfrac{1}{(4x^4)^{1/2}}(3x-5)\right]^{-2} = \left(\dfrac{3x-5}{2x^2}\right)^{-2} = \left(\dfrac{3x}{2x^2}-\dfrac{5}{2x^2}\right)^{-2} = \left(\dfrac{3}{2x}-\dfrac{5}{2x^2}\right)^{-2}$

10) $x^{-3}(x^{-4} + x^{-1})^{-1} = \left[(x^{-3})^{-1/1}(x^{-4} + x^{-1})\right]^{-1} = \left[(x^{-3})^{-1}(x^{-4} + x^{-1})\right]^{-1}$

$= [x^3(x^{-4} + x^{-1})]^{-1} = (x^{-1} + x^2)^{-1} = \frac{1}{x^{-1}+x^2} = \frac{1}{\frac{1}{x}+x^2}$

If you multiply the numerator and denominator each by x, this is equivalent to:

$\frac{1}{\frac{1}{x}+x^2} \frac{x}{x} = \frac{x}{\frac{x}{x}+x^3} = \frac{x}{1+x^3} = \frac{x}{x^3+1}$ (which is a preferred form of the answer)

Alternate solution: $x^{-3}(x^{-4} + x^{-1})^{-1} = \frac{1}{x^3}\left(\frac{1}{x^4} + \frac{1}{x}\right)^{-1} = \frac{1}{x^3}\left(\frac{1}{x^4} + \frac{x^3}{x^4}\right)^{-1}$

$= \frac{1}{x^3}\left(\frac{1+x^3}{x^4}\right)^{-1} = \frac{1}{x^3}\left(\frac{x^3+1}{x^4}\right)^{-1} = \frac{1}{x^3}\frac{x^4}{x^3+1} = \frac{x^4}{x^6+x^3}$ which is equivalent to $\frac{x}{x^3+1}$

because $\frac{x^4}{x^6+x^3} = \frac{x^4}{x^6+x^3}\frac{x^{-3}}{x^{-3}} = \frac{x^4 x^{-3}}{x^6 x^{-3}+x^3 x^{-3}} = \frac{x^1}{x^3+x^0} = \frac{x}{x^3+1}$

Exercise Set 4.12

1) Distribute on the right side: $5x - 12 = 2x + 6$

Add 12 to both sides and subtract $2x$ from both sides: $3x = 18$

Divide by 3 on both sides: $x = \frac{18}{3} = \boxed{6}$

Check: $5x - 12 = 5(6) - 12 = 30 - 12 = 18$ agrees with $2(x + 3) = 2(6 + 3) = 2(9) = 18$

2) Distribute on the left side: $12x - 30 = 16x - 50$

Add 50 to both sides and subtract $12x$ from both sides: $20 = 4x$

Divide by 4 on both sides: $\frac{20}{4} = \boxed{5} = x$

Check: $6(2x - 5) = 6[2(5) - 5] = 6(10 - 5) = 6(5) = 30$

agrees with $16x - 50 = 16(5) - 50 = 80 - 50 = 30$

3) Distribute on the left side: $-6x + 8 = x - 13$ Note: $-2(3x - 4) = -2(3x) - 2(-4)$

Add 13 to both sides and add $6x$ to both sides: $21 = 7x$

Divide by 7 on both sides: $\frac{21}{7} = \boxed{3} = x$

Check: $-2(3x - 4) = -2[3(3) - 4] = -2(9 - 4) = -2(5) = -10$

agrees with $x - 13 = 3 - 13 = -10$

4) Distribute on both sides: $24x - 48 = 15x + 24$

Add 48 to both sides and subtract $15x$ from both sides: $9x = 72$

Divide by 9 on both sides: $x = \frac{72}{9} = \boxed{8}$

Check: $4(6x - 12) = 4[6(8) - 12] = 4(48 - 12) = 4(36) = 144$

agrees with $3(5x + 8) = 3[5(8) + 8] = 3(40 + 8) = 3(48) = 144$

5 Factoring Expressions

Exercise Set 5.1

1) $10x^2 + 14x = 2x(5x + 7)$
2) $6x^3 - 9x^2 = 3x^2(2x - 3)$
3) $80x^6 + 60x^4 = 20x^4(4x^2 + 3)$
4) $18x^9 + 36x^8 = 18x^8(x + 2)$
5) $5x^5 - 3x^3 = x^3(5x^2 - 3)$
6) $32x^2 - 24 = 8(4x^2 - 3)$
7) $9x^4 + 15x^3 = 3x^3(3x + 5)$
8) $16x^7 - 20x^4 = 4x^4(4x^3 - 5)$
9) $32x^8 + 12x^3 = 4x^3(8x^5 + 3)$
10) $90x^{15} - 108x^9 = 18x^9(5x^6 - 6)$
11) $30x^3 - 60x^2 + 45 = 15(2x^3 - 4x^2 + 3)$
12) $72x^7 - 96x^6 + 144x^5 = 24x^5(3x^2 - 4x + 6)$

Exercise Set 5.2

1) $-8x^2 - 12x = \boxed{-4x(2x + 3)}$ since $-4x(2x + 3) = -4x(2x) - 4x(3)$

Tip: When factoring out a minus sign, every term in parentheses changes sign. This simple rule can help you get all of the signs correct.

2) $-15x^5 - 9x^3 = \boxed{-3x^3(5x^2 + 3)}$ since $-3x^3(5x^2 + 3) = -3x^3(5x^2) - 3x^3(3)$

3) $-2x^4 + 1 = \boxed{-(2x^4 - 1)}$ since $-(2x^4 - 1) = -(2x^4) - (-1)$

Note: It would be unconventional to include a 1 before the parentheses (Sec. 1.3).

4) $-36x^8 + 72x^4 = \boxed{-36x^4(x^4 - 2)}$ since $-36x^4(x^4 - 2) = -36x^4(x^4) - 36x^4(-2)$

5) $-12x^3 - 16x^2 + 8x = \boxed{-4x(3x^2 + 4x - 2)}$ since this is $-4x(3x^2) - 4x(4x) - 4x(-2)$

6) $-24x^2 + 18x - 21 = \boxed{-3(8x^2 - 6x + 7)}$ since this is $-3(8x^2) - 3(-6x) - 3(7)$

7) $-28x^{15} - 42x^{12} - 35x^9 = \boxed{-7x^9(4x^6 + 6x^3 + 5)}$
since this is $-7x^9(4x^6) - 7x^9(6x^3) - 7x^9(5)$

8) $-54x^{5/4} + 30x^{3/4} + 12x^{1/4} = \boxed{-6x^{1/4}\left(9x - 5x^{1/2} - 2\right)}$

Notes: $\frac{1}{4} + 1 = \frac{1}{4} + \frac{4}{4} = \frac{5}{4}$ and $\frac{1}{4} + \frac{1}{2} = \frac{1}{4} + \frac{2}{4} = \frac{3}{4}$

Exercise Set 5.3

1) $x^2 - 25 = x^2 - 5^2 = (x + 5)(x - 5)$

Note: You can verify these answers using the FOIL method (Sec. 4.3).

2) $16x^2 - 81 = (4x)^2 - 9^2 = (4x + 9)(4x - 9)$

3) $36x^2 - 1 = (6x)^2 - 1^2 = (6x + 1)(6x - 1)$

4) $225x^6 - 100 = (15x^3)^2 - 10^2 = (15x^3 + 10)(15x^3 - 10)$

5) $49x^8 - 144 = (7x^4)^2 - 12^2 = (7x^4 + 12)(7x^4 - 12)$

6) $-64x^2 + 121 = 121 - 64x^2 = 11^2 - (8x)^2 = (11 + 8x)(11 - 8x)$

Alternate answer: $(8x + 11)(-8x + 11)$

Exercise Set 5.4

1) $x^2 + 14x + 49 = x^2 + 2(x)(7) + 7^2 = \boxed{(x + 7)^2}$

Note: You can verify these answers using the FOIL method (Sec. 4.3).

2) $x^2 - 2x + 1 = x^2 + 2(x)(-1) + 1^2 = \boxed{(x - 1)^2}$

Tip: The sign in the answer matches the sign of the middle term (if the first and last terms are both positive).

3) $4x^2 + 20x + 25 = (2x)^2 + 2(2x)5 + 5^2 = \boxed{(2x + 5)^2}$

4) $16x^2 + 64x + 64 = (4x)^2 + 2(4x)8 + 8^2 = \boxed{(4x + 8)^2}$ equivalent to $16(x + 2)^2$

Note: The answer $(4x + 8)^2$ is fine. The alternate answer is given just for information.

5) $49x^2 - 126x + 81 = (7x)^2 + 2(7x)(-9) + 9^2 = \boxed{(7x - 9)^2}$

6) $100x^2 - 100x + 25 = (10x)^2 + 2(10x)(-5) + 5^2 = \boxed{(10x - 5)^2}$ equivalent to $25(2x - 1)^2$

7) $36x^2 - 144x + 144 = (6x)^2 + 2(6x)(-12) + 12^2 = \boxed{(6x - 12)^2}$ equivalent to $36(x - 2)^2$

8) $64x^2 + 96x + 36 = (8x)^2 + 2(8x)(6) + 6^2 = \boxed{(8x + 6)^2}$ equivalent to $4(4x + 3)^2$

9) $9x^2 + 54x + 81 = (3x)^2 + 2(3x)(9) + 9^2 = \boxed{(3x + 9)^2}$ equivalent to $9(x + 3)^2$

10) $49x^2 - 112x + 64 = (7x)^2 + 2(7x)(-8) + 8^2 = \boxed{(7x - 8)^2}$

11) $169x^2 - 442x + 289 = (13x)^2 + 2(13x)(-17) + 17^2 = \boxed{(13x - 17)^2}$

12) $225x^2 + 750x + 625 = (15x)^2 + 2(15x)(25) + 25^2 = \boxed{(15x + 25)^2}$ equivalent to $25(3x + 5)^2$

13) $\frac{x^2}{4} + \frac{3x}{4} + \frac{9}{16} = \left(\frac{x}{2}\right)^2 + 2\left(\frac{x}{2}\right)\left(\frac{3}{4}\right) + \left(\frac{3}{4}\right)^2 = \boxed{\left(\frac{x}{2} + \frac{3}{4}\right)^2}$ equivalent to $\frac{1}{4}\left(x + \frac{3}{2}\right)^2$

14) $x^4 + 6x^2 + 9 = (x^2)^2 + 2(x^2)(3) + 3^2 = \boxed{(x^2 + 3)^2}$

15) $49x^6 - 28x^3 + 4 = (7x^3)^2 + 2(7x^3)(-2) + 2^2 = \boxed{(7x^3 - 2)^2}$

16) $4x^{18} + 20x^9 + 25 = (2x^9)^2 + 2(2x^9)(5) + 5^2 = \boxed{(2x^9 + 5)^2}$

17) $x^{4/3} + 6x^{2/3} + 9 = \left(x^{2/3}\right)^2 + 2\left(x^{2/3}\right)(3) + 3^2 = \boxed{\left(x^{2/3} + 3\right)^2}$

Note: $\left(x^{2/3}\right)^2 = x^{(2/3)(2)} = x^{4/3}$ because $\frac{2}{3}(2) = \frac{4}{3}$. Recall that $(x^m)^n = x^{mn}$ from Chapter 3.

Exercise Set 5.5

1) $(x + 5)(x + 4)$ Check: $(x + 5)(x + 4) = x^2 + 4x + 5x + 20 = x^2 + 9x + 20$

2) $(x + 6)(x + 4)$ Check: $(x + 6)(x + 4) = x^2 + 4x + 6x + 24 = x^2 + 10x + 24$

3) $(5x + 2)(x + 7)$ Check: $(5x + 2)(x + 7) = 5x^2 + 35x + 2x + 14 = 5x^2 + 37x + 14$

4) $(3x + 5)(2x + 7)$ Check: $(3x + 5)(2x + 7) = 6x^2 + 21x + 10x + 35 = 6x^2 + 31x + 35$

5) $(5x + 9)(2x + 7)$ Check: $(5x + 9)(2x + 7) = 10x^2 + 35x + 18x + 63 = 10x^2 + 53x + 63$

6) $(8x + 1)(4x + 1)$ Check: $(8x + 1)(4x + 1) = 32x^2 + 8x + 4x + 1 = 32x^2 + 12x + 1$

7) $(9x + 5)(4x + 5)$ Check: $(9x + 5)(4x + 5) = 36x^2 + 45x + 20x + 25 = 36x^2 + 65x + 25$

8) $(9x + 5)(3x + 20)$ Check: $(9x + 5)(3x + 20)$
$= 27x^2 + 180x + 15x + 100 = 27x^2 + 195x + 100$

9) $(7x + 8)(x + 2)$ Check: $(7x + 8)(x + 2) = 7x^2 + 14x + 8x + 16 = 7x^2 + 22x + 16$

10) $(4x + 9)(4x + 1)$ Check: $(4x + 9)(4x + 1) = 16x^2 + 4x + 36x + 9 = 16x^2 + 40x + 9$

11) $(10x + 6)(2x + 7)$ Check: $(10x + 6)(2x + 7) = 20x^2 + 70x + 12x + 42$
$= 20x^2 + 82x + 42$ equivalent to $2(10x^2 + 41x + 21)$

12) $(12x + 5)(2x + 1)$ Check: $(12x + 5)(2x + 1) = 24x^2 + 12x + 10x + 5 = 24x^2 + 22x + 5$

13) $(9x + 11)(4x + 3)$ Check: $(9x + 11)(4x + 3)$
$= 36x^2 + 27x + 44x + 33 = 36x^2 + 71x + 33$

14) $(9x + 2)(x + 2)$ Check: $(9x + 2)(x + 2) = 9x^2 + 18x + 2x + 4 = 9x^2 + 20x + 4$

15) $(8x + 3)(3x + 8)$ Check: $(8x + 3)(3x + 8) = 24x^2 + 64x + 9x + 24 = 24x^2 + 73x + 24$

16) $(5x + 16)(3x + 2)$ Check: $(5x + 16)(3x + 2)$
$= 15x^2 + 10x + 48x + 32 = 15x^2 + 58x + 32$

Exercise Set 5.6

1) $(x - 7)(x - 1)$ Check: $(x - 7)(x - 1) = x^2 - x - 7x + 7 = x^2 - 8x + 7$

2) $(x + 11)(x - 1)$ Check: $(x + 11)(x - 1) = x^2 - x + 11x - 11 = x^2 + 10x - 11$

3) $(3x - 5)(x + 1)$ Check: $(3x - 5)(x + 1) = 3x^2 + 3x - 5x - 5 = 3x^2 - 2x - 5$

4) $(3x - 5)(3x - 2)$ Check: $(3x - 5)(3x - 2) = 9x^2 - 6x - 15x + 10 = 9x^2 - 21x + 10$

5) $(3x - 5)(2x + 7)$ Check: $(3x - 5)(2x + 7) = 6x^2 + 21x - 10x - 35 = 6x^2 + 11x - 35$

6) $(5x + 3)(2x - 9)$ Check: $(5x + 3)(2x - 9) = 10x^2 - 45x + 6x - 27 = 10x^2 - 39x - 27$

7) $(8x - 9)(4x - 1)$ Check: $(8x - 9)(4x - 1) = 32x^2 - 8x - 36x + 9 = 32x^2 - 44x + 9$

8) $(12x - 1)(2x + 5)$ Check: $(12x - 1)(2x + 5) = 24x^2 + 60x - 2x - 5 = 24x^2 + 58x - 5$

Exercise Set 5.7

1) $x^2 + 18x + 4 = x^2 + 18x + 81 - 77 = (x + 9)^2 - 77$

Check: $(x + 9)^2 - 77 = x^2 + 18x + 81 - 77 = x^2 + 18x + 4$

2) $x^2 - 12x + 60 = x^2 - 12x + 36 + 24 = (x - 6)^2 + 24$

Check: $(x - 6)^2 + 24 = x^2 - 12x + 36 + 24 = x^2 - 12x + 60$

3) $36x^2 + 36x - 36 = 36x^2 + 36x + 9 - 45 = (6x + 3)^2 - 45$

Check: $(6x + 3)^2 - 45 = 36x^2 + 36x + 9 - 45 = 36x^2 + 36x - 36$

4) $64x^2 - 80x + 16 = 64x^2 - 80x + 25 - 9 = (8x - 5)^2 - 9$

Check: $(8x - 5)^2 - 9 = 64x^2 - 80x + 25 - 9 = 64x^2 - 80x + 16$

5) $81x^2 + 108x + 54 = 81x^2 + 108x + 36 + 18 = (9x + 6)^2 + 18$

Check: $(9x + 6)^2 + 18 = 81x^2 + 108x + 36 + 18 = 81x^2 + 108x + 54$

6) $100x^2 - 20x - 1 = 100x^2 - 20x + 1 - 2 = (10x - 1)^2 - 2$

Check: $(10x - 1)^2 - 2 = 100x^2 - 20x + 1 - 2 = 100x^2 - 20x - 1$

7) $49x^2 - 126x + 49 = 49x^2 - 126x + 81 - 32 = (7x - 9)^2 - 32$

Check: $(7x - 9)^2 - 32 = 49x^2 - 126x + 81 - 32 = 49x^2 - 126x + 49$

8) $144x^2 + 72x = 144x^2 + 72x + 9 - 9 = (12x + 3)^2 - 9$

Check: $(12x + 3)^2 - 9 = 144x^2 + 72x + 9 - 9 = 144x^2 + 72x$

Note: Of course, we could factor $144x^2 + 72x$ as $72x(2x + 1)$, but the purpose of this section is to practice completing the square.

Exercise Set 5.8

1) $\frac{14x^5 - 42x^2}{3} = \boxed{\frac{14x^2(x^3 - 3)}{3}}$ Alternate answers: $\frac{14x^2}{3}(x^3 - 3)$ or $14x^2 \frac{x^3 - 3}{3}$

Note: You can check the answers using the distributive property (Chapter 4).

2) $\frac{5x}{24x + 30} = \boxed{\frac{5x}{6(4x + 5)}}$ Alternate answers: $\frac{5x}{6} \frac{1}{4x + 5}$ or $\frac{1}{6} \frac{5x}{4x + 5}$

3) $\frac{36x^8+63x^5}{32x^7-40x^3} = \frac{9x^5(4x^3+7)}{8x^3(4x^4-5)} = \boxed{\frac{9x^2(4x^3+7)}{8(4x^4-5)}}$ Alternate answer: $\frac{9x^2}{8}\frac{4x^3+7}{4x^4-5}$ Note: $\frac{x^5}{x^3} = x^2$

4) $\frac{48x^5-60x^4+96x^3}{36x^2} = \frac{12x^3(4x^2-5x+8)}{36x^2} = \boxed{\frac{x(4x^2-5x+8)}{3}}$ Alternate answer: $\frac{x}{3}(4x^2-5x+8)$

Note: $\frac{12x^3}{36x^2} = \frac{x}{3}$ because $\frac{x^3}{x^2} = x^{3-2} = x^1 = x$ (Sec. 3.2) and $\frac{12}{36}$ reduces to $\frac{1}{3}$.

Exercise Set 5.9

1) $(6x+9)^2 = [3(2x+3)]^2 = 3^2(2x+3)^2 = 9(2x+3)^2$

2) $(3x^4-4x^2)^3 = [x^2(3x^2-4)]^3 = (x^2)^3(3x^2-4)^3 = x^6(3x^2-4)^3$

3) $(16x^3+20x)^5 = [4x(4x^2+5)]^5 = (4x)^5(4x^2+5)^5 = 1024x^5(4x^2+5)^5$

Note: Recall from Sec. 3.8 that $(x^m y^n)^p = x^{mp} y^{np}$ and Sec. 3.5 that $(x^m)^n = x^{mn}$.

4) $(12x^4-21x^3)^6 = [3x^3(4x-7)]^6 = (3x^3)^6(4x-7)^6 = 729x^{18}(4x-7)^6$

5) $(2x^9+6x^5)^4 = [2x^5(x^4+3)]^4 = (2x^5)^4(x^4+3)^4 = 16x^{20}(x^4+3)^4$

6) $(x^{12}+x^9-x^7)^8 = [x^7(x^5+x^2-1)]^8 = (x^7)^8(x^5+x^2-1)^8 = x^{56}(x^5+x^2-1)^8$

7) $(24x^3-60x-48)^2 = [12(2x^3-5x-4)]^2 = 12^2(2x^3-5x-4)^2 = 144(2x^3-5x-4)^2$

8) $\sqrt{18x^2-27} = \sqrt{9(2x^2-3)} = \sqrt{9}\sqrt{2x^2-3} = 3\sqrt{2x^2-3}$

9) $\sqrt{25x^4+75x^3+50x^2} = \sqrt{25x^2(x^2+3x+2)} = \sqrt{25x^2}\sqrt{x^2+3x+2} = 5x\sqrt{x^2+3x+2}$

10) $(24x^8+40x^6)^{2/3} = [8x^6(3x^2+5)]^{2/3} = (8x^6)^{2/3}(3x^2+5)^{2/3} = 8^{2/3}(x^6)^{2/3}(3x^2+5)^{2/3}$

$= \left(\sqrt[3]{8}\right)^2\left(\sqrt[3]{x^6}\right)^2(3x^2+5)^{2/3} = 2^2(x^2)^2(3x^2+5)^{2/3} = 4x^4(3x^2+5)^{2/3}$

Note: It may help to review fractional exponents (Sec. 3.7).

11) $(9x^5-12x^3)^{-1} = [3x^3(3x^2-4)]^{-1} = (3x^3)^{-1}(3x^2-4)^{-1} = 3^{-1}x^{-3}(3x^2-4)^{-1}$

$= \frac{1}{3}\frac{1}{x^3}\frac{1}{(3x^2-4)^1} = \frac{1}{3x^3(3x^2-4)}$ Note: Recall from Sec. 3.4 that $y^{-n} = \frac{1}{y^n}$.

12) $(14x^7+10x^6)^{-3} = [2x^6(7x+5)]^{-3} = (2x^6)^{-3}(7x+5)^{-3} = 2^{-3}x^{-18}(7x+5)^{-3}$

$= \frac{1}{2^3x^{18}}\frac{1}{(7x+5)^3} = \frac{1}{8x^{18}(7x+5)^3}$ Note: Recall from Sec. 3.4 that $y^{-n} = \frac{1}{y^n}$.

Exercise Set 5.10

1) $\sqrt{8} = \sqrt{4(2)} = \sqrt{4}\sqrt{2} = 2\sqrt{2}$

Note: You can check these answers with a calculator. For example, $\sqrt{8} \approx 2.828427125$ agrees with $2\sqrt{2} \approx 2.828427125$.

2) $\sqrt{27} = \sqrt{9(3)} = \sqrt{9}\sqrt{3} = 3\sqrt{3}$

3) $\sqrt{20} = \sqrt{4(5)} = \sqrt{4}\sqrt{5} = 2\sqrt{5}$

4) $\sqrt{108} = \sqrt{36(3)} = \sqrt{36}\sqrt{3} = 6\sqrt{3}$

5) $\sqrt{98} = \sqrt{49(2)} = \sqrt{49}\sqrt{2} = 7\sqrt{2}$

6) $\sqrt{45} = \sqrt{9(5)} = \sqrt{9}\sqrt{5} = 3\sqrt{5}$

7) $\sqrt{24} = \sqrt{4(6)} = \sqrt{4}\sqrt{6} = 2\sqrt{6}$

8) $\sqrt{90} = \sqrt{9(10)} = \sqrt{9}\sqrt{10} = 3\sqrt{10}$

9) $\sqrt{700} = \sqrt{100(7)} = \sqrt{100}\sqrt{7} = 10\sqrt{7}$

10) $\sqrt{288} = \sqrt{144(2)} = \sqrt{144}\sqrt{2} = 12\sqrt{2}$

Note: If you first factored out a smaller perfect square than 144, you'll need to factor another number out to reach the final answer.

11) $\sqrt{192} = \sqrt{64(3)} = \sqrt{64}\sqrt{3} = 8\sqrt{3}$

12) $\sqrt{150} = \sqrt{25(6)} = \sqrt{25}\sqrt{6} = 5\sqrt{6}$

Exercise Set 5.11

1) $8x(x + 4) = 0 \rightarrow 8x = 0$ or $x + 4 = 0 \rightarrow x = \boxed{0}$ or $x = \boxed{-4}$

Check: $8x^2 + 32x = 8(0)^2 + 32(0) = 0$ and $8(-4)^2 + 32(-4) = 8(16) - 128 = 128 - 128 = 0$

2) $12x^4(x - 6) = 0 \rightarrow 12x^4 = 0$ or $x - 6 = 0 \rightarrow x = \boxed{0}$ or $x = \boxed{6}$

Check: $12x^5 - 72x^4 = 12(0)^5 - 72(0)^4 = 0$ and $12(6)^5 - 72(6^4) = 93,312 - 93,312 = 0$

3) $x^8(x - 7) = 0 \rightarrow x^8 = 0$ or $x - 7 = 0 \rightarrow x = \boxed{0}$ or $x = \boxed{7}$

Check: $x^9 - 7x^8 = 0^9 - 7(0)^8 = 0$ and $7^9 - 7(7)^8 = 7^9 - 7^9 = 0$

4) $10x^6 - 20x^5 = 0 \rightarrow 10x^5(x - 2) = 0 \rightarrow 10x^5 = 0$ or $x - 2 = 0 \rightarrow x = \boxed{0}$ or $x = \boxed{2}$

Check: $10x^6 = 10(0)^6 = 0$ agrees with $20x^5 = 20(0)^5 = 0$

and $10x^6 = 10(2)^6 = 10(64) = 640$ agrees with $20x^5 = 20(2)^5 = 20(32) = 640$

6 Quadratic Equations

Exercise Set 6.2

1) $3x^2 = 48 \rightarrow x^2 = 16 \rightarrow \boxed{x = \pm 4}$ Check: $3x^2 = 3(\pm 4)^2 = 3(16) = 48$

2) $2x^2 - 3x = 0 \rightarrow x(2x - 3) = 0 \rightarrow x = 0$ or $2x - 3 = 0 \rightarrow x = 0$ or $2x = 3 \rightarrow \boxed{x = 0}$ or $\boxed{x = \frac{3}{2}}$

Check: $2x^2 - 3x = 2(0)^2 - 3(0) = 0 - 0 = 0$ and $2\left(\frac{3}{2}\right)^2 - 3\left(\frac{3}{2}\right) = 2\left(\frac{9}{4}\right) - \frac{9}{2} = \frac{9}{2} - \frac{9}{2} = 0$

3) $3x^2 = 4x \rightarrow 3x^2 - 4x = 0 \rightarrow x(3x - 4) = 0 \rightarrow x = 0$ or $3x - 4 = 0 \rightarrow x = 0$ or $3x = 4$

$\rightarrow \boxed{x = 0}$ or $\boxed{x = \frac{4}{3}}$ Check: $3x^2 = 3(0)^2 = 0$ agrees with $4x = 4(0) = 0$

and $3x^2 = 3\left(\frac{4}{3}\right)^2 = 3\left(\frac{16}{9}\right) = \frac{48}{9} = \frac{16}{3}$ agrees with $4x = 4\left(\frac{4}{3}\right) = \frac{16}{3}$

4) $9x^2 - 225 = 0 \rightarrow 9x^2 = 225 \rightarrow x^2 = 25 \rightarrow \boxed{x = \pm 5}$

Check: $9x^2 - 225 = 9(\pm 5)^2 - 225 = 225 - 225 = 0$

5) $x^2 + 36 = 100 \rightarrow x^2 = 64 \rightarrow \boxed{x = \pm 8}$ Check: $x^2 + 36 = (\pm 8)^2 + 36 = 64 + 36 = 100$

6) $x^2 + 5x = 0 \rightarrow x(x + 5) = 0 \rightarrow x = 0$ or $x + 5 = 0 \rightarrow \boxed{x = 0}$ or $\boxed{x = -5}$

Check: $x^2 + 5x = 0^2 + 5(0) = 0$ and $x^2 + 5x = (-5)^2 + 5(-5) = 25 - 25 = 0$

7) $\frac{x^2}{2} = 3x \rightarrow x^2 = 6x \rightarrow x^2 - 6x = 0 \rightarrow x(x - 6) = 0 \rightarrow x = 0$ or $x - 6 = 0 \rightarrow \boxed{x = 0}$ or $\boxed{x = 6}$

Check: $\frac{x^2}{2} = \frac{0^2}{2} = 0$ agrees with $3x = 3(0) = 0$ and $\frac{x^2}{2} = \frac{6^2}{2} = \frac{36}{2} = 18$ agrees with $3x = 3(6) = 18$

8) $2x^2 - 48 = 50 \rightarrow 2x^2 = 98 \rightarrow x^2 = 49 \rightarrow \boxed{x = \pm 7}$

Check: $2x^2 - 48 = 2(\pm 7)^2 - 48 = 2(49) - 48 = 98 - 48 = 50$

Exercise Set 6.3

1) $a = 1, b = -7, c = 12, x = \frac{-(-7) \pm \sqrt{(-7)^2 - 4(1)(12)}}{2(1)} = \frac{7 \pm \sqrt{49 - 48}}{2} = \frac{7 \pm \sqrt{1}}{2} = \frac{7 \pm 1}{2}$

$x = \frac{7 + 1}{2} = \frac{8}{2} = \boxed{4}$ or $x = \frac{7 - 1}{2} = \frac{6}{2} = \boxed{3}$

Check: $x^2 - 7x + 12 = 4^2 - 7(4) + 12 = 16 - 28 + 12 = -12 + 12 = 0$

$x^2 - 7x + 12 = 3^2 - 7(3) + 12 = 9 - 21 + 12 = -12 + 12 = 0$

2) $a = 2, b = -8, c = -10, x = \frac{-(-8)\pm\sqrt{(-8)^2-4(2)(-10)}}{2(2)} = \frac{8\pm\sqrt{64+80}}{4} = \frac{8\pm\sqrt{144}}{4} = \frac{8\pm12}{4}$

$x = \frac{8+12}{4} = \frac{20}{4} = \boxed{5}$ or $x = \frac{8-12}{4} = \frac{-4}{4} = \boxed{-1}$

Note: If you divide both sides of the equation by 2, you could use $a = 1, b = -4, c = -5$

Check: $2x^2 - 8x - 10 = 2(5)^2 - 8(5) - 10 = 2(25) - 40 - 10 = 50 - 50 = 0$

$2x^2 - 8x - 10 = 2(-1)^2 - 8(-1) - 10 = 2(1) + 8 - 10 = 2 - 2 = 0$

3) $a = 4, b = 7, c = -15, x = \frac{-7\pm\sqrt{7^2-4(4)(-15)}}{2(4)} = \frac{-7\pm\sqrt{49+240}}{8} = \frac{-7\pm\sqrt{289}}{8} = \frac{-7\pm17}{8}$

$x = \frac{-7+17}{8} = \frac{10}{8} = \boxed{\frac{5}{4}}$ or $x = \frac{-7-17}{8} = \frac{-24}{8} = \boxed{-3}$

Check: $4x^2 + 7x - 15 = 4\left(\frac{5}{4}\right)^2 + 7\left(\frac{5}{4}\right) - 15 = 4\left(\frac{25}{16}\right) + \frac{35}{4} - 15 = \frac{100}{16} + \frac{140}{16} - \frac{240}{16} = 0$

$4x^2 + 7x - 15 = 4(-3)^2 + 7(-3) - 15 = 4(9) - 21 - 15 = 36 - 36 = 0$

4) $a = 6, b = -1, c = -40, x = \frac{-(-1)\pm\sqrt{(-1)^2-4(6)(-40)}}{2(6)} = \frac{1\pm\sqrt{1+960}}{12} = \frac{1\pm\sqrt{961}}{12} = \frac{1\pm31}{12}$

$x = \frac{1+31}{12} = \frac{32}{12} = \boxed{\frac{8}{3}}$ or $x = \frac{1-31}{12} = \frac{-30}{12} = \boxed{-\frac{5}{2}}$

Check: $6x^2 - x - 40 = 6\left(\frac{8}{3}\right)^2 - \frac{8}{3} - 40 = 6\left(\frac{64}{9}\right) - \frac{8}{3} - 40 = \frac{384}{9} - \frac{24}{9} - \frac{360}{9} = 0$

$6x^2 - x - 40 = 6\left(-\frac{5}{2}\right)^2 - \left(-\frac{5}{2}\right) - 40 = 6\left(\frac{25}{4}\right) + \frac{5}{2} - 40 = \frac{150}{4} + \frac{10}{4} - \frac{160}{4} = 0$

Exercise Set 6.4

1) Reorder the terms: $-4x^2 + 9x + 5 = 0$

Note: If you multiply by -1 on both sides, this is equivalent to $4x^2 - 9x - 5 = 0$.

2) Combine like terms: $6 - 8x + 3x^2 = 0$

Reorder the terms: $3x^2 - 8x + 6 = 0$

3) Subtract $6x^2$ from both sides and add $12x$ to both sides: $8x^2 + 3 - 6x^2 + 12x = 0$

Combine like terms: $2x^2 + 3 + 12x = 0$

Reorder the terms: $2x^2 + 12x + 3 = 0$

4) Subtract $6x$ from both sides: $5x^2 - 6x - 7 = 0$ (Be sure to put $-6x$ in the middle.)

5) Subtract $6x$ from both sides and add 8 to both sides: $4x - 9x^2 - 6x + 8 = 0$

Combine like terms: $-2x - 9x^2 + 8 = 0$

Reorder the terms: $-9x^2 - 2x + 8 = 0$ (equivalent to $9x^2 + 2x - 8 = 0$)

6) Subtract $5x$ and 9 from both sides and add $4x^2$ to both sides:

$11x - 9 + 3x^2 - 5x - 9 + 4x^2 = 0$

Combine like terms: $6x - 18 + 7x^2 = 0$

Reorder the terms: $7x^2 + 6x - 18 = 0$

7) Subtract 7 from both sides and add $2x^2$ to both sides: $3 + 8x - 7 + 2x^2 = 0$

Combine like terms: $-4 + 8x + 2x^2 = 0$

Reorder the terms: $2x^2 + 8x - 4 = 0$ (equivalent to $x^2 + 4x - 2 = 0$)

8) Add $4x$ to both sides (this term will cancel out): $5x^2 = 7$

Subtract 7 from both sides: $5x^2 - 7 = 0$

Note: This equation doesn't have a linear term. Think of it as $5x^2 + 0x - 7 = 0$.

9) Subtract $2x^2$ and 11 from both sides and add $3x$ to both sides:

$x^2 - 9x + 5 - 2x^2 + 3x - 11 = 0$

Combine like terms: $-x^2 - 6x - 6 = 0$

Note: If you multiply by -1 on both sides, this is equivalent to $x^2 + 6x + 6 = 0$.

10) Apply the FOIL method (Sec. 4.3): $x^2 + 6x + 3x + 18 = 25$

Subtract 25 from both sides: $x^2 + 6x + 3x + 18 - 25 = 0$

Combine like terms: $x^2 + 9x - 7 = 0$

Exercise Set 6.5

1) $a = 1, b = -8, c = 12, x = \frac{-b \pm \sqrt{b^2 - 4ac}}{2a} = \frac{-(-8) \pm \sqrt{(-8)^2 - 4(1)(12)}}{2(1)}$

$x = \frac{8 \pm \sqrt{64 - 48}}{2} = \frac{8 \pm \sqrt{16}}{2} = \frac{8 \pm 4}{2}$

$x = \frac{8+4}{2} = \frac{12}{2} = \boxed{6}$ or $x = \frac{8-4}{2} = \frac{4}{2} = \boxed{2}$

Check: $x^2 - 8x + 12 = 6^2 - 8(6) + 12 = 36 - 48 + 12 = -12 + 12 = 0$

and $x^2 - 8x + 12 = 2^2 - 8(2) + 12 = 4 - 16 + 12 = -12 + 12 = 0$

2) $a = 2, b = 8, c = -10, x = \frac{-b \pm \sqrt{b^2 - 4ac}}{2a} = \frac{-8 \pm \sqrt{8^2 - 4(2)(-10)}}{2(2)}$

$x = \frac{-8 \pm \sqrt{64 + 80}}{4} = \frac{-8 \pm \sqrt{144}}{4} = \frac{-8 \pm 12}{4}$ (Note: You could use $a = 1, b = 4, c = -5$ instead)

$x = \frac{-8+12}{4} = \frac{4}{4} = \boxed{1}$ or $x = \frac{-8-12}{4} = \frac{-20}{4} = \boxed{-5}$

Check: $2x^2 + 8x - 10 = 2(1)^2 + 8(1) - 10 = 2(1) + 8 - 10 = 2 - 2 = 0$

and $2x^2 + 8x - 10 = 2(-5)^2 + 8(-5) - 10 = 2(25) - 40 - 10 = 50 - 50 = 0$

3) First rewrite the equation in standard form: $2x^2 + 16x + 30 = 0$

$a = 2, b = 16, c = 30, x = \frac{-b \pm \sqrt{b^2 - 4ac}}{2a} = \frac{-16 \pm \sqrt{16^2 - 4(2)(30)}}{2(2)}$

$x = \frac{-16 \pm \sqrt{256 - 240}}{4} = \frac{-16 \pm \sqrt{16}}{4} = \frac{-16 \pm 4}{4}$ (Note: You could use $a = 1, b = 8, c = 15$ instead)

$x = \frac{-16 + 4}{4} = \frac{-12}{4} = \boxed{-3}$ or $x = \frac{-16 - 4}{4} = \frac{-20}{4} = \boxed{-5}$

Check: $16x + 30 + 2x^2 = 16(-3) + 30 + 2(-3)^2 = -48 + 30 + 2(9) = -18 + 18 = 0$

and $16x + 30 + 2x^2 = 16(-5) + 30 + 2(-5)^2 = -80 + 30 + 2(25) = -50 + 50 = 0$

4) First rewrite the equation in standard form: $-x^2 - 5x + 14 = 0$

$a = -1, b = -5, c = 14, x = \frac{-b \pm \sqrt{b^2 - 4ac}}{2a} = \frac{-(-5) \pm \sqrt{(-5)^2 - 4(-1)(14)}}{2(-1)}$

$x = \frac{5 \pm \sqrt{25 + 56}}{-2} = \frac{5 \pm \sqrt{81}}{-2} = \frac{5 \pm 9}{-2}$

$x = \frac{5 + 9}{-2} = \frac{14}{-2} = \boxed{-7}$ or $x = \frac{5 - 9}{-2} = \frac{-4}{-2} = \boxed{2}$

Check: $14 - x^2 - 5x = 14 - (-7)^2 - 5(-7) = 14 - 49 + 35 = -35 + 35 = 0$

and $14 - x^2 - 5x = 14 - 2^2 - 5(2) = 14 - 4 - 10 = 14 - 14 = 0$

5) Note that the constant term is on the wrong side.

First add 16 to both sides: $4x^2 - 16x + 16 = 0$

Tip: Each term is divisible by 4. Divide by 4 on both sides: $x^2 - 4x + 4 = 0$

$a = 1, b = -4, c = 4, x = \frac{-b \pm \sqrt{b^2 - 4ac}}{2a} = \frac{-(-4) \pm \sqrt{(-4)^2 - 4(1)(4)}}{2(1)}$

$x = \frac{4 \pm \sqrt{16 - 16}}{2} = \frac{4 \pm \sqrt{0}}{2} = \frac{4 \pm 0}{2} = \frac{4}{2} = \boxed{2}$ Note: This problem only has one answer.

Check: $4x^2 - 16x = 4(2)^2 - 16(2) = 4(4) - 32 = 16 - 32 = -16$

6) Subtract x^2 from both sides: $30 - x - x^2 = 0$

Reorder the terms to put the equation in standard form: $-x^2 - x + 30 = 0$

$a = -1, b = -1, c = 30, x = \frac{-b \pm \sqrt{b^2 - 4ac}}{2a} = \frac{-(-1) \pm \sqrt{(-1)^2 - 4(-1)(30)}}{2(-1)}$

$x = \frac{1 \pm \sqrt{1 + 120}}{-2} = \frac{1 \pm \sqrt{121}}{-2} = \frac{1 \pm 11}{-2}$

$x = \frac{1 + 11}{-2} = \frac{12}{-2} = \boxed{-6}$ or $x = \frac{1 - 11}{-2} = \frac{-10}{-2} = \boxed{5}$

Check: $30 - x = 30 - (-6) = 30 + 6 = 36$ agrees with $x^2 = (-6)^2 = 36$

and $30 - x = 30 - 5 = 25$ agrees with $x^2 = 5^2 = 25$

7) According to the reflexive property (Sec. 1.6): $x^2 + 11x + 18 = 0$

(Alternatively, you could bring every term to the left via subtraction. In that case, each constant would be negative, but the final answers would be the same. The arithmetic is a little trickier in that case, since every number would be negative.)

$a = 1, b = 11, c = 18, x = \frac{-b \pm \sqrt{b^2 - 4ac}}{2a} = \frac{-11 \pm \sqrt{11^2 - 4(1)(18)}}{2(1)}$

$x = \frac{-11 \pm \sqrt{121 - 72}}{2} = \frac{-11 \pm \sqrt{49}}{2} = \frac{-11 \pm 7}{2}$

$x = \frac{-11 + 7}{2} = \frac{-4}{2} = \boxed{-2}$ or $x = \frac{-11 - 7}{2} = \frac{-18}{2} = \boxed{-9}$

Check: $x^2 + 11x + 18 = (-2)^2 + 11(-2) + 18 = 4 - 22 + 18 = -18 + 18 = 0$

and $x^2 + 11x + 18 = (-9)^2 + 11(-9) + 18 = 81 - 99 + 18 = -18 + 18 = 0$

8) Add $3x^2$ and 6 to both sides and reorder the terms: $3x^2 + 9x + 6 = 0$

$a = 3, b = 9, c = 6, x = \frac{-b \pm \sqrt{b^2 - 4ac}}{2a} = \frac{-9 \pm \sqrt{9^2 - 4(3)(6)}}{2(3)}$

$x = \frac{-9 \pm \sqrt{81 - 72}}{6} = \frac{-9 \pm \sqrt{9}}{6} = \frac{-9 \pm 3}{6}$ (Note: You could use $a = 1, b = 3, c = 2$ instead)

$x = \frac{-9 + 3}{6} = \frac{-6}{6} = \boxed{-1}$ or $x = \frac{-9 - 3}{6} = \frac{-12}{6} = \boxed{-2}$

Check: $9x = 9(-1) = -9$ agrees with $-3x^2 - 6 = -3(-1)^2 - 6 = -3(1) - 6 = -3 - 6 = -9$

and $9x = 9(-2) = -18$ agrees with $-3x^2 - 6 = -3(-2)^2 - 6 = -3(4) - 6 = -12 - 6 = -18$

9) Bring every term to the same side. Combine like terms: $2x^2 - 10x - 100 = 0$

Tip: Each term is divisible by 2. Divide by 2 on both sides: $x^2 - 5x - 50 = 0$

$a = 1, b = -5, c = -50, x = \frac{-b \pm \sqrt{b^2 - 4ac}}{2a} = \frac{-(-5) \pm \sqrt{(-5)^2 - 4(1)(-50)}}{2(1)}$

$x = \frac{5 \pm \sqrt{25 + 200}}{2} = \frac{5 \pm \sqrt{225}}{2} = \frac{5 \pm 15}{2}$

$x = \frac{5 + 15}{2} = \frac{20}{2} = \boxed{10}$ or $x = \frac{5 - 15}{2} = \frac{-10}{2} = \boxed{-5}$

Check: $7x^2 - 10x = 7(10)^2 - 10(10) = 600$ agrees with $5x^2 + 100 = 5(10)^2 + 100 = 600$

and $7x^2 - 10x = 7(-5)^2 - 10(-5) = 7(25) + 50 = 175 + 50 = 225$ agrees with $5x^2 + 100$

$= 5(-5)^2 + 100 = 125 + 100 = 225$

10) Bring every term to the same side. Combine like terms: $-3x^2 + 21x = 0$

Note: The constant terms cancel since $7 - 9 + 2 = -2 + 2 = 0$

$a = -3, b = 21, c = 0, x = \frac{-b \pm \sqrt{b^2 - 4ac}}{2a} = \frac{-21 \pm \sqrt{21^2 - 4(-3)(0)}}{2(-3)}$

(Note: You could use $a = -1, b = 7, c = 0$ instead if you divide both sides by 3)

$$x = \frac{-21 \pm \sqrt{21^2 - 0}}{-6} = \frac{-21 \pm \sqrt{21^2}}{-6} = \frac{-21 \pm 21}{-6}$$

$$x = \frac{-21 + 21}{-6} = \frac{0}{-6} = \boxed{0} \text{ or } x = \frac{-21 - 21}{-6} = \frac{-42}{-6} = \boxed{7}$$

Note: This problem could be solved by factoring (Sec. 6.2), except that the instructions state to use the quadratic formula.

Check: $2x^2 + 21x + 7 = 2(0)^2 + 21(0) + 7 = 7$ agrees with $9 + 5x^2 - 2 = 9 + 5(0)^2 - 2 = 7$

and $2x^2 + 21x + 7 = 2(7)^2 + 21(7) + 7 = 2(49) + 147 + 7 = 98 + 154 = 252$ agrees

with $9 + 5x^2 - 2 = 7 + 5(7)^2 = 7 + 5(49) = 7 + 245 = 252$

11) Subtract x^2 from both sides to put this in standard form: $-x^2 + 10x + 24 = 0$

$$a = -1, b = 10, c = 24, x = \frac{-b \pm \sqrt{b^2 - 4ac}}{2a} = \frac{-10 \pm \sqrt{10^2 - 4(-1)(24)}}{2(-1)}$$

$$x = \frac{-10 \pm \sqrt{100 + 96}}{-2} = \frac{-10 \pm \sqrt{196}}{-2} = \frac{-10 \pm 14}{-2}$$

$$x = \frac{-10 + 14}{-2} = \frac{4}{-2} = \boxed{-2} \text{ or } x = \frac{-10 - 14}{-2} = \frac{-24}{-2} = \boxed{12}$$

Check: $10x + 24 = 10(-2) + 24 = -20 + 24 = 4$ agrees with $x^2 = (-2)^2 = 4$

and $10x + 24 = 10(12) + 24 = 120 + 24 = 144$ agrees with $x^2 = 12^2 = 144$

12) Bring every term to the same side. Combine like terms: $5x^2 + 25x - 120 = 0$

Tip: Each term is divisible by 5. Divide by 5 on both sides: $x^2 + 5x - 24 = 0$

$$a = 1, b = 5, c = -24, x = \frac{-b \pm \sqrt{b^2 - 4ac}}{2a} = \frac{-5 \pm \sqrt{5^2 - 4(1)(-24)}}{2(1)}$$

$$x = \frac{-5 \pm \sqrt{25 + 96}}{2} = \frac{-5 \pm \sqrt{121}}{2} = \frac{-5 \pm 11}{2}$$

$$x = \frac{-5 + 11}{2} = \frac{6}{2} = \boxed{3} \text{ or } x = \frac{-5 - 11}{2} = \frac{-16}{2} = \boxed{-8}$$

Check: $5x^2 + 10x - 90 = 5(3)^2 + 10(3) - 90 = 5(9) + 30 - 90 = 45 - 60 = -15$ agrees

with $30 - 15x = 30 - 15(3) = 30 - 45 = -15$

and $5x^2 + 10x - 90 = 5(-8)^2 + 10(-8) - 90 = 5(64) - 80 - 90 = 320 - 170 = 150$ agrees

with $30 - 15x = 30 - 15(-8) = 30 + 120 = 150$

13) Bring every term to the same side. Combine like terms: $3x^2 - 75 = 0$

Note: The linear terms cancel since $-4x + 4x = 0$

$$a = 3, b = 0, c = -75, x = \frac{-b \pm \sqrt{b^2 - 4ac}}{2a} = \frac{-0 \pm \sqrt{0^2 - 4(3)(-75)}}{2(3)}$$

$$x = \frac{0 \pm \sqrt{900}}{6} = \frac{\pm 30}{6} = \pm 5 \text{ The answers are } x = \boxed{-5} \text{ and } x = \boxed{5}$$

(Note: You could use $a = 1, b = 0, c = -25$ instead if you divide both sides by 3)

Note: This problem could be solved like the problems in Sec. 6.2, except that the instructions state to use the quadratic formula.

Check: $3x^2 - 4x = 3(-5)^2 - 4(-5) = 3(25) + 20 = 75 + 20 = 95$ agrees with $75 - 4x$ $= 75 - 4(-5) = 95$

and $3x^2 - 4x = 3(5)^2 - 4(5) = 75 - 20 = 55$ agrees with $75 - 4x = 75 - 4(5) = 55$

14) Bring every term to the same side. Combine like terms: $x^2 - 3x - 54 = 0$

$a = 1, b = -3, c = -54, x = \frac{-b \pm \sqrt{b^2 - 4ac}}{2a} = \frac{-(-3) \pm \sqrt{(-3)^2 - 4(1)(-54)}}{2(1)}$

$x = \frac{3 \pm \sqrt{9 + 216}}{2} = \frac{3 \pm \sqrt{225}}{2} = \frac{3 \pm 15}{2}$

$x = \frac{3 + 15}{2} = \frac{18}{2} = \boxed{9}$ or $x = \frac{3 - 15}{2} = \frac{-12}{2} = \boxed{-6}$

Check: $7x^2 + 5x + 26 = 7(9)^2 + 5(9) + 26 = 7(81) + 45 + 26 = 567 + 71 = 638$

agrees with $6x^2 + 8x + 80 = 6(9)^2 + 8(9) + 80 = 6(81) + 72 + 80 = 486 + 152 = 638$

and $7x^2 + 5x + 26 = 7(-6)^2 + 5(-6) + 26 = 7(36) - 30 + 26 = 252 - 4 = 248$ agrees

with $6x^2 + 8x + 80 = 6(-6)^2 + 8(-6) + 80 = 6(36) - 48 + 80 = 216 + 32 = 248$

15) Subtract 24 from both sides and add x to both sides: $\frac{x^2}{2} + x - 24 = 0$

(If you don't like the one-half, you could multiply both sides by 2.)

$a = \frac{1}{2}, b = 1, c = -24, x = \frac{-b \pm \sqrt{b^2 - 4ac}}{2a} = \frac{-1 \pm \sqrt{1^2 - 4\left(\frac{1}{2}\right)(-24)}}{2\left(\frac{1}{2}\right)}$

$x = \frac{-1 \pm \sqrt{1 + 48}}{1} = \frac{-1 \pm \sqrt{49}}{1} = \frac{-1 \pm 7}{1} = -1 \pm 7$

$x = -1 + 7 = \boxed{6}$ or $x = -1 - 7 = \boxed{-8}$

Check: $\frac{x^2}{2} = \frac{6^2}{2} = \frac{36}{2} = 18$ agrees with $24 - x = 24 - 6 = 18$

and $\frac{x^2}{2} = \frac{(-8)^2}{2} = \frac{64}{2} = 32$ agrees with $24 - x = 24 - (-8) = 24 + 8 = 32$

16) First apply the FOIL method (Sec. 4.3): $2x^2 - 12x - 8x + 48 = 6$

Subtract 6 from both sides and combine like terms: $2x^2 - 20x + 42 = 0$

Tip: Each term is divisible by 2. Divide by 2 on both sides: $x^2 - 10x + 21 = 0$

$a = 1, b = -10, c = 21, x = \frac{-b \pm \sqrt{b^2 - 4ac}}{2a} = \frac{-(-10) \pm \sqrt{(-10)^2 - 4(1)(21)}}{2(1)}$

$x = \frac{10 \pm \sqrt{100 - 84}}{2} = \frac{10 \pm \sqrt{16}}{2} = \frac{10 \pm 4}{2}$

$x = \frac{10 + 4}{2} = \frac{14}{2} = \boxed{7}$ or $x = \frac{10 - 4}{2} = \frac{6}{2} = \boxed{3}$

Check: $(2x - 8)(x - 6) = [2(7) - 8](7 - 6) = (14 - 8)(1) = 6(1) = 6$

and $(2x - 8)(x - 6) = [2(3) - 8](3 - 6) = (6 - 8)(-3) = (-2)(-3) = 6$

Exercise Set 6.6

1) $a = 8, b = -10, c = 3, x = \frac{-b \pm \sqrt{b^2 - 4ac}}{2a} = \frac{-(-10) \pm \sqrt{(-10)^2 - 4(8)(3)}}{2(8)}$

$x = \frac{10 \pm \sqrt{100 - 96}}{16} = \frac{10 \pm \sqrt{4}}{16} = \frac{10 \pm 2}{16}$

$x = \frac{10 + 2}{16} = \frac{12}{16} = \frac{12/4}{16/4} = \boxed{\frac{3}{4}}$ or $x = \frac{10 - 2}{16} = \frac{8}{16} = \boxed{\frac{1}{2}}$

Check: $8x^2 - 10x + 3 = 8\left(\frac{3}{4}\right)^2 - 10\left(\frac{3}{4}\right) + 3 = \frac{8(9)}{16} - \frac{30}{4} + 3 = \frac{72}{16} - \frac{120}{16} + \frac{48}{16} = 0$

and $8x^2 - 10x + 3 = 8\left(\frac{1}{2}\right)^2 - 10\left(\frac{1}{2}\right) + 3 = \frac{8}{4} - \frac{10}{2} + 3 = 2 - 5 + 3 = 0$

2) $a = 12, b = -5, c = -2, x = \frac{-b \pm \sqrt{b^2 - 4ac}}{2a} = \frac{-(-5) \pm \sqrt{(-5)^2 - 4(12)(-2)}}{2(12)}$

$x = \frac{5 \pm \sqrt{25 + 96}}{24} = \frac{5 \pm \sqrt{121}}{24} = \frac{5 \pm 11}{24}$

$x = \frac{5 + 11}{24} = \frac{16}{24} = \frac{16/8}{24/8} = \boxed{\frac{2}{3}}$ or $x = \frac{5 - 11}{24} = \frac{-6}{24} = \boxed{-\frac{1}{4}}$

Check: $12x^2 - 5x - 2 = 12\left(\frac{2}{3}\right)^2 - 5\left(\frac{2}{3}\right) - 2 = \frac{12(4)}{9} - \frac{10}{3} - 2 = \frac{48}{9} - \frac{30}{9} - \frac{18}{9} = 0$

and $12x^2 - 5x - 2 = 12\left(-\frac{1}{4}\right)^2 - 5\left(-\frac{1}{4}\right) - 2 = \frac{12}{16} + \frac{5}{4} - 2 = \frac{12}{16} + \frac{20}{16} - \frac{32}{16} = 0$

3) $a = 21, b = 47, c = 20, x = \frac{-b \pm \sqrt{b^2 - 4ac}}{2a} = \frac{-47 \pm \sqrt{47^2 - 4(21)(20)}}{2(21)}$

$x = \frac{-47 \pm \sqrt{2209 - 1680}}{42} = \frac{-47 \pm \sqrt{529}}{42} = \frac{-47 \pm 23}{42}$

$x = \frac{-47 + 23}{42} = \frac{-24}{42} = -\frac{24/6}{42/6} = \boxed{-\frac{4}{7}}$ or $x = \frac{-47 - 23}{42} = \frac{-70}{42} = -\frac{70/14}{42/14} = \boxed{-\frac{5}{3}}$

Check: $21x^2 + 47x + 20 = 21\left(-\frac{4}{7}\right)^2 + 47\left(-\frac{4}{7}\right) + 20 = \frac{21(16)}{49} - \frac{188}{7} + 20 = \frac{336}{49} - \frac{1316}{49} + \frac{980}{49} = 0$

and $21x^2 + 47x + 20 = 21\left(-\frac{5}{3}\right)^2 + 47\left(-\frac{5}{3}\right) + 20 = \frac{21(25)}{9} - \frac{235}{3} + 20 = \frac{525}{9} - \frac{705}{9} + \frac{180}{9} = 0$

4) Bring every term to the same side: $45x^2 + 52x - 32 = 0$

$a = 45, b = 52, c = -32, x = \frac{-b \pm \sqrt{b^2 - 4ac}}{2a} = \frac{-52 \pm \sqrt{52^2 - 4(45)(-32)}}{2(45)}$

$x = \frac{-52 \pm \sqrt{2704 + 5760}}{90} = \frac{-52 \pm \sqrt{8464}}{90} = \frac{-52 \pm 92}{90}$

$x = \frac{-52 + 92}{90} = \frac{40}{90} = \boxed{\frac{4}{9}}$ or $x = \frac{-52 - 92}{90} = \frac{-144}{90} = -\frac{144/18}{90/18} = \boxed{-\frac{8}{5}}$

Check: $45x^2 = 45\left(\frac{4}{9}\right)^2 = \frac{45(16)}{81} = \frac{720}{81} = \frac{720/9}{81/9} = \frac{80}{9}$ agrees with $32 - 52x = 32 - 52\left(\frac{4}{9}\right)$

$= \frac{288}{9} - \frac{208}{9} = \frac{80}{9}$ and $45x^2 = 45\left(-\frac{8}{5}\right)^2 = \frac{45(64)}{25} = \frac{2880}{25} = \frac{2880/5}{25/5} = \frac{576}{5}$ agrees with

$32 - 52x = 32 - 52\left(-\frac{8}{5}\right) = \frac{160}{5} + \frac{416}{5} = \frac{576}{5}$

5) Bring every term to the same side: $24x^2 - 2x - 15 = 0$

$a = 24, b = -2, c = -15, x = \frac{-b \pm \sqrt{b^2 - 4ac}}{2a} = \frac{-(-2) \pm \sqrt{(-2)^2 - 4(24)(-15)}}{2(24)}$

$x = \frac{2 \pm \sqrt{4 + 1440}}{48} = \frac{2 \pm \sqrt{1444}}{48} = \frac{2 \pm 38}{48}$

$x = \frac{2 + 38}{48} = \frac{40}{48} = \frac{40/8}{48/8} = \boxed{\frac{5}{6}}$ or $x = \frac{2 - 38}{48} = \frac{-36}{48} = -\frac{36/12}{48/12} = \boxed{-\frac{3}{4}}$

Check: $24x^2 - 2x = 24\left(\frac{5}{6}\right)^2 - 2\left(\frac{5}{6}\right) = \frac{24(25)}{36} - \frac{10}{6} = \frac{600}{36} - \frac{60}{36} = \frac{540}{36} = 15$

and $24x^2 - 2x = 24\left(-\frac{3}{4}\right)^2 - 2\left(-\frac{3}{4}\right) = \frac{24(9)}{16} + \frac{6}{4} = \frac{216}{16} + \frac{24}{16} = \frac{240}{16} = 15$

6) Bring every term to the same side: $24x^2 - 89x + 30 = 0$

$a = 24, b = -89, c = 30, x = \frac{-b \pm \sqrt{b^2 - 4ac}}{2a} = \frac{-(-89) \pm \sqrt{(-89)^2 - 4(24)(30)}}{2(24)}$

$x = \frac{89 \pm \sqrt{7921 - 2880}}{48} = \frac{89 \pm \sqrt{5041}}{48} = \frac{89 \pm 71}{48}$

$x = \frac{89 + 71}{48} = \frac{160}{48} = \frac{160/16}{48/16} = \boxed{\frac{10}{3}}$ or $x = \frac{89 - 71}{48} = \frac{18}{48} = \frac{18/6}{48/6} = \boxed{\frac{3}{8}}$

Check: $89x - 24x^2 = 89\left(\frac{10}{3}\right) - 24\left(\frac{10}{3}\right)^2 = \frac{890}{3} - \frac{24(100)}{9} = \frac{2670}{9} - \frac{2400}{9} = \frac{270}{9} = 30$

and $89x - 24x^2 = 89\left(\frac{3}{8}\right) - 24\left(\frac{3}{8}\right)^2 = \frac{267}{8} - \frac{24(9)}{64} = \frac{2136}{64} - \frac{216}{64} = -\frac{1920}{64} = 30$

7) Bring every term to the same side. Combine like terms: $14x^2 + 3x - 5 = 0$

$a = 14, b = 3, c = -5, x = \frac{-b \pm \sqrt{b^2 - 4ac}}{2a} = \frac{-3 \pm \sqrt{3^2 - 4(14)(-5)}}{2(14)}$

$x = \frac{-3 \pm \sqrt{9 + 280}}{28} = \frac{-3 \pm \sqrt{289}}{28} = \frac{-3 \pm 17}{28}$

$x = \frac{-3 + 17}{28} = \frac{14}{28} = \boxed{\frac{1}{2}}$ or $x = \frac{-3 - 17}{28} = \frac{-20}{28} = -\frac{20/4}{28/4} = \boxed{-\frac{5}{7}}$

Check: $8x^2 + 3x = 8\left(\frac{1}{2}\right)^2 + 3\left(\frac{1}{2}\right) = \frac{8}{4} + \frac{6}{4} = \frac{14}{4}$ agrees with $5 - 6x^2 = 5 - 6\left(\frac{1}{2}\right)^2 =$

$\frac{20}{4} - \frac{6}{4} = \frac{14}{4}$ and $8x^2 + 3x = 8\left(-\frac{5}{7}\right)^2 + 3\left(-\frac{5}{7}\right) = \frac{8(25)}{49} - \frac{15}{7} = \frac{200}{49} - \frac{105}{49} = \frac{95}{49}$ agrees

with $5 - 6x^2 = 5 - 6\left(-\frac{5}{7}\right)^2 = 5 - \frac{6(25)}{49} = \frac{245}{49} - \frac{150}{49} = \frac{95}{49}$

8) Bring every term to the same side. Combine like terms: $27x^2 + 42x + 16 = 0$

$a = 27, b = 42, c = 16, x = \frac{-b \pm \sqrt{b^2 - 4ac}}{2a} = \frac{-42 \pm \sqrt{42^2 - 4(27)(16)}}{2(27)}$

$x = \frac{-42 \pm \sqrt{1764 - 1728}}{54} = \frac{-42 \pm \sqrt{36}}{54} = \frac{-42 \pm 6}{54}$

$x = \frac{-42 + 6}{54} = \frac{-36}{54} = -\frac{36/18}{54/18} = \boxed{-\frac{2}{3}}$ or $x = \frac{-42 - 6}{54} = \frac{-48}{54} = -\frac{48/6}{54/6} = \boxed{-\frac{8}{9}}$

Check: $27x^2 + 18x = 27\left(-\frac{2}{3}\right)^2 + 18\left(-\frac{2}{3}\right) = \frac{27(4)}{9} - \frac{36}{3} = \frac{108}{9} - \frac{108}{9} = 0$ agrees with

$-16 - 24x = -16 - 24\left(-\frac{2}{3}\right) = -16 + \frac{48}{3} = -16 + 16 = 0$

and $27x^2 + 18x = 27\left(-\frac{8}{9}\right)^2 + 18\left(-\frac{8}{9}\right) = \frac{27(64)}{81} - \frac{144}{9} = \frac{1728}{81} - \frac{1296}{81} = \frac{432}{81}$ agrees

with $-16 - 24x = -16 - 24\left(-\frac{8}{9}\right) = -\frac{144}{9} + \frac{192}{9} = \frac{48}{9} = \frac{48 \cdot 9}{9 \cdot 9} = \frac{432}{81}$

9) Bring every term to the same side. Combine like terms: $10x^2 - 29x - 72 = 0$

$a = 10, b = -29, c = -72, x = \frac{-b \pm \sqrt{b^2 - 4ac}}{2a} = \frac{-(-29) \pm \sqrt{(-29)^2 - 4(10)(-72)}}{2(10)}$

$x = \frac{29 \pm \sqrt{841 + 2880}}{20} = \frac{29 \pm \sqrt{2721}}{20} = \frac{29 \pm 61}{20}$

$x = \frac{29 + 61}{20} = \frac{90}{20} = \boxed{\frac{9}{2}}$ or $x = \frac{29 - 61}{20} = \frac{-32}{20} = -\frac{32/4}{20/4} = \boxed{-\frac{8}{5}}$

Check: $10x^2 - 17x - 54 = 10\left(\frac{9}{2}\right)^2 - 17\left(\frac{9}{2}\right) - 54 = \frac{10(81)}{4} - \frac{153}{2} - 54 = \frac{810}{4} - \frac{306}{4} - \frac{216}{4}$

$= \frac{288}{4} = 72$ agrees with $12x + 18 = 12\left(\frac{9}{2}\right) + 18 = \frac{108}{2} + 18 = 54 + 18 = 72$

and $10x^2 - 17x - 54 = 10\left(-\frac{8}{5}\right)^2 - 17\left(-\frac{8}{5}\right) - 54 = \frac{10(64)}{25} + \frac{136}{5} - 54 = \frac{640}{25} + \frac{680}{25} - \frac{1350}{25}$

$= -\frac{30}{25}$ agrees with $12x + 18 = 12\left(-\frac{8}{5}\right) + 18 = -\frac{96}{5} + \frac{90}{5} = -\frac{6}{5} = -\frac{6 \cdot 5}{5 \cdot 5} = -\frac{30}{25}$

10) Bring every term to the same side. Combine like terms: $-21x^2 - 58x - 21 = 0$

Multiply both sides of the equation by -1 to get $21x^2 + 58x + 21 = 0$

$a = 21, b = 58, c = 21, x = \frac{-b \pm \sqrt{b^2 - 4ac}}{2a} = \frac{-58 \pm \sqrt{58^2 - 4(21)(21)}}{2(21)}$

$x = \frac{-58 \pm \sqrt{3364 - 1764}}{42} = \frac{-58 \pm \sqrt{1600}}{42} = \frac{-58 \pm 40}{42}$

$x = \frac{-58 + 40}{42} = \frac{-18}{42} = -\frac{18/6}{42/6} = \boxed{-\frac{3}{7}}$ or $x = \frac{-58 - 40}{42} = \frac{-98}{42} = -\frac{98/14}{42/14} = \boxed{-\frac{7}{3}}$

Check: $24 - 9x^2 - 29x = 24 - 9\left(-\frac{3}{7}\right)^2 - 29\left(-\frac{3}{7}\right) = 24 - \frac{9(9)}{49} + \frac{87}{7} = \frac{1176}{49} - \frac{81}{49} + \frac{609}{49}$

$= \frac{1704}{49}$ agrees with $12x^2 + 45 + 29x = 12\left(-\frac{3}{7}\right)^2 + 45 + 29x = 12\left(\frac{9}{49}\right) + 45 + 29\left(-\frac{3}{7}\right)$

$$= \frac{108}{49} + \frac{2205}{49} - \frac{609}{49} = \frac{1704}{49}$$

and $24 - 9x^2 - 29x = 24 - 9\left(-\frac{7}{3}\right)^2 - 29\left(-\frac{7}{3}\right) = 24 - \frac{9(49)}{9} + \frac{203}{3} = \frac{216}{9} - \frac{441}{9} + \frac{609}{9}$

$= \frac{384}{9}$ agrees with $12x^2 + 45 + 29x = 12\left(-\frac{7}{3}\right)^2 + 45 + 29\left(-\frac{7}{3}\right) = \frac{12(49)}{9} + 45 - \frac{203}{3}$

$= \frac{588}{9} + \frac{405}{9} - \frac{609}{9} = \frac{384}{9}$

Exercise Set 6.7

1) $a = 1, b = 5, c = 2, x = \frac{-b \pm \sqrt{b^2 - 4ac}}{2a} = \frac{-5 \pm \sqrt{5^2 - 4(1)(2)}}{2(1)} = \frac{-5 \pm \sqrt{25 - 8}}{2} = \boxed{\frac{-5 \pm \sqrt{17}}{2}}$

If you use a calculator: $x \approx -0.438$ or $x \approx -4.56$

Check: $x^2 + 5x + 2 \approx (-0.438)^2 + 5(-0.438) + 2 \approx 0.192 - 2.19 + 2 \approx 0.002$

and $x^2 + 5x + 2 \approx (-4.56)^2 + 5(-4.56) + 2 \approx 20.8 - 22.8 + 2 = 0$

2) $a = 3, b = 6, c = -5, x = \frac{-b \pm \sqrt{b^2 - 4ac}}{2a} = \frac{-6 \pm \sqrt{6^2 - 4(3)(-5)}}{2(3)} = \frac{-6 \pm \sqrt{36 + 60}}{6} = \frac{-6 \pm \sqrt{96}}{6}$

Note that $\sqrt{96} = \sqrt{16(6)} = \sqrt{16}\sqrt{6} = 4\sqrt{6}$, such that $x = \frac{-6 \pm 4\sqrt{6}}{6} = \frac{2(-3 \pm 2\sqrt{6})}{2(3)} = \boxed{\frac{-3 \pm 2\sqrt{6}}{3}}$

If you use a calculator: $x \approx 0.633$ or $x \approx -2.63$

Check: $3x^2 + 6x - 5 \approx 3(0.633)^2 + 6(0.633) - 5 \approx 1.20 + 3.80 - 5 = 0$

and $3x^2 + 6x - 5 \approx 3(-2.63)^2 + 6(-2.63) - 5 \approx 20.8 - 15.8 - 5 = 0$

3) Bring every term to the same side: $4x^2 - 9x + 4 = 0$

$a = 4, b = -9, c = 4, x = \frac{-b \pm \sqrt{b^2 - 4ac}}{2a} = \frac{-(-9) \pm \sqrt{(-9)^2 - 4(4)(4)}}{2(4)} = \frac{9 \pm \sqrt{81 - 64}}{8} = \boxed{\frac{9 \pm \sqrt{17}}{8}}$

If you use a calculator: $x \approx 1.64$ or $x \approx 0.610$ (if you round 0.6096 up)

Check: $4x^2 + 4 \approx 4(1.64)^2 + 4 \approx 10.8 + 4 = 14.8$ agrees with $9x \approx 9(1.64) \approx 14.8$

and $4x^2 + 4 \approx 4(0.610)^2 + 4 \approx 1.49 + 4 = 5.49$ agrees with $9x \approx 9(0.610) \approx 5.49$

4) Bring every term to the same side: $5x^2 - 2x - 2 = 0$

$a = 5, b = -2, c = -2, x = \frac{-b \pm \sqrt{b^2 - 4ac}}{2a} = \frac{-(-2) \pm \sqrt{(-2)^2 - 4(5)(-2)}}{2(5)} = \frac{2 \pm \sqrt{4 + 40}}{10} = \frac{2 \pm \sqrt{44}}{10}$

Note that $\sqrt{44} = \sqrt{4(11)} = \sqrt{4}\sqrt{11} = 2\sqrt{11}$, such that $x = \frac{2 \pm 2\sqrt{11}}{10} = \frac{2(1 \pm \sqrt{11})}{2(5)} = \boxed{\frac{1 \pm \sqrt{11}}{5}}$

If you use a calculator: $x \approx 0.863$ or $x \approx -0.463$

Check: $5x^2 - 2x \approx 5(0.863)^2 - 2(0.863) \approx 3.72 - 1.73 = 1.99 \approx 2$

and $5x^2 - 2x \approx 5(-0.463)^2 - 2(-0.463) \approx 1.07 + 0.926 \approx 2$

5) Bring every term to the same side: $-3x^2 + 2x + 6 = 0$

$a = -3, b = 2, c = 6, x = \frac{-b \pm \sqrt{b^2 - 4ac}}{2a} = \frac{-2 \pm \sqrt{2^2 - 4(-3)(6)}}{2(-3)} = \frac{-2 \pm \sqrt{4 + 72}}{-6} = \frac{-2 \pm \sqrt{76}}{-6}$

Note that $\sqrt{76} = \sqrt{4(19)} = \sqrt{4}\sqrt{19} = 2\sqrt{19}$, such that $x = \frac{-2 \pm 2\sqrt{19}}{-6} = \frac{-2(1 \pm \sqrt{19})}{-2(3)} = \boxed{\frac{1 \pm \sqrt{19}}{3}}$

Note that $-2(1 \mp \sqrt{19}) = -2(1 \pm \sqrt{19})$ since they give the same two answers.

If you use a calculator: $x \approx 1.79$ or $x \approx -1.12$

Check: $2x + 6 \approx 2(1.79) + 6 \approx 9.58$ agrees with $3x^2 \approx 3(1.79)^2 \approx 9.61$

and $2x + 6 \approx 2(-1.12) + 6 \approx 3.76$ agrees with $3x^2 \approx 3(-1.12)^2 \approx 3.76$

Note: There is a little rounding error. You can reduce this by keeping more digits.

6) Bring every term to the same side: $2x^2 - 8x + 7 = 0$

$a = 2, b = -8, c = 7, x = \frac{-b \pm \sqrt{b^2 - 4ac}}{2a} = \frac{-(-8) \pm \sqrt{(-8)^2 - 4(2)(7)}}{2(2)} = \frac{8 \pm \sqrt{64 - 56}}{4} = \frac{8 \pm \sqrt{8}}{4}$

Note that $\sqrt{8} = \sqrt{4(2)} = \sqrt{4}\sqrt{2} = 2\sqrt{2}$, such that $x = \frac{8 \pm 2\sqrt{2}}{4} = \frac{2(4 \pm \sqrt{2})}{2(2)} = \boxed{\frac{4 \pm \sqrt{2}}{2}}$

If you use a calculator: $x \approx 2.71$ or $x \approx 1.29$

Check: $8x - 2x^2 \approx 8(2.71) - 2(2.71)^2 \approx 21.7 - 14.7 = 7$

and $8x - 2x^2 \approx 8(1.29) - 2(1.29)^2 \approx 10.3 - 3.33 \approx 6.97 \approx 7$

7) Bring every term to the same side. Combine like terms: $4x^2 - 10x - 5 = 0$

$a = 4, b = -10, c = -5, x = \frac{-b \pm \sqrt{b^2 - 4ac}}{2a} = \frac{-(-10) \pm \sqrt{(-10)^2 - 4(4)(-5)}}{2(4)} = \frac{10 \pm \sqrt{100 + 80}}{8} = \frac{10 \pm \sqrt{180}}{8}$

Note that $\sqrt{180} = \sqrt{36(5)} = \sqrt{36}\sqrt{5} = 6\sqrt{5}$, such that $x = \frac{10 \pm 6\sqrt{5}}{8} = \frac{2(5 \pm 3\sqrt{5})}{2(4)} = \boxed{\frac{5 \pm 3\sqrt{5}}{4}}$

If you use a calculator: $x \approx 2.93$ or $x \approx -0.427$

Check: $9x^2 - 10x \approx 9(2.93)^2 - 10(2.93) \approx 77.3 - 29.3 = 48$ agrees with

$5 + 5x^2 \approx 5 + 5(2.93)^2 \approx 5 + 42.9 = 47.9$

and $9x^2 - 10x \approx 9(-0.427)^2 - 10(-0.427) \approx 1.64 + 4.27 = 5.91$

$5 + 5x^2 \approx 5 + 5(-0.427)^2 \approx 5 + 0.912 \approx 5.91$

8) Bring every term to the same side. Combine like terms: $5x^2 + 7x + 1 = 0$

$a = 5, b = 7, c = 1, x = \frac{-b \pm \sqrt{b^2 - 4ac}}{2a} = \frac{-7 \pm \sqrt{7^2 - 4(5)(1)}}{2(5)} = \frac{-7 \pm \sqrt{49 - 20}}{10} = \boxed{\frac{-7 \pm \sqrt{29}}{10}}$

If you use a calculator: $x \approx -0.161$ or $x \approx -1.24$

Check: $5x^2 + 15x \approx 5(-0.161)^2 + 15(-0.161) \approx 0.130 - 2.42 = -2.30$

agrees with $8x - 1 = 8(-0.161) - 1 = -2.29$

and $5x^2 + 15 \approx 5(-1.24)^2 + 15(-1.24) \approx 7.69 - 18.6 \approx -10.9$

agrees with $8x - 1 = 8(-1.24) - 1 = -10.9$

Exercise Set 6.8

1) $a = 1, b = 2, c = 3, b^2 - 4ac = 2^2 - 4(1)(3) = 4 - 12 = \boxed{-8}$ (two complex answers)

2) $a = 2, b = -5, c = 3, b^2 - 4ac = (-5)^2 - 4(2)(3) = 25 - 24 = \boxed{1}$ (two real answers)

3) $a = 5, b = -7, c = -3, b^2 - 4ac = (-7)^2 - 4(5)(-3) = 49 + 60 = \boxed{109}$ (two real answers)

4) $a = -4, b = 8, c = -4, b^2 - 4ac = 8^2 - 4(-4)(-4) = 64 - 64 = \boxed{0}$ (one real answer)

Note: Three minus signs multiplying make a minus sign (in contrast to Problem 3, where two minus signs multiplying make a plus sign.)

5) $a = \frac{1}{2}, b = -\frac{3}{4}, c = \frac{3}{8}, b^2 - 4ac = \left(-\frac{3}{4}\right)^2 - 4\left(\frac{1}{2}\right)\left(\frac{3}{8}\right) = \frac{9}{16} - \frac{12}{16} = \boxed{-\frac{3}{16}}$ (two complex answers)

6) First bring all of the terms to the same side: $4x^2 - 10x + 6 = 0$

$a = 4, b = -10, c = 6, b^2 - 4ac = (-10)^2 - 4(4)(6) = 100 - 96 = \boxed{4}$ (two real answers)

7) First put the equation in standard form: $-9x^2 + 12x - 4 = 0$

$a = -9, b = 12, c = -4, b^2 - 4ac = 12^2 - 4(-9)(-4) = 144 - 144 = \boxed{0}$ (one real answer)

8) First put the equation in standard form: $-2x^2 - 15x + 25 = 0$

$a = -2, b = -15, c = 25, b^2 - 4ac = (-15)^2 - 4(-2)(25) = 225 + 200 = \boxed{425}$ (two real answers)

9) Combine like terms and put in standard form: $5x^2 + 11x + 6 = 0$

$a = 5, b = 11, c = 6, b^2 - 4ac = 11^2 - 4(5)(6) = 121 - 120 = \boxed{1}$ (two real answers)

10) Combine like terms and put in standard form: $8x^2 - 12x + 5 = 0$

$a = 8, b = -12, c = 5, b^2 - 4ac = (-12)^2 - 4(8)(5) = 144 - 160 = \boxed{-16}$ (two complex answers)

Exercise Set 6.9

1) $(x + 7)(x + 1) = 0 \rightarrow x + 7 = 0$ or $x + 1 = 0 \rightarrow x = \boxed{-7}$ or $x = \boxed{-1}$

Check: $x^2 + 8x + 7 = (-7)^2 + 8(-7) + 7 = 49 - 56 + 7 = 0$

and $x^2 + 8x + 7 = (-1)^2 + 8(-1) + 7 = 1 - 8 + 7 = 0$

2) $(2x + 3)(x - 5) = 0 \rightarrow 2x + 3 = 0$ or $x - 5 = 0 \rightarrow 2x = -3$ or $x = 5 \rightarrow x = \boxed{-\frac{3}{2}}$ or $x = \boxed{5}$

Check: $2x^2 - 7x - 15 = 2\left(-\frac{3}{2}\right)^2 - 7\left(-\frac{3}{2}\right) - 15 = 2\left(\frac{9}{4}\right) + \frac{21}{2} - 15 = \frac{18}{4} + \frac{42}{4} - \frac{60}{4} = 0$

and $2x^2 - 7x - 15 = 2(5)^2 - 7(5) - 15 = 2(25) - 35 - 15 = 50 - 50 = 0$

3) First put the equation in standard form: $7x^2 - 11x - 6 = 0$

$(7x + 3)(x - 2) = 0 \rightarrow 7x + 3 = 0$ or $x - 2 = 0 \rightarrow 7x = -3$ or $x = 2 \rightarrow x = \boxed{-\dfrac{3}{7}}$ or $x = \boxed{2}$

Check: $7x^2 - 6 = 7\left(-\dfrac{3}{7}\right)^2 - 6 = 7\left(\dfrac{9}{49}\right) - 6 = \dfrac{63}{49} - \dfrac{294}{49} = -\dfrac{231}{49} = -\dfrac{33}{7}$

agrees with $11x = 11\left(-\dfrac{3}{7}\right) = -\dfrac{33}{7}$

and $7x^2 - 6 = 7(2)^2 - 6 = 7(4) - 6 = 28 - 6 = 22$ agrees with $11(x) = 11(2) = 22$

4) First put the equation in standard form: $x^2 - 12x + 36 = 0$

$(x - 6)(x - 6) = 0 \rightarrow x - 6 = 0$ or $x - 6 = 0 \rightarrow x = \boxed{6}$ or $x = \boxed{6}$ (just one answer)

Check: $x^2 - 12x = (6)^2 - 12(6) = 36 - 72 = -36$

5) First put the equation in standard form: $6x^2 - 5x - 4 = 0$

$(3x - 4)(2x + 1) = 0 \rightarrow 3x - 4 = 0$ or $2x + 1 = 0 \rightarrow 3x = 4$ or $2x = -1 \rightarrow x = \boxed{\dfrac{4}{3}}$ or $x = \boxed{-\dfrac{1}{2}}$

Check: $6x^2 = 6\left(\dfrac{4}{3}\right)^2 = 6\left(\dfrac{16}{9}\right) = \dfrac{96}{9} = \dfrac{32}{3}$ agrees with $5x + 4 = 5\left(\dfrac{4}{3}\right) + 4 = \dfrac{20}{3} + \dfrac{12}{3} = \dfrac{32}{3}$

and $6x^2 = 6\left(-\dfrac{1}{2}\right)^2 = 6\left(\dfrac{1}{4}\right) = \dfrac{6}{4} = \dfrac{3}{2}$ agrees with $5x + 4 = 5\left(-\dfrac{1}{2}\right) + 4 = -\dfrac{5}{2} + \dfrac{8}{2} = \dfrac{3}{2}$

6) First put the equation in standard form: $-3x^2 + 26x - 35 = 0$ or $3x^2 - 26x + 35 = 0$

$(3x - 5)(x - 7) = 0 \rightarrow 3x - 5 = 0$ or $x - 7 = 0 \rightarrow 3x = 5$ or $x = 7 \rightarrow x = \boxed{\dfrac{5}{3}}$ or $x = \boxed{7}$

Check: $26x - 35 = 26\left(\dfrac{5}{3}\right) - 35 = \dfrac{130}{3} - \dfrac{105}{3} = \dfrac{25}{3}$ agrees with $3x^2 = 3\left(\dfrac{5}{3}\right)^2 = 3\left(\dfrac{25}{9}\right) = \dfrac{25}{3}$

and $26x - 35 = 26(7) - 35 = 182 - 35 = 147$ agrees with $3x^2 = 3(7)^2 = 3(49) = 147$

7) First put the equation in standard form: $4x^2 + 37x + 40 = 0$

$(4x + 5)(x + 8) = 0 \rightarrow 4x + 5 = 0$ or $x + 8 = 0 \rightarrow 4x = -5$ or $x = -8 \rightarrow x = \boxed{-\dfrac{5}{4}}$ or $x = \boxed{-8}$

Check: $4x^2 + 45x = 4\left(-\dfrac{5}{4}\right)^2 + 45\left(-\dfrac{5}{4}\right) = 4\left(\dfrac{25}{16}\right) - \dfrac{225}{4} = \dfrac{100}{16} - \dfrac{900}{16} = -\dfrac{800}{16} = -50$

agrees with $8x - 40 = 8\left(-\dfrac{5}{4}\right) - 40 = -\dfrac{40}{4} - 40 = -10 - 40 = -50$

and $4x^2 + 45x = 4(-8)^2 + 45(-8) = 4(64) - 360 = 256 - 360 = -104$

agrees with $8x - 40 = 8(-8) - 40 = -64 - 40 = -104$

8) First put the equation in standard form: $9x^2 - 56x + 12 = 0$

$(x - 6)(9x - 2) = 0 \rightarrow x - 6 = 0$ or $9x - 2 = 0 \rightarrow x = 6$ or $9x = 2 \rightarrow x = \boxed{6}$ or $x = \boxed{\dfrac{2}{9}}$

Check: $6x^2 + 12 = 6(6)^2 + 12 = 6(36) + 12 = 216 + 12 = 228$ agrees with

$56x - 3x^2 = 56(6) - 3(6)^2 = 336 - 3(36) = 336 - 108 = 228$

and $6x^2 + 12 = 6\left(\dfrac{2}{9}\right)^2 + 12 = 6\left(\dfrac{4}{81}\right) + 12 = \dfrac{24}{81} + \dfrac{972}{81} = \dfrac{996}{81}$ agrees with

$56x - 3x^2 = 56\left(\dfrac{2}{9}\right) - 3\left(\dfrac{2}{9}\right)^2 = \dfrac{112}{9} - 3\left(\dfrac{4}{81}\right) = \dfrac{1008}{81} - \dfrac{12}{81} = \dfrac{996}{81}$

Exercise Set 6.10

1) x = the number and x^2 = the number squared

$x^2 = 10x - 24$ First put the equation in standard form: $x^2 - 10x + 24 = 0$

$a = 1, b = -10, c = 24, x = \frac{-b \pm \sqrt{b^2 - 4ac}}{2a} = \frac{-(-10) \pm \sqrt{(-10)^2 - 4(1)(24)}}{2(1)} = \frac{10 \pm \sqrt{100 - 96}}{2} = \frac{10 \pm \sqrt{4}}{2} = \frac{10 \pm 2}{2}$

$x = \frac{10 + 2}{2} = \frac{12}{2} = \boxed{6}$ or $x = \frac{10 - 2}{2} = \frac{8}{2} = \boxed{4}$

Check: $6^2 = 36$ is 24 less than $6(10) = 60$ because $60 - 24 = 36$

and $4^2 = 16$ is 24 less than $4(10) = 40$ because $40 - 24 = 16$

2) x and $(x + 2)$ are the two consecutive even numbers

$x(x + 2) = 360$ Distribute (Chapter 4): $x^2 + 2x = 360$

Put the equation in standard form: $x^2 + 2x - 360 = 0$

$a = 1, b = 2, c = -360, x = \frac{-b \pm \sqrt{b^2 - 4ac}}{2a} = \frac{-2 \pm \sqrt{2^2 - 4(1)(-360)}}{2(1)} = \frac{-2 \pm \sqrt{4 + 1440}}{2} = \frac{-2 \pm \sqrt{1444}}{2} = \frac{-2 \pm 38}{2}$

$x = \frac{-2 + 38}{2} = \frac{36}{2} = 18$ or $x = \frac{-2 - 38}{2} = \frac{-40}{2} = -20$

Two possibilities: The two numbers are $\boxed{18}$ and $\boxed{20}$ or

the two numbers are $\boxed{-20}$ and $\boxed{-18}$ Check: $(18)(20) = 360$ and $(-20)(-18) = 360$

3) x = the smaller number and $x + 17$ = the larger number (You could instead call x the larger number and $x - 17$ the smaller number. The answers would be the same.)

$x(x + 17) = 200$ Distribute (Chapter 4): $x^2 + 17x = 200$

Put the equation in standard form: $x^2 + 17x - 200 = 0$

$a = 1, b = 17, c = -200, x = \frac{-b \pm \sqrt{b^2 - 4ac}}{2a} = \frac{-17 \pm \sqrt{17^2 - 4(1)(-200)}}{2(1)} = \frac{-17 \pm \sqrt{289 + 800}}{2} = \frac{-17 \pm \sqrt{1089}}{2}$

$x = \frac{-17 \pm 33}{2}$

$x = \frac{-17 + 33}{2} = \frac{16}{2} = 8$ or $x = \frac{-17 - 33}{2} = \frac{-50}{2} = -25$

Two possibilities: The two numbers are $\boxed{8}$ and $\boxed{25}$ or the two numbers are $\boxed{-25}$ and $\boxed{-8}$

Check: $(8)(25) = 200$ and $(-25)(-8) = 200$

7 Variables in the Denominator

Exercise Set 7.1

1) $x = \frac{5}{3}$

2) $x = \frac{1}{6}$

3) $x = 9$

4) $x = \frac{5}{8}$

5) $\frac{6}{5} = x$

6) $\frac{1}{2} = x$

7) $x = -\frac{7}{2}$

8) $x = -8$

9) $-\frac{1}{5} = x$

10) $10 = x$

11) $x = -\frac{4}{9}$

12) $x = \frac{1}{6}$

Note: For Problems 11-12, it may help to multiply both sides by -1 (Sec. 2.8).

Exercise Set 7.2

1) $\frac{5}{x} = \frac{2}{3} \rightarrow 5(3) = 2x \rightarrow 15 = 2x \rightarrow \boxed{\frac{15}{2}} = x$

2) $\frac{6}{x} = \frac{3}{4} \rightarrow 6(4) = 3x \rightarrow 24 = 3x \rightarrow \frac{24}{3} = x \rightarrow \boxed{8} = x$

3) $\frac{4}{x} = \frac{1}{7} \rightarrow 4(7) = 1x \rightarrow \boxed{28} = x$

4) $\frac{3}{x} = \frac{4}{3} \rightarrow 3(3) = 4x \rightarrow 9 = 4x \rightarrow \boxed{\frac{9}{4}} = x$

5) $\frac{12}{x} = \frac{3}{5} \rightarrow 12(5) = 3x \rightarrow 60 = 3x \rightarrow \frac{60}{3} = x \rightarrow \boxed{20} = x$

6) $\frac{7}{x} = 3 \rightarrow \frac{7}{x} = \frac{3}{1} \rightarrow 7(1) = 3x \rightarrow 7 = 3x \rightarrow \boxed{\frac{7}{3}} = x$

7) $\frac{8}{x} = -\frac{6}{5} \rightarrow 8(5) = -6x \rightarrow 40 = -6x \rightarrow \frac{40}{-6} = x \rightarrow \boxed{-\frac{20}{3}} = x$ since $-\frac{40}{6} = -\frac{40/2}{6/2} = -\frac{20}{3}$

8) $\frac{2}{3} = \frac{10}{x} \rightarrow 2x = 3(10) \rightarrow 2x = 30 \rightarrow x = \frac{30}{2} \rightarrow x = \boxed{15}$

9) $\frac{6}{x} = -2 \rightarrow \frac{6}{x} = -\frac{2}{1} \rightarrow 6(1) = -2x \rightarrow 6 = -2x \rightarrow \frac{6}{-2} = x \rightarrow \boxed{-3} = x$

10) $-\frac{4}{x} = -\frac{1}{8} \rightarrow \frac{4}{x} = \frac{1}{8} \rightarrow 4(8) = 1x \rightarrow \boxed{32} = x$

Note: In the first step, we multiplied by -1 on both sides.

11) $-\frac{24}{x} = \frac{18}{5} \rightarrow -24(5) = 18x \rightarrow -120 = 18x \rightarrow -\frac{120}{18} = x \rightarrow \boxed{-\frac{20}{3}} = x$ since $-\frac{120}{18} = -\frac{120/6}{18/6} = -\frac{20}{3}$

12) $\frac{1}{x} = 6 \rightarrow \frac{1}{x} = \frac{6}{1} \rightarrow 1(1) = 6x \rightarrow 1 = 6x \rightarrow \boxed{\frac{1}{6}} = x$

13) $\frac{25}{x} = \frac{1}{3} \rightarrow 3(25) = 1x \rightarrow \boxed{75} = x$

14) $\frac{22}{5} = \frac{11}{x} \rightarrow 22x = 5(11) \rightarrow 22x = 55 \rightarrow x = \frac{55}{22} = \boxed{\frac{5}{2}}$ since $\frac{55}{22} = \frac{55/11}{22/11} = \frac{5}{2}$

15) $-\frac{7}{9} = -\frac{3}{x} \rightarrow \frac{7}{9} = \frac{3}{x} \rightarrow 7x = 9(3) \rightarrow 7x = 27 \rightarrow x = \boxed{\frac{27}{7}}$

Note: In the first step, we multiplied by -1 on both sides.

16) $-\frac{36}{x} = \frac{3}{4} \rightarrow -36(4) = 3x \rightarrow -144 = 3x \rightarrow -\frac{144}{3} = x \rightarrow \boxed{-48} = x$

Exercise Set 7.3

1) $\frac{x^2}{5} = \frac{25}{x} \rightarrow x^3 = 125 \rightarrow x = \sqrt[3]{125} = \boxed{5}$

Check: $\frac{x^2}{5} = \frac{5^2}{5} = \frac{25}{5} = 5$ agrees with $\frac{25}{x} = \frac{25}{5} = 5$

2) $\frac{6}{x^2} = \frac{2}{3} \rightarrow 18 = 2x^2 \rightarrow \frac{18}{2} = x^2 \rightarrow 9 = x^2 \rightarrow \pm\sqrt{9} = x \rightarrow \boxed{\pm 3} = x$

Check: $\frac{6}{x^2} = \frac{6}{(\pm 3)^2} = \frac{6}{9} = \frac{2}{3}$

3) $\frac{x^6}{4} = \frac{x^5}{7} \rightarrow 7x^6 = 4x^5 \rightarrow \frac{7x^6}{x^5} = 4 \rightarrow 7x = 4 \rightarrow x = \boxed{\frac{4}{7}}$ or $x = \boxed{0}$ (see Sec. 5.11)

Check: $\frac{x^6}{4} = \frac{(4/7)^6}{4} = \frac{4^6}{7^6}\frac{1}{4} = \frac{4096}{117,649(4)} = \frac{1024}{117,649}$ agrees with $\frac{x^5}{7} = \frac{(4/7)^5}{7} = \frac{4^5}{7^5}\frac{1}{7} = \frac{1024}{16,807(7)} = \frac{1024}{117,649}$

4) $\frac{x}{4} = \frac{9}{x} \rightarrow x^2 = 36 \rightarrow x = \pm\sqrt{36} \rightarrow x = \boxed{\pm 6}$

Check: $\frac{x}{4} = \frac{\pm 6}{4} = \pm\frac{6/2}{4/2} = \pm\frac{3}{2}$ agrees with $\frac{9}{x} = \frac{9}{\pm 6} = \pm\frac{9/3}{6/3} = \pm\frac{3}{2}$

5) $\frac{12}{x^2} = 3 \rightarrow \frac{12}{x^2} = \frac{3}{1} \rightarrow 12 = 3x^2 \rightarrow \frac{12}{3} = x^2 \rightarrow 4 = x^2 \rightarrow \pm\sqrt{4} = x \rightarrow \boxed{\pm 2} = x$

Check: $\frac{12}{x^2} = \frac{12}{(\pm 2)^2} = \frac{12}{4} = 3$

6) $\frac{x^6}{2} = \frac{x^8}{50} \rightarrow 50x^6 = 2x^8 \rightarrow 50 = \frac{2x^8}{x^6} \rightarrow 50 = 2x^2 \rightarrow 25 = x^2 \rightarrow \pm\sqrt{25} = x \rightarrow \boxed{\pm 5} = x$ or $x = \boxed{0}$

Check: $\frac{x^6}{2} = \frac{(\pm 5)^6}{2} = \frac{15,625}{2}$ agrees with $\frac{x^8}{50} = \frac{(\pm 5)^8}{50} = \frac{390,625}{50} = \frac{390,625/25}{50/25} = \frac{15,625}{2}$

7) $\frac{x^2}{2} = \frac{128}{x^2} \rightarrow x^4 = 256 \rightarrow x = 256^{1/4} \rightarrow x = \boxed{\pm 4}$

Check: $\frac{x^2}{2} = \frac{(\pm 4)^2}{2} = \frac{16}{2} = 8$ agrees with $\frac{128}{x^2} = \frac{128}{(\pm 4)^2} = \frac{128}{16} = 8$

8) $\frac{54}{x} = \frac{x^2}{4} \rightarrow 216 = x^3 \rightarrow \sqrt[3]{216} = x \rightarrow \boxed{6} = x$

Check: $\frac{54}{x} = \frac{54}{6} = 9$ agrees with $\frac{x^2}{4} = \frac{6^2}{4} = \frac{36}{4} = 9$

9) $\frac{4}{x} = -\frac{6}{x^2} \to 4x^2 = -6x \to \frac{4x^2}{x} = -6 \to 4x = -6 \to x = -\frac{6}{4} \to x = \boxed{-\frac{3}{2}}$

Check: $\frac{4}{x} = \frac{4}{-3/2} = 4\left(-\frac{2}{3}\right) = -\frac{8}{3}$ agrees with $-\frac{6}{x^2} = -\frac{6}{\left(-\frac{3}{2}\right)^2} = -\frac{6}{9/4} = -6\left(\frac{4}{9}\right) = -\frac{24}{9} = -\frac{8}{3}$

Recall from Sec. 1.7 that the way to divide by a fraction is to multiply by its reciprocal.

Note that $-\frac{24}{9} = -\frac{24/3}{9/3} = -\frac{8}{3}$

10) $\frac{4}{x^2} = 196 \to \frac{4}{x^2} = \frac{196}{1} \to 4 = 196x^2 \to \frac{4}{196} = x^2 \to \frac{1}{49} = x^2 \to \pm\sqrt{\frac{1}{49}} = x \to \boxed{\pm\frac{1}{7}} = x$

Check: $\frac{4}{x^2} = \frac{4}{(\pm1/7)^2} = \frac{4}{1/49} = 4\left(\frac{49}{1}\right) = \frac{196}{1} = 196$

Recall from Sec. 1.7 that the way to divide by a fraction is to multiply by its reciprocal.

11) $\frac{x}{16} = -\frac{32}{x^2} \to x^3 = -512 \to x = \sqrt[3]{-512} \to x = \boxed{-8}$

Check: $\frac{x}{16} = \frac{-8}{16} = -\frac{1}{2}$ agrees with $-\frac{32}{x^2} = -\frac{32}{(-8)^2} = -\frac{32}{64} = -\frac{32/32}{64/32} = -\frac{1}{2}$

12) $\frac{x^2}{6} = \frac{x^5}{6000} \to 6000x^2 = 6x^5 \to 6000 = \frac{6x^5}{x^2} \to 6000 = 6x^3$

$\frac{6000}{6} = x^3 \to 1000 = x^3 \to \sqrt[3]{1000} = x \to \boxed{10} = x$ or $x = \boxed{0}$ (see Sec. 5.11)

Check: $\frac{x^2}{6} = \frac{10^2}{6} = \frac{100}{6} = \frac{100/2}{6/2} = \frac{50}{3}$ agrees with $\frac{x^5}{6000} = \frac{10^5}{6000} = \frac{100,000}{6000} = \frac{100,000/2000}{6000/2000} = \frac{50}{3}$

13) $-\frac{27}{x^2} = -\frac{x^2}{3} \to -81 = -x^4 \to 81 = x^4 \to \pm\sqrt[4]{81} = x \to \boxed{\pm3} = x$

Check: $-\frac{27}{x^2} = -\frac{27}{(\pm3)^2} = -\frac{27}{9} = -3$ agrees with $-\frac{x^2}{3} = -\frac{(\pm3)^2}{3} = -\frac{9}{3} = -3$

14) $\frac{18}{x^5} = \frac{2}{x^7} \to 18x^7 = 2x^5 \to \frac{18x^7}{x^5} = 2 \to 18x^2 = 2 \to x^2 = \frac{2}{18} \to x^2 = \frac{1}{9} \to x = \pm\sqrt{\frac{1}{9}} \to x = \boxed{\pm\frac{1}{3}}$

Check: $\frac{18}{x^5} = \frac{18}{(\pm1/3)^5} = \frac{18}{\pm1/243} = \pm18\left(\frac{243}{1}\right) = \pm\frac{4374}{1} = \pm4374$

agrees with $\frac{2}{x^7} = \frac{2}{(\pm1/3)^7} = \frac{2}{\pm1/2187} = \pm2\left(\frac{2187}{1}\right) = \pm4374$

Recall from Sec. 1.7 that the way to divide by a fraction is to multiply by its reciprocal.

15) $\frac{x^5}{18} = \frac{x^7}{8} \to 8x^5 = 18x^7 \to 8 = \frac{18x^7}{x^5} \to 8 = 18x^2 \to \frac{8}{18} = x^2 \to \frac{4}{9} = x^2 \to \pm\sqrt{\frac{4}{9}} = x \to \boxed{\pm\frac{2}{3}} = x$

Check: $\frac{x^5}{18} = \frac{(\pm2/3)^5}{18} = \pm\frac{2^5}{3^5}\frac{1}{18} = \pm\frac{32}{243(18)} = \pm\frac{32}{4374} = \pm\frac{16}{2187}$ or $x = \boxed{0}$ (see Sec. 5.11)

agrees with $\frac{x^7}{8} = \frac{(\pm2/3)^7}{8} = \pm\frac{2^7}{3^7}\frac{1}{8} = \pm\frac{128}{2187(8)} = \pm\frac{128}{17,496} = \pm\frac{128\div8}{17,496\div8} = \pm\frac{16}{2187}$

Recall from Sec. 1.7 that the way to divide by a fraction is to multiply by its reciprocal.

16) $-\frac{4}{x^{2/3}} = \frac{36}{x} \rightarrow -4x = 36x^{2/3} \rightarrow -\frac{4x}{x^{2/3}} = 36 \rightarrow -4x^{1/3} = 36$

$\rightarrow x^{1/3} = \frac{36}{-4} \rightarrow x^{1/3} = -9 \rightarrow x^3 = (-9)^3 \rightarrow x^3 = \boxed{-729}$

Check: $-\frac{4}{2^{2/3}} = -\frac{4}{(-729)^{2/3}} = -\frac{4}{\left(\sqrt[3]{-729}\right)^2} = -\frac{4}{(-9)^2} = -\frac{4}{81}$

agrees with $\frac{36}{x} = \frac{36}{-729} = -\frac{36/9}{729/9} = -\frac{4}{81}$

Exercise Set 7.4

1) $\frac{x+8}{12} = \frac{5}{4} \rightarrow 4(x+8) = 5(12) \rightarrow 4x + 32 = 60 \rightarrow 4x = 28 \rightarrow x = \frac{28}{4} \rightarrow x = \boxed{7}$

Check: $\frac{x+8}{12} = \frac{7+8}{12} = \frac{15}{12} = \frac{15/3}{12/3} = \frac{5}{4}$

2) $\frac{2}{3} = \frac{5x-9}{9} \rightarrow 2(9) = 3(5x-9) \rightarrow 18 = 15x - 27 \rightarrow 45 = 15x \rightarrow \frac{45}{15} = x \rightarrow \boxed{3} = x$

Check: $\frac{5x-9}{9} = \frac{5(3)-9}{9} = \frac{15-9}{9} = \frac{6}{9} = \frac{6/3}{9/3} = \frac{2}{3}$

3) $\frac{3x-5}{6} = \frac{2x^2}{4x+7} \rightarrow (3x-5)(4x+7) = 6(2x^2) \rightarrow 12x^2 + 21x - 20x - 35 = 12x^2$

$\rightarrow 12x^2 + x - 35 = 12x^2 \rightarrow x - 35 = 0 \rightarrow x = \boxed{35}$

Check: $\frac{3x-5}{6} = \frac{3(35)-5}{6} = \frac{105-5}{6} = \frac{100}{6} = \frac{100/2}{6/2} = \frac{50}{3}$

agrees with $\frac{2x^2}{4x+7} = \frac{2(35)^2}{4(35)+7} = \frac{2(1225)}{140+7} = \frac{2450}{147} = \frac{2450/49}{147/49} = \frac{50}{3}$

4) $\frac{x^2+1}{6} = \frac{5}{3} \rightarrow 3(x^2+1) = 5(6) \rightarrow 3x^2 + 3 = 30 \rightarrow 3x^2 = 27$

$\rightarrow x^2 = \frac{27}{3} \rightarrow x^2 = 9 \rightarrow x = \pm\sqrt{9} \rightarrow x = \boxed{\pm 3}$

Check: $\frac{x^2+1}{6} = \frac{(\pm 3)^2+1}{6} = \frac{9+1}{6} = \frac{10}{6} = \frac{10/2}{6/2} = \frac{5}{3}$

5) $\frac{5}{3x+4} = \frac{2}{3} \rightarrow 5(3) = 2(3x+4) \rightarrow 15 = 6x + 8 \rightarrow 7 = 6x \rightarrow \boxed{\frac{7}{6}} = x$

Check: $\frac{5}{3x+4} = \frac{5}{3\left(\frac{7}{6}\right)+4} = \frac{5}{\frac{21}{6}+\frac{24}{6}} = \frac{5/1}{45/6} = \frac{5}{1}\frac{6}{45} = \frac{30}{45} = \frac{30/15}{45/15} = \frac{2}{3}$

6) $\frac{x}{x-4} = \frac{x+9}{5} \rightarrow 5x = (x+9)(x-4) \rightarrow 5x = x^2 - 4x + 9x - 36 \rightarrow 5x = x^2 + 5x - 36$

$\rightarrow 0 = x^2 - 36 \rightarrow 36 = x^2 \rightarrow \pm\sqrt{36} = x \rightarrow \boxed{\pm 6} = x$

Check: $\frac{x}{x-4} = \frac{6}{6-4} = \frac{6}{2} = 3$ agrees with $\frac{x+9}{5} = \frac{6+9}{5} = \frac{15}{5} = 3$

and $\frac{x}{x-4} = \frac{-6}{-6-4} = \frac{-6}{-10} = \frac{6}{10} = \frac{6/2}{10/2} = \frac{3}{5}$ agrees with $\frac{x+9}{5} = \frac{-6+9}{5} = \frac{3}{5}$

7) $\frac{28}{3x-2} = 4 \rightarrow 28 = 4(3x-2) \rightarrow 28 = 12x - 8 \rightarrow 36 = 12x \rightarrow \frac{36}{12} = x \rightarrow \boxed{3} = x$

Check: $\frac{28}{3x-2} = \frac{28}{3(3)-2} = \frac{28}{9-2} = \frac{28}{7} = 4$

8) $\frac{x+5}{x+2} = \frac{x+9}{x+5} \rightarrow (x+5)(x+5) = (x+9)(x+2)$

$\rightarrow x^2 + 5x + 5x + 25 = x^2 + 2x + 9x + 18 \rightarrow x^2 + 10x + 25 = x^2 + 11x + 18$

$\rightarrow 10x + 25 = 11x + 18 \rightarrow 25 = x + 18 \rightarrow \boxed{7} = x$

Check: $\frac{x+5}{x+2} = \frac{7+5}{7+2} = \frac{12}{9} = \frac{12/3}{9/3} = \frac{4}{3}$ agrees with $\frac{x+9}{x+5} = \frac{7+9}{7+5} = \frac{16}{12} = \frac{16/4}{12/4} = \frac{4}{3}$

Exercise Set 7.5

1) $\frac{3}{4x} - \frac{2}{3x} = \frac{3}{4x}\frac{3}{3} - \frac{2}{3x}\frac{4}{4} = \frac{9}{12x} - \frac{8}{12x} = \frac{9-8}{12x} = \boxed{\frac{1}{12x}}$

2) $\frac{x}{6} + \frac{4}{x} = \frac{x}{6}\frac{x}{x} + \frac{4}{x}\frac{6}{6} = \frac{x^2}{6x} + \frac{24}{6x} = \boxed{\frac{x^2+24}{6x}}$

3) $\frac{2}{x} + \frac{1}{x+1} = \frac{2}{x}\frac{x+1}{x+1} + \frac{1}{x+1}\frac{x}{x} = \frac{2(x+1)}{x(x+1)} + \frac{1x}{x(x+1)} = \frac{2x+2}{x^2+x} + \frac{x}{x^2+x} = \frac{2x+2+x}{x^2+x} = \boxed{\frac{3x+2}{x^2+x}}$

4) $\frac{8}{x-3} - \frac{6}{x+3} = \frac{8}{x-3}\frac{x+3}{x+3} - \frac{6}{x+3}\frac{x-3}{x-3} = \frac{8(x+3)}{(x-3)(x+3)} - \frac{6(x-3)}{(x-3)(x+3)} = \frac{8x+24}{x^2+3x-3x-9} - \frac{6x-18}{x^2+3x-3x-9}$

$= \frac{8x+24-(6x-18)}{x^2-9} = \frac{8x+24-6x-(-18)}{x^2-9} = \frac{8x+24-6x+18}{x^2-9} = \boxed{\frac{2x+42}{x^2-9}}$ Alternate: $\frac{2(x+21)}{(x-3)(x+3)}$

5) $\frac{5}{3x^2} - \frac{2}{5x^4} = \frac{5}{3x^2}\frac{5x^2}{5x^2} - \frac{2}{5x^4}\frac{3}{3} = \frac{25x^2}{15x^4} - \frac{6}{15x^4} = \boxed{\frac{25x^2-6}{15x^4}}$

6) $\frac{x}{2} + \frac{4}{3x+5} = \frac{x}{2}\frac{3x+5}{3x+5} + \frac{4}{3x+5}\frac{2}{2} = \frac{x(3x+5)}{2(3x+5)} + \frac{8}{2(3x+5)} = \frac{3x^2+5x}{6x+10} + \frac{8}{6x+10} = \boxed{\frac{3x^2+5x+8}{6x+10}}$

7) $\frac{5}{x-8} + \frac{3}{x+5} = \frac{5}{x-8}\frac{x+5}{x+5} + \frac{3}{x+5}\frac{x-8}{x-8} = \frac{5(x+5)}{(x-8)(x+5)} + \frac{3(x-8)}{(x-8)(x+5)} = \frac{5x+25}{x^2+5x-8x-40} + \frac{3x-24}{x^2+5x-8x-40}$

$= \frac{5x+25+3x-24}{x^2-3x-40} = \boxed{\frac{8x+1}{x^2-3x-40}}$

8) $\frac{6}{2x-1} - \frac{3}{x+7} = \frac{6}{2x-1}\frac{x+7}{x+7} - \frac{3}{x+7}\frac{2x-1}{2x-1} = \frac{6(x+7)}{(2x-1)(x+7)} - \frac{3(2x-1)}{(2x-1)(x+7)} = \frac{6x+42}{2x^2+14x-x-7} - \frac{6x-3}{2x^2+14x-x-7}$

$= \frac{6x+42-(6x-3)}{2x^2+13x-7} = \frac{6x+42-6x-(-3)}{2x^2+13x-7} = \frac{6x+42-6x+3}{2x^2+13x-7} = \boxed{\frac{45}{2x^2+13x-7}}$

Exercise Set 7.6

1) $\frac{3}{8} + \frac{7}{4x} = \frac{2}{3}$ Multiply by $24x$ on both sides: $\frac{72x}{8} + \frac{168x}{4x} = \frac{48x}{3}$ Simplify: $9x + 42 = 16x$

Subtract $9x$ from both sides: $42 = 7x$ Divide by 7 on both sides: $\boxed{6} = x$

Check: $\frac{3}{8} + \frac{7}{4x} = \frac{3}{8} + \frac{7}{4(6)} = \frac{9}{24} + \frac{7}{24} = \frac{16}{24} = \frac{16/8}{24/8} = \frac{2}{3}$

2) $\frac{5}{4x} = \frac{3}{4} - \frac{5}{2x}$ Multiply by $4x$ on both sides: $\frac{20x}{4x} = \frac{12x}{4} - \frac{20x}{2x}$ Simplify: $5 = 3x - 10$

Add 10 to both sides: $15 = 3x$ Divide by 3 on both sides: $\boxed{5} = x$

Check: $\frac{5}{4x} = \frac{5}{4(5)} = \frac{5}{20} = \frac{1}{4}$ agrees with $\frac{3}{4} - \frac{5}{2x} = \frac{3}{4} - \frac{5}{2(5)} = \frac{3}{4} - \frac{5}{10} = \frac{3}{4} - \frac{1}{2} = \frac{3}{4} - \frac{2}{4} = \frac{1}{4}$

3) $\frac{16}{3x^2} + \frac{5}{3} = 3$ Multiply by $3x^2$ on both sides: $\frac{48x^2}{3x^2} + \frac{15x^2}{3} = 9x^2$ Simplify: $16 + 5x^2 = 9x^2$

Subtract $5x^2$ from both sides: $16 = 4x^2$ Divide by 4 on both sides: $4 = x^2$

Square root both sides: $\pm\sqrt{4} = x$ Simplify: $\boxed{\pm 2} = x$

Check: $\frac{16}{3x^2} + \frac{5}{3} = \frac{16}{3(\pm 2)^2} + \frac{5}{3} = \frac{16}{3(4)} + \frac{5}{3} = \frac{16}{12} + \frac{20}{12} = \frac{36}{12} = 3$

4) $\frac{4}{x} + \frac{3}{x^2} = \frac{13}{3x}$ Multiply by $3x^2$ on both sides: $\frac{12x^2}{x} + \frac{9x^2}{x^2} = \frac{39x^2}{3x}$ Simplify: $12x + 9 = 13x$

Subtract $12x$ from both sides: $\boxed{9} = x$

Check: $\frac{4}{x} + \frac{3}{x^2} = \frac{4}{9} + \frac{3}{9^2} = \frac{36}{81} + \frac{3}{81} = \frac{39}{81} = \frac{39/3}{81/3} = \frac{13}{27}$ agrees with $\frac{13}{3x} = \frac{13}{3(9)} = \frac{13}{27}$

5) $\frac{5x}{6} = \frac{5x}{8} - \frac{5}{4}$ Multiply by 24 on both sides: $\frac{120x}{6} = \frac{120x}{8} - \frac{120}{4}$ Simplify: $20x = 15x - 30$

Subtract $15x$ from both sides: $5x = -30$ Divide by 5 on both sides: $x = \boxed{-6}$

Check: $\frac{5x}{6} = \frac{5(-6)}{6} = -\frac{30}{6} = -5$ agrees with $\frac{5x}{8} - \frac{5}{4} = \frac{5(-6)}{8} - \frac{5}{4} = -\frac{30}{8} - \frac{10}{8} = -\frac{40}{8} = -5$

6) $\frac{4}{x-4} + \frac{7}{6} = \frac{5}{2}$ Multiply by $6(x-4)$ on both sides: $\frac{24(x-4)}{x-4} + \frac{42(x-4)}{6} = \frac{30(x-4)}{2}$

Simplify the above equation: $24 + 7(x-4) = 15(x-4)$

Distribute: $24 + 7x - 28 = 15x - 60$ Add 60 to both sides: $24 + 7x - 28 + 60 = 15x$

Subtract $7x$ from both sides: $24 - 28 + 60 = 8x$ Simplify: $56 = 8x$

Divide by 8 on both sides: $\frac{56}{8} = \boxed{7} = x$

Check: $\frac{4}{x-4} + \frac{7}{6} = \frac{4}{7-3} + \frac{7}{6} = \frac{4}{3} + \frac{7}{6} = \frac{8}{6} + \frac{7}{6} = \frac{15}{6} = \frac{15/3}{6/3} = \frac{5}{2}$

7) $\frac{1}{x+3} + \frac{1}{x-3} = \frac{18}{x^2-9}$ Multiply by $(x+3)(x-3)$ on both sides: $\frac{(x+3)(x-3)}{x+3} + \frac{(x+3)(x-3)}{x-3} = 18$

Note that $(x+3)(x-3) = x^2 - 9$ such that $\frac{18(x+3)(x-3)}{x^2-9} = \frac{18(x^2-9)}{x^2-9} = 18$

Simplify the top right equation: $x - 3 + x + 3 = 18$ (Note that $\frac{x+3}{x+3} = 1$ and $\frac{x-3}{x-3} = 1$)

Combine like terms: $2x + 0 = 18$ Divide by 2 on both sides: $x = \boxed{9}$

Check: $\frac{1}{x+3} + \frac{1}{x-3} = \frac{1}{9+3} + \frac{1}{9-3} = \frac{1}{12} + \frac{1}{6} = \frac{1}{12} + \frac{2}{12} = \frac{3}{12} = \frac{3/3}{12/3} = \frac{1}{4}$

agrees with $\frac{18}{x^2-9} = \frac{18}{9^2-9} = \frac{18}{81-9} = \frac{18}{72} = \frac{18/18}{72/18} = \frac{1}{4}$

8) $\frac{3}{x+8} - \frac{2}{3x} = \frac{5}{6x}$ Multiply by $6x(x+8)$ on both sides: $\frac{18x(x+8)}{x+8} - \frac{12x(x+8)}{3x} = \frac{30x(x+8)}{6x}$

Simplify: $18x - 4(x+8) = 5(x+8)$

Distribute: $18x - 4x - 32 = 5x + 40$

Subtract $5x$ from both sides: $18x - 4x - 32 - 5x = 40$

Add 32 to both sides: $18x - 4x - 5x = 40 + 32$

Combine like terms: $9x = 72$

Divide by 9 on both sides: $x = \frac{72}{9} = \boxed{8}$

Check: $\frac{3}{x+8} - \frac{2}{3x} = \frac{3}{8+8} - \frac{2}{3(8)} = \frac{3}{16} - \frac{2}{24} = \frac{9}{48} - \frac{4}{48} = \frac{5}{48}$

agrees with $\frac{5}{6x} = \frac{5}{6(8)} = \frac{5}{48}$

Exercise Set 7.7

1) $\frac{x-16}{3} = -\frac{21}{x}$ Cross multiply: $x(x-16) = -3(21)$ Distribute: $x^2 - 16x = -63$

Put in standard form: $x^2 - 16x + 63 = 0$ Identify: $a = 1, b = -16, c = 63$

$x = \frac{-b \pm \sqrt{b^2-4ac}}{2a} = \frac{-(-16)\pm\sqrt{(-16)^2-4(1)(63)}}{2(1)} = \frac{16\pm\sqrt{256-252}}{2} = \frac{16\pm\sqrt{4}}{2} = \frac{16\pm2}{2}$

$x = \frac{16+2}{2} = \frac{18}{2} = \boxed{9}$ or $x = \frac{16-2}{2} = \frac{14}{2} = \boxed{7}$

Check: $\frac{x-16}{3} = \frac{9-16}{3} = -\frac{7}{3}$ agrees with $-\frac{21}{x} = -\frac{21}{9} = -\frac{21/3}{9/3} = -\frac{7}{3}$

and $\frac{x-16}{3} = \frac{7-16}{3} = -\frac{9}{3} = -3$ agrees with $-\frac{21}{x} = -\frac{21}{7} = -3$

2) $\frac{5}{x} = \frac{x}{x+10}$ Cross multiply: $5(x + 10) = x^2$ Distribute: $5x + 50 = x^2$

Put in standard form: $-x^2 + 5x + 50 = 0$ Identify: Identify: $a = -1, b = 5, c = 50$

$$x = \frac{-b\pm\sqrt{b^2-4ac}}{2a} = \frac{-5\pm\sqrt{5^2-4(-1)(50)}}{2(-1)} = \frac{-5\pm\sqrt{25+200}}{-2} = \frac{-5\pm\sqrt{225}}{-2} = \frac{-5\pm15}{-2}$$

$$x = \frac{-5+15}{-2} = \frac{10}{-2} = \boxed{-5} \text{ or } x = \frac{-5-15}{-2} = \frac{-20}{-2} = \boxed{10}$$

Check: $\frac{5}{x} = \frac{5}{-5} = -1$ agrees with $\frac{x}{x+10} = \frac{-5}{-5+10} = \frac{-5}{5} = -1$

and $\frac{5}{x} = \frac{5}{10} = \frac{1}{2}$ agrees with $\frac{x}{x+10} = \frac{10}{10+10} = \frac{10}{20} = \frac{1}{2}$

3) $\frac{1}{4} - \frac{2}{x^2} = \frac{7}{4x}$ Multiply by $4x^2$ on both sides: $\frac{4x^2}{4} - \frac{8x^2}{x^2} = \frac{28x^2}{4x}$ Simplify: $x^2 - 8 = 7x$

Put in standard form: $x^2 - 7x - 8 = 0$ Identify: Identify: $a = 1, b = -7, c = -8$

$$x = \frac{-b\pm\sqrt{b^2-4ac}}{2a} = \frac{-(-7)\pm\sqrt{(-7)^2-4(1)(-8)}}{2(1)} = \frac{7\pm\sqrt{49+32}}{2} = \frac{7\pm\sqrt{81}}{2} = \frac{7\pm9}{2}$$

$$x = \frac{7+9}{2} = \frac{16}{2} = \boxed{8} \text{ or } x = \frac{7-9}{2} = \frac{-2}{2} = \boxed{-1}$$

Check: $\frac{1}{4} - \frac{2}{x^2} = \frac{1}{4} - \frac{2}{8^2} = \frac{16}{64} - \frac{2}{64} = \frac{14}{64} = \frac{7}{32}$ agrees with $\frac{7}{4x} = \frac{7}{4(8)} = \frac{7}{32}$

and $\frac{1}{4} - \frac{2}{x^2} = \frac{1}{4} - \frac{2}{(-1)^2} = \frac{1}{4} - \frac{2}{1} = \frac{1}{4} - \frac{8}{4} = -\frac{7}{4}$ agrees with $\frac{7}{4x} = \frac{7}{4(-1)} = -\frac{7}{4}$

4) $\frac{x}{x-5} = \frac{x-12}{7x+1}$ Cross multiply: $x(7x + 1) = (x - 5)(x - 12)$

Distribute: $7x^2 + x = x^2 - 12x - 5x + 60$ Subtract $7x^2$ and x from both sides:

$0 = x^2 - 12x - 5x + 60 - 7x^2 - x$ Combine like terms: $0 = -6x^2 - 18x + 60$

Identify: $a = -6, b = -18, c = 60$

$$x = \frac{-b\pm\sqrt{b^2-4ac}}{2a} = \frac{-(-18)\pm\sqrt{(-18)^2-4(-6)(60)}}{2(-6)} = \frac{18\pm\sqrt{324+1440}}{-12} = \frac{18\pm\sqrt{1764}}{-12} = \frac{18\pm42}{-12}$$

$$x = \frac{18+42}{-12} = \frac{60}{-12} = \boxed{-5} \text{ or } x = \frac{18-42}{-12} = \frac{-24}{-12} = \boxed{2}$$

Check: $\frac{x}{x-5} = \frac{-5}{-5-5} = \frac{-5}{-10} = \frac{1}{2}$ agrees with $\frac{x-12}{7x+1} = \frac{-5-12}{7(-5)+1} = \frac{-17}{-35+1} = \frac{-17}{-34} = \frac{1}{2}$

and $\frac{x}{x-5} = \frac{2}{2-5} = \frac{2}{-3} = -\frac{2}{3}$ agrees with $\frac{x-12}{7x+1} = \frac{2-12}{7(2)+1} = \frac{-10}{14+1} = -\frac{10}{15} = -\frac{2}{3}$

5) $\frac{1}{2} + 3x = \frac{1}{2x}$ Multiply by $2x$ on both sides: $\frac{2x}{2} + 3x(2x) = \frac{2x}{2x}$ Simplify: $x + 6x^2 = 1$

Put in standard form: $6x^2 + x - 1 = 0$ Identify: $a = 6, b = 1, c = -1$

$$x = \frac{-b \pm \sqrt{b^2 - 4ac}}{2a} = \frac{-1 \pm \sqrt{1^2 - 4(6)(-1)}}{2(6)} = \frac{-1 \pm \sqrt{1+24}}{12} = \frac{-1 \pm \sqrt{25}}{12} = \frac{-1 \pm 5}{12}$$

$$x = \frac{-1+5}{12} = \frac{4}{12} = \boxed{\frac{1}{3}} \text{ or } x = \frac{-1-5}{12} = \frac{-6}{12} = \boxed{-\frac{1}{2}}$$

Check: $\frac{1}{2} + 3x = \frac{1}{2} + 3\left(\frac{1}{3}\right) = \frac{1}{2} + 1 = \frac{1}{2} + \frac{2}{2} = \frac{3}{2}$ agrees with $\frac{1}{2x} = \frac{1}{2\left(\frac{1}{3}\right)} = \frac{1}{2}\left(\frac{3}{1}\right) = \frac{3}{2}$

and $\frac{1}{2} + 3x = \frac{1}{2} + 3\left(-\frac{1}{2}\right) = \frac{1}{2} - \frac{3}{2} = -\frac{2}{2} = -1$ agrees with $\frac{1}{2x} = \frac{1}{2\left(-\frac{1}{2}\right)} = \frac{1}{-1} = -1$

Note: $\frac{1}{2\left(\frac{1}{3}\right)} = \frac{1}{2} \div \frac{1}{3} = \frac{1}{2}\frac{3}{1} = \frac{3}{2}$ (To divide by a fraction, multiply by its reciprocal.)

6) $\frac{24}{x^5} + \frac{1}{x^3} = \frac{11}{x^4}$ Multiply by x^5 on both sides: $\frac{24x^5}{x^5} + \frac{x^5}{x^3} = \frac{11x^5}{x^4}$ Simplify: $24 + x^2 = 11x$

Put in standard form: $x^2 - 11x + 24 = 0$ Identify: $a = 1, b = -11, c = 24$

$$x = \frac{-b \pm \sqrt{b^2 - 4ac}}{2a} = \frac{-(-11) \pm \sqrt{(-11)^2 - 4(1)(24)}}{2(1)} = \frac{11 \pm \sqrt{121 - 96}}{2} = \frac{11 \pm \sqrt{25}}{2} = \frac{11 \pm 5}{2}$$

$$x = \frac{11+5}{2} = \frac{16}{2} = \boxed{8} \text{ or } x = \frac{11-5}{2} = \frac{6}{2} = \boxed{3}$$

Check: $\frac{24}{x^5} + \frac{1}{x^3} = \frac{24}{8^5} + \frac{1}{8^3} = \frac{3(8)}{8^5} + \frac{1}{8^3} = \frac{3}{8^4} + \frac{8}{8^4} = \frac{11}{8^4}$ agrees with $\frac{11}{x^4} = \frac{11}{8^4}$

(or use a calculator to see that both sides equal approximately 0.0026855)

and $\frac{24}{x^5} + \frac{1}{x^3} = \frac{24}{3^5} + \frac{1}{3^3} = \frac{8(3)}{3^5} + \frac{1}{3^3} = \frac{8}{3^4} + \frac{3}{3^4} = \frac{11}{3^4}$ agrees with $\frac{11}{x^4} = \frac{11}{3^4}$

(or use a calculator to see that both sides equal approximately 0.13580)

7) $\frac{x+6}{x+8} = \frac{5x}{x-6}$ Cross multiply: $(x+6)(x-6) = 5x(x+8)$

Distribute: $x^2 - 6x + 6x - 36 = 5x^2 + 40x$ Simplify: $x^2 - 36 = 5x^2 + 40x$

Put in standard form: $-4x^2 - 40x - 36 = 0$

Optional step: Divide by -4 on both sides: $x^2 + 10x + 9 = 0$

Identify: $a = 1, b = 10, c = 9$

$$x = \frac{-b \pm \sqrt{b^2 - 4ac}}{2a} = \frac{-10 \pm \sqrt{10^2 - 4(1)(9)}}{2(1)} = \frac{-10 \pm \sqrt{100 - 36}}{2} = \frac{-10 \pm \sqrt{64}}{2} = \frac{-10 \pm 8}{2}$$

$$x = \frac{-10+8}{2} = \frac{-2}{2} = \boxed{-1} \text{ or } x = \frac{-10-8}{2} = \frac{-18}{2} = \boxed{-9}$$

Check: $\frac{x+6}{x+8} = \frac{-1+6}{-1+8} = \frac{5}{7}$ agrees with $\frac{5x}{x-6} = \frac{5(-1)}{-1-6} = \frac{-5}{-7} = \frac{5}{7}$

and $\frac{x+6}{x+8} = \frac{-9+6}{-9+8} = \frac{-3}{-1} = 3$ agrees with $\frac{5x}{x-6} = \frac{5(-9)}{-9-6} = \frac{-45}{-15} = 3$

8) $\frac{15-x^2}{4x^2} = \frac{1}{2x}$ Cross multiply: $2x(15 - x^2) = 4x^2$ Distribute: $30x - 2x^3 = 4x^2$

Subtract $4x^2$ from both sides and reorder the terms: $-2x^3 - 4x^2 + 30x = 0$

Factor out $-2x$ like we did in Chapter 5: $-2x(x^2 + 2x - 15) = 0$

There are two possibilities: Either $-2x = 0$ or $x^2 + 2x - 15 = 0$ (We applied similar reasoning in Sec. 5.11). One solution is $x = 0$, but this causes a problem with division by zero (which is undefined) in the original equation.

For the other solutions, we need to solve the quadratic equation $x^2 + 2x - 15 = 0$.
Identify $a = 1, b = 2, c = -15$

$$x = \frac{-b \pm \sqrt{b^2 - 4ac}}{2a} = \frac{-2 \pm \sqrt{2^2 - 4(1)(-15)}}{2(1)} = \frac{-2 \pm \sqrt{4 + 60}}{2} = \frac{-2 \pm \sqrt{64}}{2} = \frac{-2 \pm 8}{2}$$

$$x = \frac{-2+8}{2} = \frac{6}{2} = \boxed{3} \text{ or } x = \frac{-2-8}{2} = \frac{-10}{2} = \boxed{-5}$$

Check: $\frac{15-x^2}{4x^2} = \frac{15-3^2}{4(3)^2} = \frac{15-9}{4(9)} = \frac{6}{36} = \frac{1}{6}$ agrees with $\frac{1}{2x} = \frac{1}{2(3)} = \frac{1}{6}$

and $\frac{15-x^2}{4x^2} = \frac{15-(-5)^2}{4(-5)^2} = \frac{15-25}{4(25)} = \frac{-10}{100} = -\frac{1}{10}$ agrees with $\frac{1}{2x} = \frac{1}{2(-5)} = -\frac{1}{10}$

Exercise Set 7.8

1) $x = $ the number of oranges in the barrel

The 9 and 45 correspond to apples, while the 5 and x correspond to oranges.

Put the apples in the numerator: $\frac{9}{5} = \frac{45}{x}$ Cross multiply: $9x = 5(45)$ Simplify: $9x = 225$

Divide by 9 on both sides: $x = \frac{225}{9} = \boxed{25}$ oranges

Check: $\frac{45}{x} = \frac{45}{25} = \frac{45/5}{25/5} = \frac{9}{5}$ It should make sense that there are fewer oranges.

2) $x = $ the number of children's tickets sold

The parts add up to the whole: $3 + 7 = 10$ The ratio of children's tickets to the total number of tickets is 3:10.

The 3 and x correspond to children's tickets, while the 10 and 150 correspond to total tickets. Put the children's tickets in the numerator: $\frac{3}{10} = \frac{x}{150}$ Cross multiply: $450 = 10x$

Divide by 10 on both sides: $\boxed{45} = x$ is the number of adult tickets sold.

The number of adult tickets is $150 - 45 = \boxed{105}$. Check: $\frac{x}{150} = \frac{45}{150} = \frac{45/15}{150/15} = \frac{3}{10}$

Note that the ratio of children's tickets to adult tickets is $\frac{45}{105} = \frac{45/15}{105/15} = \frac{3}{7}$.

3) x = the number of pennies in the jar

Since a nickel is worth 5 cents, which is $0.05, divide $3 by $0.05 to determine how many nickels are in the jar: $\frac{\$3}{\$0.05} = \frac{3}{0.05} = \frac{3(100)}{0.05(100)} = \frac{300}{5} = 60$ nickels

The 11 and x correspond to pennies, while the 4 and 60 correspond to nickels. Put the pennies in the numerator: $\frac{11}{4} = \frac{x}{60}$. Cross multiply: $660 = 4x$

Divide by 4 on both sides: $\frac{660}{4} = 165 = x$ Since a penny is worth 1 cent, which is $0.01, the value of the pennies is $165(\$0.01) = \$1.65$. Add the value of the nickels ($3) to the value of the pennies ($1.65) to determine the total value: $\$3 + \$1.65 = \boxed{\$4.65}$

Check: The ratio of pennies to nickels is $\frac{165}{60} = \frac{165/15}{60/15} = \frac{11}{4}$

4) x = the number of red buttons in the box

The ratio of red buttons to white buttons to blue buttons is 5:3:4. These are all parts. The parts add up to the whole: $5 + 3 + 4 = 12$.

The 5 and x correspond to red buttons, while the 12 and 480 correspond to the total.

Put the red buttons in the numerator: $\frac{5}{12} = \frac{x}{480}$ Cross multiply: $5(480) = 12x$

Simplify: $2400 = 12x$ Divide by 12 on both sides: $\boxed{200} = x$ are red

y = the number of white buttons in the box

The 5 and 200 correspond to red buttons, while the 3 and y correspond to yellow buttons. Put the red buttons in the numerator: $\frac{5}{3} = \frac{200}{y}$. Cross multiply: $5y = 600$

Divide by 5 on both sides: $y = \frac{600}{5} = \boxed{120}$ are white

z = the number of blue buttons in the box

The 5 and 200 correspond to red buttons, while the 4 and z correspond to blue buttons. Put the red buttons in the numerator: $\frac{5}{4} = \frac{200}{z}$. Cross multiply: $5z = 800$

Divide by 5 on both sides: $z = \frac{800}{5} = \boxed{160}$ are blue

Check: The total number of buttons is $200 + 120 + 160 = 480$.

The ratio of red buttons to white buttons is $\frac{200}{120} = \frac{200/40}{120/40} = \frac{5}{3}$.

The ratio of red buttons to blue buttons is $\frac{200}{160} = \frac{200/40}{160/40} = \frac{5}{4}$.

Exercise Set 7.9

1) First convert the time from 30 minutes to 0.5 hours so that it matches the units of the rate (mph stands for "miles per hour"). Plug $r = 48$ mph and $t = 0.5$ hr into the rate equation: $r = \frac{d}{t}$ becomes $48 = \frac{d}{0.5}$. Multiply by 0.5 on both sides: $\boxed{24}$ miles $= d$

2) We're given $d = 75$ in. and $t = 5$ s. Use the rate equation: $r = \frac{d}{t} = \frac{75}{5} = \boxed{15}$ in./s

3) We're given $d = 24$ km and $r = 16$ km/hr. Plug these into the rate equation. $r = \frac{d}{t}$ becomes $16 = \frac{24}{t}$ Multiply by t on both sides: $16t = 24$

Divide by 16 on both sides: $t = \frac{24}{16} = \frac{24/8}{16/8} = \frac{3}{2} = \boxed{1.5}$ hr.

8 Systems of Equations

Exercise Set 8.1

1) Subtract $7x$ from both sides of the top equation: $3y = 13 - 7x$

Divide by 3 on both sides: $y = \frac{13}{3} - \frac{7x}{3}$

Replace y with $\frac{13}{3} - \frac{7x}{3}$ in the bottom equation: $9x + 8\left(\frac{13}{3} - \frac{7x}{3}\right) = -4$

Distribute: $9x + \frac{104}{3} - \frac{56x}{3} = -4$ Multiply by 3 on both sides: $27x + 104 - 56x = -12$

Subtract 104 from both sides: $27x - 56x = -12 - 104$ Combine like terms: $-29x = -116$

Divide by -29 on both sides: $x = \frac{-116}{-29} = \boxed{4}$ (this is x)

Plug $x = 4$ into $y = \frac{13}{3} - \frac{7x}{3} = \frac{13}{3} - \frac{7(4)}{3} = \frac{13}{3} - \frac{28}{3} = -\frac{15}{3} = \boxed{-5}$ (this is y)

Check: $7x + 3y = 7(4) + 3(-5) = 28 - 15 = 13$

and $9x + 8y = 9(4) + 8(-5) = 36 - 40 = -4$

2) Add y to both sides of the top equation: $6x = 48 + y$ Subtract 48 from both sides:

$6x - 48 = y$ Replace y with $6x - 48$ in the bottom equation: $5x + 3(6x - 48) = 63$

Distribute: $5x + 18x - 144 = 63$ Combine like terms: $23x - 144 = 63$

Add 144 to both sides: $23x = 207$ Divide by 23 on both sides: $x = \frac{207}{23} = \boxed{9}$ (this is x)

Plug $x = 9$ into $y = 6x - 48 = 6(9) - 48 = 54 - 48 = \boxed{6}$ (this is y)

Check: $6x - y = 6(9) - 6 = 54 - 6 = 48$ and $5x + 3y = 5(9) + 3(6) = 45 + 18 = 63$

3) Subtract $5y$ from both sides of the top equation: $8x = 11 - 5y$

Divide by 8 on both sides: $x = \frac{11}{8} - \frac{5y}{8}$

Replace x with $\frac{11}{8} - \frac{5y}{8}$ in the bottom equation: $5\left(\frac{11}{8} - \frac{5y}{8}\right) - 3y = 13$

Distribute: $\frac{55}{8} - \frac{25y}{8} - 3y = 13$ Multiply by 8 on both sides: $55 - 25y - 24y = 104$

Combine like terms: $55 - 49y = 104$ Subtract 55 from both sides: $-49y = 49$

Divide by -49 on both sides: $y = \frac{49}{-49} = \boxed{-1}$ (this is y)

Plug $y = -1$ into $x = \frac{11}{8} - \frac{5y}{8} = \frac{11}{8} - \frac{5(-1)}{8} = \frac{11}{8} + \frac{5}{8} = \frac{16}{8} = \boxed{2}$ (this is x)

Check: $8x + 5y = 8(2) + 5(-1) = 16 - 5 = 11$

and $5x - 3y = 5(2) - 3(-1) = 10 + 3 = 13$

4) Add $5y$ to both sides of the top equation: $6x = 5y - 2$

Divide by 6 on both sides: $x = \frac{5y}{6} - \frac{2}{6}$ which simplifies to $x = \frac{5y}{6} - \frac{1}{3}$

Replace x with $\frac{5y}{6} - \frac{1}{3}$ in the bottom equation: $7\left(\frac{5y}{6} - \frac{1}{3}\right) - 9y = -34$

Distribute: $\frac{35y}{6} - \frac{7}{3} - 9y = -34$ Multiply by 6 on both sides: $35y - \frac{42}{3} - 54y = -204$

Simplify and combine like terms: $-19y - 14 = -204$ Multiply by -1 on both sides: $19y + 14 = 204$ Subtract 14 from both sides: $19y = 190$

Divide by 10 on both sides: $y = \frac{190}{19} = \boxed{10}$ (this is y)

Plug $y = 10$ into $x = \frac{5y}{6} - \frac{2}{6} = 5\left(\frac{10}{6}\right) - \frac{2}{6} = \frac{50}{6} - \frac{2}{6} = \frac{48}{6} = \boxed{8}$ (this is x)

Check: $6x - 5y = 6(8) - 5(10) = 48 - 50 = -2$

and $7x - 9y = 7(8) - 9(10) = 56 - 90 = -34$

5) Add $3y$ to both sides of the top equation: $7x = 3y + 35$

Divide by 7 on both sides: $x = \frac{3y}{7} + \frac{35}{7}$ which simplifies to $x = \frac{3y}{7} + 5$

Replace x with $\frac{3y}{7} + 5$ in the bottom equation: $-6\left(\frac{3y}{7} + 5\right) + 5y = -47$

Distribute (remember to distribute the $-$ sign): $-\frac{18y}{7} - 30 + 5y = -47$

Multiply by 7 on both sides: $-18y - 210 + 35y = -329$

Combine like terms: $17y - 210 = -329$ Add 210 to both sides: $17y = -119$

Divide by 17 on both sides: $y = -\frac{119}{17} = \boxed{-7}$ (this is y)

Plug $y = -7$ into $x = \frac{3y}{7} + 5 = \frac{3(-7)}{7} + 5 = -\frac{21}{7} + 5 = -3 + 5 = \boxed{2}$ (this is x)

Check: $7x - 3y = 7(2) - 3(-7) = 14 + 21 = 35$

and $-6x + 5y = -6(2) + 5(-7) = -12 - 35 = -47$

6) Add $7y$ to both sides of the top equation: $5x = 7y + 20$

Divide by 5 on both sides: $x = \frac{7y}{5} + \frac{20}{5}$ which simplifies to $x = \frac{7y}{5} + 4$

Replace x with $\frac{7y}{5} + 4$ in the bottom equation: $2\left(\frac{7y}{5} + 4\right) + 3y = 95$

Distribute: $\frac{14y}{5} + 8 + 3y = 95$ Multiply by 5 on both sides: $14y + 40 + 15y = 475$

Combine like terms: $29y + 40 = 475$ Subtract 40 from both sides: $29y = 435$

Divide by 29 on both sides: $y = \frac{435}{29} = \boxed{15}$ (this is y)

Plug $y = 15$ into $x = \frac{7y}{5} + 4 = \frac{7(15)}{5} + 4 = \frac{105}{5} + 4 = 21 + 4 = \boxed{25}$ (this is x)

Check: $5x - 7y = 5(25) - 7(15) = 125 - 105 = 20$

and $2x + 3y = 2(25) + 3(15) = 50 + 45 = 95$

7) Add $6x$ to the bottom equation: $9y = 6x + 6$ Divide by 9 on both sides: $y = \frac{6x}{9} + \frac{6}{9}$

which simplifies to $y = \frac{2x}{3} + \frac{2}{3}$ Replace y with $\frac{2x}{3} + \frac{2}{3}$ in the top equation:

$8x + 6\left(\frac{2x}{3} + \frac{2}{3}\right) = 22$ Distribute: $8x + \frac{12x}{3} + \frac{12}{3} = 22$ Simplify: $8x + 4x + 4 = 22$

Combine like terms: $12x + 4 = 22$ Subtract 4 from both sides: $12x = 18$

Divide by 12 on both sides and reduce the fraction: $x = \frac{18}{12} = \frac{18/6}{12/6} = \boxed{\frac{3}{2}}$ (this is x)

Plug $x = \frac{3}{2}$ into $y = \frac{2x}{3} + \frac{2}{3} = \frac{2}{3}\left(\frac{3}{2}\right) + \frac{2}{3} = \frac{6}{6} + \frac{4}{6} = \frac{10}{6} = \frac{10/2}{6/2} = \boxed{\frac{5}{3}}$ (this is y)

Check: $8x + 6y = 8\left(\frac{3}{2}\right) + 6\left(\frac{5}{3}\right) = \frac{24}{2} + \frac{30}{3} = 12 + 10 = 22$

and $-6x + 9y = -6\left(\frac{3}{2}\right) + 9\left(\frac{5}{3}\right) = -\frac{18}{2} + \frac{45}{3} = -9 + 15 = 6$

Exercise Set 8.2

1) Subtract $5x$ from both sides of the top equation: $2z = 24 - 5x$

Divide by 2 on both sides: $z = 12 - \frac{5x}{2}$ Replace z with $12 - \frac{5x}{2}$ in the other equations:

z only appears in one of these: $3y - 7\left(12 - \frac{5x}{2}\right) = 7$ Distribute: $3y - 84 - 7\left(-\frac{5x}{2}\right) = 7$

Simplify: $3y - 84 + \frac{35x}{2} = 7$ Multiply by 2 on both sides: $6y - 168 + 35x = 14$

Add 168 to both sides: $6y + 35x = 182$ Subtract $35x$ from both sides: $6y = 182 - 35x$

Divide by 6 on both sides: $y = \frac{182}{6} - \frac{35x}{6}$ which simplifies to $y = \frac{91}{3} - \frac{35x}{6}$

Replace y with $\frac{91}{3} - \frac{35x}{6}$ in the middle equation: $3x + 4\left(\frac{91}{3} - \frac{35x}{6}\right) = 40$

Distribute: $3x + \frac{364}{3} - \frac{140x}{6} = 40$ which simplifies to $3x + \frac{364}{3} - \frac{70x}{3} = 40$

Multiply by 3 on both sides: $9x + 364 - 70x = 120$ Combine like terms: $-61x = -244$

Divide by -61 on both sides: $x = \frac{-244}{-61} = \boxed{4}$ (this is x)

Plug $x = 4$ into $y = \frac{91}{3} - \frac{35x}{6} = \frac{91}{3} - \frac{35(4)}{6} = \frac{91}{3} - \frac{140}{6} = \frac{91}{3} - \frac{70}{3} = \frac{21}{3} = \boxed{7}$ (this is y)

Plug $x = 4$ into $z = 12 - \frac{5x}{2} = 12 - \frac{5(4)}{2} = 12 - \frac{20}{2} = 12 - 10 = \boxed{2}$ (this is z)

Check: $5x + 2z = 5(4) + 2(2) = 20 + 4 = 24$

and $3x + 4y = 3(4) + 4(7) = 12 + 28 = 40$

and $3y - 7z = 3(7) - 7(2) = 21 - 14 = 7$

2) Subtract $5x$ and $4y$ from both sides of the top equation: $3z = 90 - 5x - 4y$

Divide by 3 on both sides: $z = 30 - \frac{5x}{3} - \frac{4y}{3}$ Replace z with $30 - \frac{5x}{3} - \frac{4y}{3}$ in the other

equations: $10x - 3y - 6\left(30 - \frac{5x}{3} - \frac{4y}{3}\right) = 5$ and $9x - 4y - 5\left(30 - \frac{5x}{3} - \frac{4y}{3}\right) = 2$

Distribute: $10x - 3y - 180 + \frac{30x}{3} + \frac{24y}{3} = 5$ and $9x - 4y - 150 + \frac{25x}{3} + \frac{20y}{3} = 2$

Simplify the left equation: $10x - 3y - 180 + 10x + 8y = 5$

Multiply by 3 on both sides of the right equation: $27x - 12y - 450 + 25x + 20y = 6$

Combine like terms: $20x + 5y = 185$ and $52x + 8y = 456$

Subtract $20x$ from both sides of the left equation: $5y = 185 - 20x$ Divide by 5 on

both sides: $y = 37 - 4x$ Replace y with $37 - 4x$ in the right equation above:

$52x + 8(37 - 4x) = 456$ Distribute: $52x + 296 - 32x = 456$

Combine like terms: $20x = 160$ Divide by 20 on both sides: $x = \frac{160}{20} = \boxed{8}$ (this is x)

Plug $x = 8$ into $y = 37 - 4x = 37 - 4(8) = 37 - 32 = \boxed{5}$ (this is y)

Plug $x = 8$ and $y = 5$ into $z = 30 - \frac{5x}{3} - \frac{4y}{3} = 30 - \frac{5(8)}{3} - \frac{4(5)}{3} = 30 - \frac{40}{3} - \frac{20}{3}$

$= 30 - \frac{60}{3} = 30 - 20 = \boxed{10}$ (this is z)

Check: $5x + 4y + 3z = 5(8) + 4(5) + 3(10) = 40 + 20 + 30 = 90$

and $10x - 3y - 6z = 10(8) - 3(5) - 6(10) = 80 - 15 - 60 = 5$

and $9x - 4y - 5z = 9(8) - 4(5) - 5(10) = 72 - 20 - 50 = 2$

3) Add $5z$ to both sides of the top equation: $4x = 5z + 13$ Divide by 4 on both sides:

$x = \frac{5z}{4} + \frac{13}{4}$ Replace x with $\frac{5z}{4} + \frac{13}{4}$ in the other equations:

$5\left(\frac{5z}{4} + \frac{13}{4}\right) + 4y + 3z = 2$ and $9\left(\frac{5z}{4} + \frac{13}{4}\right) - 3y - 6z = -21$ Distribute:

$\frac{25z}{4} + \frac{65}{4} + 4y + 3z = 2$ and $\frac{45z}{4} + \frac{117}{4} - 3y - 6z = -21$ Multiply by 4 on both sides:

$25z + 65 + 16y + 12z = 8$ and $45z + 117 - 12y - 24z = -84$ Combine like terms:

$37z + 16y = -57$ and $21z - 12y = -201$ Add $12y$ to both sides of the right equation:

$21z = 12y - 201$ Divide by 21 on both sides: $z = \frac{12y}{21} - \frac{201}{21}$ Reduce: $z = \frac{4y}{7} - \frac{67}{7}$

Replace z with $\frac{4y}{7} - \frac{67}{7}$ in the equation left equation above: $37\left(\frac{4y}{7} - \frac{67}{7}\right) + 16y = -57$

Distribute: $\frac{148y}{7} - \frac{2479}{7} + 16y = -57$ Multiply by 7 on both sides:

$148y - 2479 + 112y = -399$ Combine like terms: $260y = 2080$

Divide by 260 on both sides: $y = \frac{2080}{260} = \boxed{8}$ (this is y)

Plug $y = 8$ into $z = \frac{4y}{7} - \frac{67}{7} = \frac{4(8)}{7} - \frac{67}{7} = \frac{32}{7} - \frac{67}{7} = -\frac{35}{7} = \boxed{-5}$ (this is z)

Plug $z = -5$ into $x = \frac{5z}{4} + \frac{13}{4} = \frac{5(-5)}{4} + \frac{13}{4} = -\frac{25}{4} + \frac{13}{4} = -\frac{12}{4} = \boxed{-3}$ (this is x)

Check: $4x - 5z = 4(-3) - 5(-5) = -12 + 25 = 13$

and $5x + 4y + 3z = 5(-3) + 4(8) + 3(-5) = -15 + 32 - 15 = 2$

and $9x - 3y - 6z = 9(-3) - 3(8) - 6(-5) = -27 - 24 + 30 = -21$

4) Add $-9z$ to and subtract $8y$ from both sides of the top equation: $4x = 9z - 8y - 2$

Divide by 4 on both sides: $x = \frac{9z}{4} - 2y - \frac{1}{2}$ Replace x with $\frac{9z}{4} - 2y - \frac{1}{2}$ in the other

equations: $-6\left(\frac{9z}{4} - 2y - \frac{1}{2}\right) - 4y + 3z = 10$ and $7\left(\frac{9z}{4} - 2y - \frac{1}{2}\right) + 6y + 6z = -3$

Distribute: $-\frac{54z}{4} + 12y + \frac{6}{2} - 4y + 3z = 10$ and $\frac{63z}{4} - 14y - \frac{7}{2} + 6y + 6z = -3$

Simplify the left equation: $-\frac{27z}{2} + 12y + 3 - 4y + 3z = 10$ Multiply by 2 on both

sides: $-27z + 24y + 6 - 8y + 6z = 20$ Multiply by 4 on both sides of the right

equation: $63z - 56y - 14 + 24y + 24z = -12$ Combine like terms in each equation:

$-21z + 16y = 14$ and $87z - 32y = 2$ Add $21z$ to both sides of the left equation:

$16y = 21z + 14$ Divide by 16 on both sides: $y = \frac{21z}{16} + \frac{14}{16}$ Simplify: $y = \frac{21z}{16} + \frac{7}{8}$

Replace y with $\frac{21z}{16} + \frac{7}{8}$ in the right equation from above:

$87z - 32\left(\frac{21z}{16} + \frac{7}{8}\right) = 2$ Distribute: $87z - 42z - 28 = 2$ Combine like terms:

$45z = 30$ Divide by 45 on both sides: $z = \frac{30}{45} = \frac{30/15}{45/15} = \boxed{\frac{2}{3}}$ (this is z)

Plug $z = \frac{2}{3}$ into $y = \frac{21z}{16} + \frac{7}{8} = \frac{21}{16}\left(\frac{2}{3}\right) + \frac{7}{8} = \frac{42}{48} + \frac{7}{8} = \frac{42/6}{48/6} + \frac{7}{8} = \frac{7}{8} + \frac{7}{8} = \frac{14}{8} = \boxed{\frac{7}{4}}$ (this is y)

Plug $z = \frac{2}{3}$ and $y = \frac{7}{4}$ into $x = \frac{9z}{4} - 2y - \frac{1}{2} = \frac{9}{4}\left(\frac{2}{3}\right) - 2\left(\frac{7}{4}\right) - \frac{1}{2} = \frac{18}{12} - \frac{14}{4} - \frac{1}{2}$

$= \frac{18/6}{12/6} - \frac{14/2}{4/2} - \frac{1}{2} = \frac{3}{2} - \frac{7}{2} - \frac{1}{2} = \frac{3-7-1}{2} = \boxed{-\frac{5}{2}}$ (this is x)

Check: $4x + 8y - 9z = 4\left(-\frac{5}{2}\right) + 8\left(\frac{7}{4}\right) - 9\left(\frac{2}{3}\right) = -10 + 14 - 6 = -2$

and $-6x - 4y + 3z = -6\left(-\frac{5}{2}\right) - 4\left(\frac{7}{4}\right) + 3\left(\frac{2}{3}\right) = 15 - 7 + 2 = 10$

and $7x + 6y + 6z = 7\left(-\frac{5}{2}\right) + 6\left(\frac{7}{4}\right) + 6\left(\frac{2}{3}\right) = -\frac{35}{2} + \frac{21}{2} + 4 = -\frac{14}{2} + 4 = -7 + 4 = -3$

Exercise Set 8.3

1) $D = 2R \rightarrow R = \frac{D}{2}$ such that $A = \pi R^2 = \pi \left(\frac{D}{2}\right)^2 = \boxed{\frac{\pi D^2}{4}}$

Recall the rule $(ax)^n = a^n x^n$ from Chapter 3 (where $a = \frac{1}{2}$).

2) $V = IR \rightarrow I = \frac{V}{R}$ such that $P = IV = \left(\frac{V}{R}\right)V = \frac{V^2}{R}$

for which $P = \frac{V^2}{R} \rightarrow PR = V^2 \rightarrow \boxed{R = \frac{V^2}{P}}$

3) $S = 6L^2 \rightarrow \frac{S}{6} = L^2 \rightarrow \sqrt{\frac{S}{6}} = L$ such that $V = L^3 = \left(\sqrt{\frac{S}{6}}\right)^3 = \left[\left(\frac{S}{6}\right)^{\frac{1}{2}}\right]^3 = \left(\frac{S}{6}\right)^{3/2} = \boxed{\frac{S^{3/2}}{6^{3/2}}}$

Recall from Sec. 1.11 that $\sqrt{x} = x^{1/2}$. Alternate answers: $V = \frac{S^{3/2}}{6\sqrt{6}}$ or $V = \frac{S\sqrt{S}}{6\sqrt{6}}$

4) $E = mgh$ and $E = \frac{1}{2}kx^2 \rightarrow mgh = \frac{1}{2}kx^2 \rightarrow 2mgh = kx^2 \rightarrow \frac{2mgh}{k} = x^2 \rightarrow \boxed{\sqrt{\frac{2mgh}{k}} = x}$

5) $\theta = \omega t \rightarrow \frac{\theta}{\omega} = t$ such that $v = \frac{s}{t} \rightarrow vt = s \rightarrow v\left(\frac{\theta}{\omega}\right) = s \rightarrow \boxed{\frac{v\theta}{\omega} = s}$

6) $q = 3p \rightarrow \frac{q}{p} = 3 \rightarrow \frac{1}{p} = \frac{3}{q}$ such that $\frac{1}{p} + \frac{1}{q} = \frac{1}{f} \rightarrow \frac{3}{q} + \frac{1}{q} = \frac{1}{f} \rightarrow \frac{4}{q} = \frac{1}{f}$ Cross multiply

(Sec. 7.2): $4f = q \rightarrow \boxed{f = \frac{q}{4}}$ (If you got $f = \frac{3p}{4}$, you eliminated the wrong variable.)

7) $v_f = v_i + at \rightarrow v_f - v_i = at \rightarrow \frac{v_f - v_i}{a} = t$ such that

$x = v_i t + \frac{1}{2}at^2 = v_i \left(\frac{v_f - v_i}{a}\right) + \frac{1}{2}a\left(\frac{v_f - v_i}{a}\right)^2 = \frac{v_i v_f - v_i^2}{a} + \frac{a}{2}\frac{v_f^2 - 2v_i v_f + v_i^2}{a^2}$

$x = \frac{2v_i v_f - 2v_i^2}{2a} + \frac{v_f^2 - 2v_i v_f + v_i^2}{2a} = \frac{2v_i v_f - 2v_i^2 + v_f^2 - 2v_i v_f + v_i^2}{2a} = \frac{-2v_i^2 + v_f^2 + v_i^2}{2a} = \frac{v_f^2 - v_i^2}{2a}$

for which $2ax = v_f^2 - v_i^2 \rightarrow v_i^2 + 2ax = v_f^2 \rightarrow \boxed{\sqrt{v_i^2 + 2ax} = v_f}$

- We applied the FOIL method (Sec. 4.3) to expand $\left(v_f - v_i\right)^2$.
- Note that $\frac{a}{2}\frac{v_f^2 - 2v_i v_f + v_i^2}{a^2} = \frac{v_f^2 - 2v_i v_f + v_i^2}{2a}$ because $\frac{a}{2}\frac{1}{a^2} = \frac{a}{2a^2} = \frac{1}{2a}$.
- On the third line, $2v_i v_f$ cancels out and $-2v_i^2 + v_i^2 = v_i^2$ (just like $-2x + x = -x$).
- The result, which can be expressed as $v_f^2 = v_i^2 + 2ax$, is one of the classic equations of uniform acceleration which can be found in a standard physics textbook.

8) $v = \frac{2\pi R}{T}$ such that $G\frac{m}{R^2} = \frac{v^2}{R} \rightarrow \frac{Gm}{R^2} = \frac{1}{R}\left(\frac{2\pi R}{T}\right)^2 \rightarrow \frac{Gm}{R^2} = \frac{1}{R}\frac{4\pi^2 R^2}{T^2} \rightarrow \frac{Gm}{R^2} = \frac{4\pi^2 R}{T^2}$ (since $\frac{R^2}{R} = R$)

Cross multiply: $GmT^2 = 4\pi^2 R^3 \rightarrow T^2 = \frac{4\pi^2 R^3}{Gm} \rightarrow T = 2\pi\sqrt{\frac{R^3}{Gm}}$

This is a well-known formula for the period of a satellite in a circular orbit (which can be found in a standard physics textbook). In the last step, we used $\sqrt{4\pi^2} = 2\pi$, which follows from $(2\pi)^2 = 2^2\pi^2 = 4\pi^2$.

Exercise Set 8.4

1) $8x - 3y = 11$ Multiply by 3 on both sides: $24x - 9y = 33$

Add $24x - 9y = 33$ to $5x + 9y = 83$ to get $24x - 9y + 5x + 9y = 33 + 83$

Combine like terms: $29x = 116 \rightarrow x = \frac{116}{29} = \boxed{4}$ (this is x) Plug $x = 4$ into $8x - 3y = 11$

$\rightarrow 8(4) - 3y = 11 \rightarrow 32 - 3y = 11 \rightarrow -3y = -21 \rightarrow y = \frac{-21}{-3} = \boxed{7}$ (this is y)

Check: $8x - 3y = 8(4) - 3(7) = 32 - 21 = 11$

and $5x + 9y = 5(4) + 9(7) = 20 + 63 = 83$

2) $4x + 5y = 57$ Multiply by 3 on both sides: $12x + 15y = 171$

$-3x + 8y = 16$ Multiply by 4 on both sides: $-12x + 32y = 64$

Add the equations on the right together: $12x + 15y - 12x + 32y = 171 + 64$

Combine like terms: $47y = 235 \rightarrow y = \frac{235}{47} = \boxed{5}$ (this is y) Plug $y = 5$ into $4x + 5y = 57$

$\rightarrow 4x + 5(5) = 57 \rightarrow 4x + 25 = 57 \rightarrow 4x = 32 \rightarrow x = \frac{32}{4} = \boxed{8}$ (this is x)

Check: $4x + 5y = 4(8) + 5(5) = 32 + 25 = 57$

and $-3x + 8y = -3(8) + 8(5) = -24 + 40 = 16$

3) $7x + 5y = 85$ Multiply by 6 on both sides: $42x + 30y = 510$

$4x + 6y = 58$ Multiply by -5 on both sides: $-20x - 30y = -290$

Add the equations on the right together: $42x + 30y - 20x - 30y = 510 - 290$

Combine like terms: $22x = 220 \rightarrow x = \frac{220}{22} = \boxed{10}$ (this is x) Plug $x = 10$ into $7x + 5y = 85$

$\rightarrow 7(10) + 5y = 85 \rightarrow 70 + 5y = 85 \rightarrow 5y = 15 \rightarrow y = \frac{15}{5} = \boxed{3}$ (this is y)

Check: $7x + 5y = 7(10) + 5(3) = 70 + 15 = 85$

and $4x + 6y = 4(10) + 6(3) = 40 + 18 = 58$

4) $6x - 3y = 75$ Multiply by 2 on both sides: $12x - 6y = 150$

$4x - 8y = 92$ Multiply by -3 on both sides: $-12x + 24y = -276$

Add the equations on the right together: $12x - 6y - 12x + 24y = 150 - 276$

Combine like terms: $18y = -126 \rightarrow y = -\frac{126}{18} = \boxed{-7}$ (this is y) Plug $y = -7$ into $6x - 3y = 75$

$\rightarrow 6x - 3(-7) = 75 \rightarrow 6x + 21 = 75 \rightarrow 6x = 54 \rightarrow x = \frac{54}{6} = \boxed{9}$ (this is x)

Check: $6x - 3y = 6(9) - 3(-7) = 54 + 21 = 75$

and $4x - 8y = 4(9) - 8(-7) = 36 + 56 = 92$

5) $8x - 5y = -13$ Multiply by 3 on both sides: $24x - 15y = -39$

$-9x + 3y = 33$ Multiply by 5 on both sides: $-45x + 15y = 165$

Add the equations on the right together: $24x - 15y - 45x + 15y = -39 + 165$

Combine like terms: $-21x = 126 \rightarrow x = \frac{126}{-21} = \boxed{-6}$ (this is x) Plug $x = -6$ into $-9x + 3y = 33$

$\rightarrow -9(-6) + 3y = 33 \rightarrow 54 + 3y = 33 \rightarrow 3y = -21 \rightarrow \boxed{y = -7}$ (this is y)

Check: $8x - 5y = 8(-6) - 5(-7) = -48 + 35 = -13$

and $-9x + 3y = -9(-6) + 3(-7) = 54 - 21 = 33$

6) $-8x + 5y = 68$ Multiply by 4 on both sides: $-32x + 20y = 272$

$-3x + 4y = 51$ Multiply by -5 on both sides: $15x - 20y = -255$

Add the equations on the right together: $-32x + 20y + 15x - 20y = 272 - 255$

Combine like terms: $-17x = 17 \rightarrow x = \frac{17}{-17} = \boxed{-1}$ (this is x) Plug $x = -1$ into $-3x + 4y = 51$

$\rightarrow -3(-1) + 4y = 51 \rightarrow 3 + 4y = 51 \rightarrow 4y = 48 \rightarrow y = \frac{48}{4} = \boxed{12}$ (this is y)

Check: $-8x + 5y = -8(-1) + 5(12) = 8 + 60 = 68$

and $-3x + 4y = -3(-1) + 4(12) = 3 + 48 = 51$

7) $-4x + 3y = 2$ Multiply by 2 on both sides: $-8x + 6y = 4$

Add $8x + 9y = 16$ to $-8x + 6y = 4$ to get $8x + 9y - 8x + 6y = 16 + 4$

Combine like terms: $15y = 20 \rightarrow y = \frac{20}{15} = \boxed{\frac{4}{3}}$ (this is y) Plug $y = \frac{4}{3}$ into $-4x + 3y = 2$

$\rightarrow -4x + 3\left(\frac{4}{3}\right) = 2 \rightarrow -4x + 4 = 2 \rightarrow -4x = -2 \rightarrow x = \frac{-2}{-4} = \boxed{\frac{1}{2}}$ (this is x)

Check: $8x + 9y = 8\left(\frac{1}{2}\right) + 9\left(\frac{4}{3}\right) = 4 + 12 = 16$

and $-4x + 3y = -4\left(\frac{1}{2}\right) + 3\left(\frac{4}{3}\right) = -2 + 4 = 2$

8) $-10x - 12y = 3$ Multiply by 2 on both sides: $-20x - 24y = 6$

$-15x - 8y = -3$ Multiply by -3 on both sides: $45x + 24y = 9$

Add the equations on the right together: $-20x - 24y + 45x + 24y = 6 + 9$

Combine like terms: $25x = 15 \rightarrow x = \frac{15}{25} = \boxed{\frac{3}{5}}$ (this is x) Plug $x = \frac{3}{5}$ into $-10x - 12y = 3$

$\rightarrow -10\left(\frac{3}{5}\right) - 12y = 3 \rightarrow -6 - 12y = 3 \rightarrow -12y = 9 \rightarrow y = \frac{9}{-12} = \boxed{-\frac{3}{4}}$ (this is y)

Check: $-10x - 12y = -10\left(\frac{3}{5}\right) - 12\left(-\frac{3}{4}\right) = -6 + 9 = 3$

and $-15x - 8y = -15\left(\frac{3}{5}\right) - 8\left(-\frac{3}{4}\right) = -9 + 6 = -3$

Exercise Set 8.5

1) $\begin{vmatrix} 8 & 5 \\ 7 & 6 \end{vmatrix} = 8(6) - 5(7) = 48 - 35 = 13$

2) $\begin{vmatrix} -3 & 6 \\ 4 & 9 \end{vmatrix} = -3(9) - 6(4) = -27 - 24 = -51$

3) $\begin{vmatrix} 4 & -5 \\ 9 & 8 \end{vmatrix} = 4(8) - (-5)(9) = 32 + 45 = 77$

4) $\begin{vmatrix} 7 & -6 \\ -3 & 7 \end{vmatrix} = 7(7) - (-6)(-3) = 49 - 18 = 31$

5) $\begin{vmatrix} 8 & 12 \\ 6 & 9 \end{vmatrix} = 8(9) - 12(6) = 72 - 72 = 0$

6) $\begin{vmatrix} -6 & -7 \\ 8 & -9 \end{vmatrix} = -6(-9) - (-7)(8) = 54 + 56 = 110$

7) $\begin{vmatrix} 4 & 3 & 8 \\ 9 & 5 & 1 \\ 2 & 7 & 6 \end{vmatrix} = \begin{matrix} 4 & 3 & 8 & 4 & 3 \\ 9 & 5 & 1 & 9 & 5 \\ 2 & 7 & 6 & 2 & 7 \end{matrix} = 4(5)(6) + 3(1)(2) + 8(9)(7) - 8(5)(2) - 4(1)(7) - 3(9)(6)$

$= 120 + 6 + 504 - 80 - 28 - 162 = 630 - 270 = 360$

8) $\begin{vmatrix} 5 & -2 & 7 \\ 8 & 4 & 1 \\ -6 & 0 & 3 \end{vmatrix} = \begin{matrix} 5 & -2 & 7 & 5 & -2 \\ 8 & 4 & 1 & 8 & 4 \\ -6 & 0 & 3 & -6 & 0 \end{matrix}$

$= 5(4)(3) + (-2)(1)(-6) + 7(8)(0) - 7(4)(-6) - 5(1)(0) - (-2)(8)(3)$

$= 60 + 12 + 0 + 168 - 0 + 48 = 288$

9) $\begin{vmatrix} 2 & 2 & 2 \\ 2 & 2 & -2 \\ 2 & -2 & -2 \end{vmatrix} = \begin{matrix} 2 & 2 & 2 & 2 & 2 \\ 2 & 2 & -2 & 2 & 2 \\ 2 & -2 & -2 & 2 & -2 \end{matrix}$

$= 2(2)(-2) + 2(-2)(2) + 2(2)(-2) - 2(2)(2) - 2(-2)(-2) - 2(2)(-2)$

$= -8 - 8 - 8 - 8 - 8 + 8 = -40 + 8 = -32$

$$10) \begin{vmatrix} 9 & 4 & 1 \\ 3 & 8 & -13 \\ -5 & 2 & -9 \end{vmatrix} = \begin{matrix} 9 & 4 & 1 \\ 3 & 8 & -13 \\ -5 & 2 & -9 \end{matrix} \begin{matrix} 9 & 4 \\ 3 & 8 \\ -5 & 2 \end{matrix}$$

$$= 9(8)(-9) + 4(-13)(-5) + 1(3)(2) - 1(8)(-5) - 9(-13)(2) - 4(3)(-9)$$

$$= -648 + 260 + 6 + 40 + 234 + 108 = -648 + 648 = 0$$

Exercise Set 8.6

1) $D_c = \begin{vmatrix} a_1 & b_1 \\ a_2 & b_2 \end{vmatrix} = \begin{vmatrix} 4 & -6 \\ 3 & -5 \end{vmatrix} = 4(-5) - (-6)(3) = -20 + 18 = -2$

$D_x = \begin{vmatrix} c_1 & b_1 \\ c_2 & b_2 \end{vmatrix} = \begin{vmatrix} 2 & -6 \\ -1 & -5 \end{vmatrix} = 2(-5) - (-6)(-1) = -10 - 6 = -16$

$D_y = \begin{vmatrix} a_1 & c_1 \\ a_2 & c_2 \end{vmatrix} = \begin{vmatrix} 4 & 2 \\ 3 & -1 \end{vmatrix} = 4(-1) - 2(3) = -4 - 6 = -10$

$x = \dfrac{D_x}{D_c} = \dfrac{-16}{-2} = \boxed{8}$, $y = \dfrac{D_y}{D_c} = \dfrac{-10}{-2} = \boxed{5}$

Check: $4x - 6y = 4(8) - 6(5) = 32 - 30 = 2$

and $3x - 5y = 3(8) - 5(5) = 24 - 25 = -1$

2) $D_c = \begin{vmatrix} a_1 & b_1 \\ a_2 & b_2 \end{vmatrix} = \begin{vmatrix} 7 & 3 \\ 3 & 7 \end{vmatrix} = 7(7) - 3(3) = 49 - 9 = 40$

$D_x = \begin{vmatrix} c_1 & b_1 \\ c_2 & b_2 \end{vmatrix} = \begin{vmatrix} 15 & 3 \\ -45 & 7 \end{vmatrix} = 15(7) - 3(-45) = 105 + 135 = 240$

$D_y = \begin{vmatrix} a_1 & c_1 \\ a_2 & c_2 \end{vmatrix} = \begin{vmatrix} 7 & 15 \\ 3 & -45 \end{vmatrix} = 7(-45) - 15(3) = -315 - 45 = -360$

$x = \dfrac{D_x}{D_c} = \dfrac{240}{40} = \boxed{6}$, $y = \dfrac{D_y}{D_c} = \dfrac{-360}{40} = \boxed{-9}$

Check: $7x + 3y = 7(6) + 3(-9) = 42 - 27 = 15$

and $3x + 7y = 3(6) + 7(-9) = 18 - 63 = -45$

3) $D_c = \begin{vmatrix} a_1 & b_1 \\ a_2 & b_2 \end{vmatrix} = \begin{vmatrix} 5 & 9 \\ 6 & -2 \end{vmatrix} = 5(-2) - 9(6) = -10 - 54 = -64$

$D_x = \begin{vmatrix} c_1 & b_1 \\ c_2 & b_2 \end{vmatrix} = \begin{vmatrix} 7 & 9 \\ -30 & -2 \end{vmatrix} = 7(-2) - 9(-30) = -14 + 270 = 256$

$D_y = \begin{vmatrix} a_1 & c_1 \\ a_2 & c_2 \end{vmatrix} = \begin{vmatrix} 5 & 7 \\ 6 & -30 \end{vmatrix} = 5(-30) - 7(6) = -150 - 42 = -192$

$x = \dfrac{D_x}{D_c} = \dfrac{256}{-64} = \boxed{-4}$, $y = \dfrac{D_y}{D_c} = \dfrac{-192}{-64} = \boxed{3}$

Check: $5x + 9y = 5(-4) + 9(3) = -20 + 27 = 7$

and $6x - 2y = 6(-4) - 2(3) = -24 - 6 = -30$

4) $D_c = \begin{vmatrix} a_1 & b_1 \\ a_2 & b_2 \end{vmatrix} = \begin{vmatrix} 8 & 1 \\ -3 & 5 \end{vmatrix} = 8(5) - 1(-3) = 40 + 3 = 43$ Note: The coefficient of $+y$ is 1

$D_x = \begin{vmatrix} c_1 & b_1 \\ c_2 & b_2 \end{vmatrix} = \begin{vmatrix} 70 & 1 \\ 49 & 5 \end{vmatrix} = 70(5) - 1(49) = 350 - 49 = 301$

$D_y = \begin{vmatrix} a_1 & c_1 \\ a_2 & c_2 \end{vmatrix} = \begin{vmatrix} 8 & 70 \\ -3 & 49 \end{vmatrix} = 8(49) - 70(-3) = 392 + 210 = 602$

$x = \frac{D_x}{D_c} = \frac{301}{43} = \boxed{7}$, $y = \frac{D_y}{D_c} = \frac{602}{43} = \boxed{14}$

Check: $8x + y = 8(7) + 14 = 56 + 14 = 70$

and $-3x + 5y = -3(7) + 5(14) = -21 + 70 = 49$

5) $D_c = \begin{vmatrix} a_1 & b_1 \\ a_2 & b_2 \end{vmatrix} = \begin{vmatrix} 6 & -8 \\ 5 & 2 \end{vmatrix} = 6(2) - (-8)(5) = 12 + 40 = 52$

$D_x = \begin{vmatrix} c_1 & b_1 \\ c_2 & b_2 \end{vmatrix} = \begin{vmatrix} -1 & -8 \\ 10 & 2 \end{vmatrix} = -1(2) - (-8)(10) = -2 + 80 = 78$

$D_y = \begin{vmatrix} a_1 & c_1 \\ a_2 & c_2 \end{vmatrix} = \begin{vmatrix} 6 & -1 \\ 5 & 10 \end{vmatrix} = 6(10) - (-1)(5) = 60 + 5 = 65$

$x = \frac{D_x}{D_c} = \frac{78}{52} = \frac{78/26}{52/26} = \boxed{\frac{3}{2}}$, $y = \frac{D_y}{D_c} = \frac{65}{52} = \frac{65/13}{52/13} = \boxed{\frac{5}{4}}$

Check: $6x - 8y = 6\left(\frac{3}{2}\right) - 8\left(\frac{5}{4}\right) = 9 - 10 = -1$

and $5x + 2y = 5\left(\frac{3}{2}\right) + 2\left(\frac{5}{4}\right) = \frac{15}{2} + \frac{10}{4} = \frac{30}{4} + \frac{10}{4} = \frac{40}{4} = 10$

6) $D_c = \begin{vmatrix} a_1 & b_1 \\ a_2 & b_2 \end{vmatrix} = \begin{vmatrix} 4 & -6 \\ -5 & 3 \end{vmatrix} = 4(3) - (-6)(-5) = 12 - 30 = -18$

$D_x = \begin{vmatrix} c_1 & b_1 \\ c_2 & b_2 \end{vmatrix} = \begin{vmatrix} 68 & -6 \\ -67 & 3 \end{vmatrix} = 68(3) - (-6)(-67) = 204 - 402 = -198$

$D_y = \begin{vmatrix} a_1 & c_1 \\ a_2 & c_2 \end{vmatrix} = \begin{vmatrix} 4 & 68 \\ -5 & -67 \end{vmatrix} = 4(-67) - 68(-5) = -268 + 340 = 72$

$x = \frac{D_x}{D_c} = \frac{-198}{-18} = \boxed{11}$, $y = \frac{D_y}{D_c} = \frac{72}{-18} = \boxed{-4}$

Check: $4x - 6y = 4(11) - 6(-4) = 44 + 24 = 68$

and $-5x + 3y = -5(11) + 3(-4) = -55 - 12 = -67$

7) $D_c = \begin{vmatrix} a_1 & b_1 \\ a_2 & b_2 \end{vmatrix} = \begin{vmatrix} -8 & 9 \\ -2 & -3 \end{vmatrix} = (-8)(-3) - 9(-2) = 24 + 18 = 42$

$D_x = \begin{vmatrix} c_1 & b_1 \\ c_2 & b_2 \end{vmatrix} = \begin{vmatrix} -12 & 9 \\ -17 & -3 \end{vmatrix} = (-12)(-3) - 9(-17) = 36 + 153 = 189$

$D_y = \begin{vmatrix} a_1 & c_1 \\ a_2 & c_2 \end{vmatrix} = \begin{vmatrix} -8 & -12 \\ -2 & -17 \end{vmatrix} = (-8)(-17) - (-12)(-2) = 136 - 24 = 112$

$$x = \frac{D_x}{D_c} = \frac{189}{42} = \frac{189/21}{42/21} = \boxed{\frac{9}{2}} \quad , \quad y = \frac{D_y}{D_c} = \frac{112}{42} = \frac{112/14}{42/14} = \boxed{\frac{8}{3}}$$

Check: $-8x + 9y = -8\left(\frac{9}{2}\right) + 9\left(\frac{8}{3}\right) = -36 + 24 = -12$

and $-2x - 3y = -2\left(\frac{9}{2}\right) - 3\left(\frac{8}{3}\right) = -9 - 8 = -17$

Exercise Set 8.7

1) $D_c = \begin{vmatrix} a_1 & b_1 & c_1 \\ a_2 & b_2 & c_2 \\ a_3 & b_3 & c_3 \end{vmatrix} = \begin{vmatrix} 3 & 2 & 2 \\ 2 & 8 & 3 \\ 1 & 5 & 2 \end{vmatrix} = \begin{matrix} 3 & 2 & 2 \\ 2 & 8 & 3 \\ 1 & 5 & 2 \end{matrix} \begin{matrix} 3 & 2 \\ 2 & 8 \\ 1 & 5 \end{matrix}$

$D_c = 3(8)(2) + 2(3)(1) + 2(2)(5) - 2(8)(1) - 3(3)(5) - 2(2)(2)$

$D_c = 48 + 6 + 20 - 16 - 45 - 8 = 74 - 69 = 5$

$D_x = \begin{vmatrix} d_1 & b_1 & c_1 \\ d_2 & b_2 & c_2 \\ d_3 & b_3 & c_3 \end{vmatrix} = \begin{vmatrix} 20 & 2 & 2 \\ 25 & 8 & 3 \\ 15 & 5 & 2 \end{vmatrix} = \begin{matrix} 20 & 2 & 2 \\ 25 & 8 & 3 \\ 15 & 5 & 2 \end{matrix} \begin{matrix} 20 & 2 \\ 25 & 8 \\ 15 & 5 \end{matrix}$

$D_x = 20(8)(2) + 2(3)(15) + 2(25)(5) - 2(8)(15) - 20(3)(5) - 2(25)(2)$

$D_x = 320 + 90 + 250 - 240 - 300 - 100 = 660 - 640 = 20$

$D_y = \begin{vmatrix} a_1 & d_1 & c_1 \\ a_2 & d_2 & c_2 \\ a_3 & d_3 & c_3 \end{vmatrix} = \begin{vmatrix} 3 & 20 & 2 \\ 2 & 25 & 3 \\ 1 & 15 & 2 \end{vmatrix} = \begin{matrix} 3 & 20 & 2 \\ 2 & 25 & 3 \\ 1 & 15 & 2 \end{matrix} \begin{matrix} 3 & 20 \\ 2 & 25 \\ 1 & 15 \end{matrix}$

$D_y = 3(25)(2) + 20(3)(1) + 2(2)(15) - 2(25)(1) - 3(3)(15) - 20(2)(2)$

$D_y = 150 + 60 + 60 - 50 - 135 - 80 = 270 - 265 = 5$

$D_z = \begin{vmatrix} a_1 & b_1 & d_1 \\ a_2 & b_2 & d_2 \\ a_3 & b_3 & d_3 \end{vmatrix} = \begin{vmatrix} 3 & 2 & 20 \\ 2 & 8 & 25 \\ 1 & 5 & 15 \end{vmatrix} = \begin{matrix} 3 & 2 & 20 \\ 2 & 8 & 25 \\ 1 & 5 & 15 \end{matrix} \begin{matrix} 3 & 2 \\ 2 & 8 \\ 1 & 5 \end{matrix}$

$D_z = 3(8)(15) + 2(25)(1) + 20(2)(5) - 20(8)(1) - 3(25)(5) - 2(2)(15)$

$D_z = 360 + 50 + 200 - 160 - 375 - 60 = 610 - 495 = 15$

$$x = \frac{D_x}{D_c} = \frac{20}{5} = \boxed{4} \quad , \quad y = \frac{D_y}{D_c} = \frac{5}{5} = \boxed{1} \quad , \quad z = \frac{D_z}{D_c} = \frac{15}{5} = \boxed{3}$$

Check: $3x + 2y + 2z = 3(4) + 2(1) + 2(3) = 12 + 2 + 6 = 20$

and $2x + 8y + 3z = 2(4) + 8(1) + 3(3) = 8 + 8 + 9 = 25$

and $x + 5y + 2z = 4 + 5(1) + 2(3) = 4 + 5 + 6 = 15$

$$2) \ D_c = \begin{vmatrix} a_1 & b_1 & c_1 \\ a_2 & b_2 & c_2 \\ a_3 & b_3 & c_3 \end{vmatrix} = \begin{vmatrix} 5 & 4 & 9 \\ 10 & 0 & -3 \\ 3 & -10 & 5 \end{vmatrix} = \begin{matrix} 5 & 4 & 9 \\ 10 & 0 & -3 \\ 3 & -10 & 5 \end{matrix} \begin{matrix} 5 & 4 \\ 10 & 0 \\ 3 & -10 \end{matrix}$$

Note: In the middle equation, the coefficient of y is zero.

$D_c = 5(0)(5) + 4(-3)(3) + 9(10)(-10) - 9(0)(3) - 5(-3)(-10) - 4(10)(5)$

$D_c = 0 - 36 - 900 - 0 - 150 - 200 = -1286$

$$D_x = \begin{vmatrix} d_1 & b_1 & c_1 \\ d_2 & b_2 & c_2 \\ d_3 & b_3 & c_3 \end{vmatrix} = \begin{vmatrix} 150 & 4 & 9 \\ 50 & 0 & -3 \\ 24 & -10 & 5 \end{vmatrix} = \begin{matrix} 150 & 4 & 9 \\ 50 & 0 & -3 \\ 24 & -10 & 5 \end{matrix} \begin{matrix} 150 & 4 \\ 50 & 0 \\ 24 & -10 \end{matrix}$$

$D_x = 150(0)(5) + 4(-3)(24) + 9(50)(-10) - 9(0)(24) - 150(-3)(-10) - 4(50)(5)$

$D_x = 0 - 288 - 4500 - 0 - 4500 - 1000 = -10,288$

$$D_y = \begin{vmatrix} a_1 & d_1 & c_1 \\ a_2 & d_2 & c_2 \\ a_3 & d_3 & c_3 \end{vmatrix} = \begin{vmatrix} 5 & 150 & 9 \\ 10 & 50 & -3 \\ 3 & 24 & 5 \end{vmatrix} = \begin{matrix} 5 & 150 & 9 \\ 10 & 50 & -3 \\ 3 & 24 & 5 \end{matrix} \begin{matrix} 5 & 150 \\ 10 & 50 \\ 3 & 24 \end{matrix}$$

$D_y = 5(50)(5) + 150(-3)(3) + 9(10)(24) - 9(50)(3) - 5(-3)(24) - 150(10)(5)$

$D_y = 1250 - 1350 + 2160 - 1350 + 360 - 7500 = 3770 - 10,200 = -6430$

$$D_z = \begin{vmatrix} a_1 & b_1 & d_1 \\ a_2 & b_2 & d_2 \\ a_3 & b_3 & d_3 \end{vmatrix} = \begin{vmatrix} 5 & 4 & 150 \\ 10 & 0 & 50 \\ 3 & -10 & 24 \end{vmatrix} = \begin{matrix} 5 & 4 & 150 \\ 10 & 0 & 50 \\ 3 & -10 & 24 \end{matrix} \begin{matrix} 5 & 4 \\ 10 & 0 \\ 3 & -10 \end{matrix}$$

$D_z = 5(0)(24) + 4(50)(3) + 150(10)(-10) - 150(0)(3) - 5(50)(-10) - 4(10)(24)$

$D_z = 0 + 600 - 15,000 - 0 + 2500 - 960 = 3100 - 15,960 = -12,860$

$$x = \frac{D_x}{D_c} = \frac{-10,288}{-1286} = \boxed{8} \ , \quad y = \frac{D_y}{D_c} = \frac{-6430}{-1286} = \boxed{5} \ , \quad z = \frac{D_z}{D_c} = \frac{-12,860}{-1286} = \boxed{10}$$

Check: $5x + 4y + 9z = 5(8) + 4(5) + 9(10) = 40 + 20 + 90 = 150$

and $10x - 3z = 10(8) - 3(10) = 80 - 30 = 50$

and $3x - 10y + 5z = 3(8) - 10(5) + 5(10) = 24 - 50 + 50 = 24$

$$3) \ D_c = \begin{vmatrix} a_1 & b_1 & c_1 \\ a_2 & b_2 & c_2 \\ a_3 & b_3 & c_3 \end{vmatrix} = \begin{vmatrix} 3 & 4 & 3 \\ 4 & -3 & -5 \\ -9 & 5 & 7 \end{vmatrix} = \begin{matrix} 3 & 4 & 3 \\ 4 & -3 & -5 \\ -9 & 5 & 7 \end{matrix} \begin{matrix} 3 & 4 \\ 4 & -3 \\ -9 & 5 \end{matrix}$$

$D_c = 3(-3)(7) + 4(-5)(-9) + 3(4)(5) - 3(-3)(-9) - 3(-5)(5) - 4(4)(7)$

$D_c = -63 + 180 + 60 - 81 + 75 - 112 = 315 - 256 = 59$

$$D_x = \begin{vmatrix} d_1 & b_1 & c_1 \\ d_2 & b_2 & c_2 \\ d_3 & b_3 & c_3 \end{vmatrix} = \begin{vmatrix} 15 & 4 & 3 \\ -10 & -3 & -5 \\ 28 & 5 & 7 \end{vmatrix} = \begin{matrix} 15 & 4 & 3 \\ -10 & -3 & -5 \\ 28 & 5 & 7 \end{matrix} \begin{matrix} 15 & 4 \\ -10 & -3 \\ 28 & 5 \end{matrix}$$

$D_x = 15(-3)(7) + 4(-5)(28) + 3(-10)(5) - 3(-3)(28) - 15(-5)(5) - 4(-10)(7)$

$$D_x = -315 - 560 - 150 + 252 + 375 + 280 = 907 - 1025 = -118$$

$$D_y = \begin{vmatrix} a_1 & d_1 & c_1 \\ a_2 & d_2 & c_2 \\ a_3 & d_3 & c_3 \end{vmatrix} = \begin{vmatrix} 3 & 15 & 3 \\ 4 & -10 & -5 \\ -9 & 28 & 7 \end{vmatrix} = \begin{matrix} 3 & 15 & 3 \\ 4 & -10 & -5 \\ -9 & 28 & 7 \end{matrix} \begin{matrix} 3 & 15 \\ 4 & -10 \\ -9 & 28 \end{matrix}$$

$$D_y = 3(-10)(7) + 15(-5)(-9) + 3(4)(28) - 3(-10)(-9) - 3(-5)(28) - 15(4)(7)$$

$$D_y = -210 + 675 + 336 - 270 + 420 - 420 = 1431 - 900 = 531$$

$$D_z = \begin{vmatrix} a_1 & b_1 & d_1 \\ a_2 & b_2 & d_2 \\ a_3 & b_3 & d_3 \end{vmatrix} = \begin{vmatrix} 3 & 4 & 15 \\ 4 & -3 & -10 \\ -9 & 5 & 28 \end{vmatrix} = \begin{matrix} 3 & 4 & 15 \\ 4 & -3 & -10 \\ -9 & 5 & 28 \end{matrix} \begin{matrix} 3 & 4 \\ 4 & -3 \\ -9 & 5 \end{matrix}$$

$$D_z = 3(-3)(28) + 4(-10)(-9) + 15(4)(5) - 15(-3)(-9) - 3(-10)(5) - 4(4)(28)$$

$$D_z = -252 + 360 + 300 - 405 + 150 - 448 = 810 - 1105 = -295$$

$$x = \frac{D_x}{D_c} = \frac{-118}{59} = \boxed{-2} \quad , \quad y = \frac{D_y}{D_c} = \frac{531}{59} = \boxed{9} \quad , \quad z = \frac{D_z}{D_c} = \frac{-295}{59} = \boxed{-5}$$

Check: $3x + 4y + 3z = 3(-2) + 4(9) + 3(-5) = -6 + 36 - 15 = 15$

and $4x - 3y - 5z = 4(-2) - 3(9) - 5(-5) = -8 - 27 + 25 = -10$

and $-9x + 5y + 7z = -9(-2) + 5(9) + 7(-5) = 18 + 45 - 35 = 28$

Exercise Set 8.8

1) indeterminate: Multiply $2x - y = 8$ by -1 to get $-2x + y = -8$ (the two equations are not independent)

2) no solution: Multiply $3x + 2y = 6$ by 2 to get $6x + 4y = 12$, which contradicts $6x + 4y = 7$ (since $6x + 4y$ can't equal 12 and also equal 7)

3) indeterminate: You need 3 independent equations to solve for 3 unknowns.

4) all real numbers: Subtract $2x$ and $3y$ from both sides of the first equation to get $0 = 0$, and subtract x and y from both sides of the second equation to get $3 = 3$ (both variables canceled out each time, and the resulting equation is always true in each case)

5) no solution: There are 3 independent equations, but only 2 unknowns. The third equation contradicts the first two equations. From the first equation: $5x + 2y = 20$ $\rightarrow 2y = 20 - 5x \rightarrow y = 10 - \frac{5x}{2}$ Plug this into the second equation: $8x - 3\left(10 - \frac{5x}{2}\right) = 1$ $\rightarrow 8x - 30 + \frac{15x}{2} = 1 \rightarrow 16x - 60 + 15x = 2 \rightarrow 31x = 62 \rightarrow x = 2$ Plug this into the first equation: $5(2) + 2y = 20 \rightarrow 10 + 2y = 20 \rightarrow 2y = 10 \rightarrow y = 5$ Plug $x = 2$ and $y = 5$ into the third equation: $-3x + 7y = -3(2) + 7(5) = -6 + 35 = 29$ doesn't equal 15 (which is a contradiction)

6) no solution: Subtract $4x$ and add $3y$ to both sides of the first equation to get $0 = 1$, and subtract $2x$ and $5y$ from both sides of the second equation to get $7 = 0$. Both equations are false regardless of the values of the variables. (Even if just one equation were false, there would be no solution.)

Exercise Set 8.9

1) x = the smaller number and y = the larger number, $x + y = 21$ and $2x = y - 3$
Subtract x from both sides of the first equation: $y = 21 - x$ Replace y with $21 - x$ in the second equation: $2x = 21 - x - 3$ Combine like terms: $3x = 18$ Divide by 3 on both sides: $x = \frac{18}{3} = \boxed{6}$ Plug $x = 6$ into $y = 21 - x = 21 - 6 = \boxed{15}$
Check: The sum is $6 + 15 = 21$. When the smaller number is doubled, we get $2(6) = 12$, which is 3 less than 15.

2) B = Brian's age and K = Kevin's age, $B + 2K = 30$ and $2B - K = 15$
Subtract $2K$ from the first equation: $B = 30 - 2K$ Replace B with $30 - 2K$ in the second equation: $2(30 - 2K) - K = 15$ Distribute: $60 - 4K - K = 15$ Combine like terms: $-5K = -45$ Divide by -5 on both sides: $K = \frac{-45}{-5} = \boxed{9}$ Plug $K = 9$ into $B = 30 - 2K = 30 - 2(9) = 30 - 18 = \boxed{12}$ Brian is $B = 12$ and Kevin is $K = 9$ years old
Check: Brian's age plus twice Kevin's age is $12 + 2(9) = 12 + 18 = 30$ and twice Brian's age minus Kevin's age is $2(12) - 9 = 24 - 9 = 15$

3) p = the cost of one pencil and e = the cost of one eraser
$4p + 9e = 4.25$ and $8p + 3e = 4.75$ Multiply the second equation by -3
$-24p - 9e = -14.25$ Add this to the first equation: $4p + 9e - 24p - 9e = 4.25 - 14.25$
Combine like terms. Note that e cancels out: $-20p = -10$ Divide by -20 on both sides: $p = \frac{-10}{-20} = \frac{1}{2} = \boxed{\$0.50}$ Plug $p = 0.5$ into the first equation: $4(0.5) + 9e = 4.25$
$\rightarrow 2 + 9e = 4.25 \rightarrow 9e = 2.25 \rightarrow e = \frac{2.25}{9} = \boxed{\$0.25} = \frac{1}{4}$ Each pencil costs $0.50 (or 50 cents or a half dollar) and each eraser costs $0.25 (or 25 cents or a quarter dollar).
Note: You can avoid decimals by multiplying both sides of the original equations by 100 (Sec. 1.8), or even by multiplying by 4. Alternatively, express all of the amounts in cents rather than dollars.
Check: 4 pencils and 9 erasers cost $4(\$0.50) + 9(\$0.25) = \$2.00 + \$2.25 = \$4.25$ and 8 pencils and 3 erasers cost $8(\$0.50) + 3(\$0.25) = \$4.00 + \$0.75 = \$4.75$

4) $L = $ length and $W = $ width, $P = 2L + 2W = 35$ and $A = LW = 75$

Divide both sides by L in the second equation: $W = \frac{75}{L}$

Replace W with $\frac{75}{L}$ in the first equation: $2L + 2\left(\frac{75}{L}\right) = 35$

Distribute: $2L + \frac{150}{L} = 35$

Multiply by L on both sides: $2L^2 + 150 = 35L$

Subtract $35L$ from both sides: $2L^2 - 35L + 150 = 0$

This is a quadratic equation (Chapter 6). Either use the quadratic formula or note that this equation may be factored as $(2L - 15)(L - 10) = 0$

As discussed in Sec. 6.9, this leads to either $2L - 15 = 0$ or $L - 10 = 0$

The first case gives $L = \frac{15}{2} = 7.5$ cm and the second case gives $L = \boxed{10}$ cm

Plug $L = 10$ cm into $LW = 75$ to see that $10W = 75 \rightarrow W = \frac{75}{10} = \boxed{7.5}$ cm

The length is 10 cm and the width is 7.5 cm.

Check: The perimeter is $P = 2L + 2W = 2(10) + 2(7.5) = 20 + 15 = 35$ cm and the area is $A = LW = 10(7.5) = 75$ square cm

9 Inequalities

Exercise Set 9.1

1) $-5, -3, 0$, and 3 are less than 4 $(x < 4)$

2) Only 4.5 is greater than 3 $(x > 3)$

3) $-6, -4, 0, 4$, and 5 are less than or equal to 5 $(x \leq 5)$

4) 2 and 3 are greater than or equal to 2 $(x \geq 2)$

5) Only 7 is greater than 6 $(6 < x)$

6) $-8.2, -6.4$, and 6.4 are less than 7 $(7 > x)$

7) Only -2 is less than -1 $(x < -1)$

Note: -2 is smaller than -1 because it is more negative. That is, -2 is farther from 0.

8) $-8, -7, 7, 8$, and 9 are greater than or equal to -8 $(x \geq -8)$

Note: -7 is greater than -8 because it is less negative. That is, -7 is closer to 0.

Exercise Set 9.2

1) Add 8 to both sides: $\boxed{x > 15}$

Check: 15.1 is a little greater than 15: $x - 8 = 15.1 - 8 = 7.1$ is a little more than 7

2) Divide by 3 on both sides: $\boxed{x \leq 9}$

Check: 8.9 is a little less than 9: $3x = 3(8.9) = 26.7$ is a little less than 27

3) Add 9 to both sides: $6x < 24$ Divide by 6 on both sides: $\boxed{x < 4}$

Check: 3.9 is a little less than 4: $6x - 9 = 6(3.9) - 9 = 14.4$ is a little less than 15

4) Multiply by 6 on both sides: $\boxed{x \leq -42}$

Check: -42.1 is a little less than -42 (because -42.1 is more negative):

$\frac{x}{6} = -\frac{42.1}{6} \approx -7.02$ is a little less than -7 (because -7.02 is more negative; it is farther from zero)

5) Add x to both sides: $8x \geq 40$ Divide by 8 on both sides: $\boxed{x \geq 5}$

Check: 5.1 is a little more than 5: $7x = 7(5.1) = 35.7$ is a little more than $40 - x = 40 - 5.1 = 34.9$

6) Subtract 8 from both sides: $9x > 24 + 6x$ Subtract $6x$ from both sides: $3x > 24$

Divide by 3 on both sides: $\boxed{x > 8}$

Check: 8.1 is a little more than 8: $9x + 8 = 9(8.1) + 8 = 80.9$ is a little more than $32 + 6x = 32 + 6(8.1) = 80.6$

7) Subtract 9 from both sides: $\frac{x}{4} < 7$ Multiply by 4 on both sides: $\boxed{x < 28}$

Check: 27.9 is a little less than 28: $\frac{x}{4} + 9 = \frac{27.9}{4} + 9 = 15.975$ is a little less than 16

8) Add 25 to both sides: $50 - 2x \leq 3x$ Add $2x$ to both sides: $50 \leq 5x$

Divide by 5 on both sides: $\boxed{10 \leq x}$

Check: 10.1 is a little more than 10: $25 - 2x = 25 - 2(10.1) = 4.8$ is a little less than $3x - 25 = 3(10.1) - 25 = 5.3$

9) Subtract $3x$ from both sides: $4x + 2 > -2$ Subtract 2 from both sides: $4x > -4$

Divide by 4 on both sides: $\boxed{x > -1}$

Check: -0.9 is a little more than -1 (because -0.9 is less negative): $7x + 2 = 7(-0.9) + 2 = -4.3$ is a little more than $3x - 2 = 3(-0.9) - 2 = -4.7$ (because -4.3 is less negative; it is closer to zero)

10) Subtract $5x$ from both sides: $6 < 4x + 4$ Subtract 4 from both sides: $2 < 4x$

Divide by 4 on both sides: $\boxed{\frac{1}{2} < x}$

Check: 0.6 is a little more than one-half: $5x + 6 = 5(0.6) + 6 = 9$ is a little less than $9x + 4 = 9(0.6) + 4 = 9.4$

11) Subtract x from both sides: $7x - 3 \leq -7$ Add 3 to both sides: $7x \leq -4$

Divide by 7 on both sides: $\boxed{x \leq -\frac{4}{7}}$

Check: -0.6 is a little less than $-\frac{4}{7}$ (because -0.6 is more negative than $-\frac{4}{7} \approx -0.5714$): $8x - 3 = 8(-0.6) - 3 = -7.8$ is a little less than $x - 7 = -0.6 - 7 = -7.6$ (because -7.8 is more negative; it is farther from zero)

12) Subtract 6 from both sides: $5x \geq 3x$ Subtract $3x$ from both sides: $2x \geq 0$

Divide by 2 on both sides: $\boxed{x \geq 0}$

Check: 0.1 is a little more than zero: $5x + 6 = 5(0.1) + 6 = 6.5$ is a little more than $3x + 6 = 3(0.1) + 6 = 6.3$

Exercise Set 9.3

1) Multiply by -1 on both sides and reverse the inequality: $\boxed{x > -1}$ (equivalent to $-1 < x$)
Check: -0.9 is a little more than -1 (because -0.9 is less negative):
$-x = -(-0.9) = 0.9$ is a little less than 1

2) Divide by -2 on both sides and reverse the inequality: $\boxed{x < 8}$ (equivalent to $8 > x$)
Check: 7.9 is a little less than 8: $-2x = -2(7.9) = -15.8$ is a little more than -16
(because -15.8 is less negative; it is closer to zero)

3) Subtract 3 from both sides: $-6x \geq 18$ Divide by -6 on both sides and reverse the
inequality: $\boxed{x \leq -3}$ (equivalent to $-3 \geq x$)
Check: -3.1 is a little less than -3 (because -3.1 is more negative):
$3 - 6x = 3 - 6(-3.1) = 3 + 18.6 = 21.6$ is a little more than 21

4) Multiply by -7 on both sides and reverse the inequality: $\boxed{x > -56}$ (equivalent to
$-56 < x$) Check: -55.9 is a little more than -56 (because -55.9 is less negative):
$-\frac{x}{7} = -\frac{(-55.9)}{7} \approx 7.99$ is a little less than 8

5) Add 10 to both sides: $-8x > -40$ Divide by -8 on both sides and reverse the
inequality: $\boxed{x < 5}$ (equivalent to $5 > x$)
Check: 4.9 is a little less than 5: $-8x - 10 = -8(4.9) - 10 = -49.2$ is a little more
than -50 (because -49.2 is less negative; it is closer to zero)

6) Subtract 14 from both sides: $56 \leq -7x$ Divide by -7 on both sides and reverse
the inequality: $\boxed{-8 \geq x}$ (equivalent to $x \leq -8$)
Check: -8.1 is a little less than -8 (because -8.1 is more negative): 70 is a little less
than $14 - 7x = 14 - 7(-8.1) = 14 + 56.7 = 70.7$

7) Subtract 12 from both sides: $-36 < -9x$ Divide by -9 on both sides and reverse
the inequality: $\boxed{4 > x}$ (equivalent to $x < 4$)
Check: 3.9 is a little less than 4: -24 is a little less than $-9x + 12 = -9(3.9) + 12 =$
-23.1 (because -24 is more negative; it is farther from zero)

8) Add 52 to both sides: $-4x > -48$ Divide by -4 on both sides and reverse the
inequality: $\boxed{x < 12}$ (equivalent to $12 > x$)
Check: 11.9 is a little less than 12: $-52 - 4x = -52 - 4(11.9) = -99.6$ is a little
more than -100 (because -99.6 is less negative; it is closer to zero)

Exercise Set 9.4

1) Both sides are positive, so the inequality reverses: $x < \frac{2}{3}$ Since x can't be negative, we can improve our solution to $\boxed{0 < x < \frac{2}{3}}$.

Check: 0.6 is a little less than $\frac{2}{3}$ (since $\frac{2}{3} \approx 0.667$), consistent with $x < \frac{2}{3}$, and $\frac{1}{x} = \frac{1}{0.6} \approx$ 1.667 is a little more than $\frac{3}{2}$ (since $\frac{3}{2} = 1.5$), consistent with $\frac{1}{x} > \frac{3}{2}$.

2) Both sides are negative, so the inequality reverses: $x > -3$ Since x can't be positive, we can improve our solution to $\boxed{0 > x > -3}$, which is equivalent to $\boxed{-3 < x < 0}$.

Check: -2.9 is a little more than -3 (since -2.9 is less negative), consistent with $x > -3$, and $\frac{1}{x} = \frac{1}{-2.9} \approx -0.345$ is a little less than $-\frac{1}{3}$ (since $\frac{1}{3} \approx 0.333$ and since -0.345 is more negative than -0.333), consistent with $\frac{1}{x} < -\frac{1}{3}$

3) Any negative value of x will satisfy this inequality, so one solution is $x < 0$. If x is positive, both sides are positive, so the inequality reverses when we reciprocate: $x \geq \frac{5}{4}$. The two solutions are $\boxed{x < 0 \text{ or } x \geq \frac{5}{4}}$, which is equivalent to $\boxed{0 > x \text{ or } \frac{5}{4} \leq x}$.

Check: 1.3 is a little more than $\frac{5}{4}$ (since $\frac{5}{4} = 1.25$), consistent with $x \geq \frac{5}{4}$, and $\frac{1}{x} = \frac{1}{1.3} \approx$ 0.769 is a little less than $\frac{4}{5}$ (since $\frac{4}{5} = 0.8$), consistent with $\frac{1}{x} \leq \frac{4}{5}$; also, any negative value of x satisfies $\frac{1}{x} \leq \frac{4}{5}$ since a negative number is less than a positive number

4) Both sides are positive, so the inequality reverses: $x < \frac{1}{4}$ Since x can't be negative, we can improve our solution to $\boxed{0 < x < \frac{1}{4}}$.

Check: 0.2 is a little less than $\frac{1}{4}$ (since $\frac{1}{4} = 0.25$), consistent with $x < \frac{1}{4}$, and $\frac{1}{x} = \frac{1}{0.2} = 5$ is a little more than 4, consistent with $\frac{1}{x} > 4$

5) Any positive value of x will satisfy this inequality, so one solution is $x > 0$. If x is negative, both sides are negative, so the inequality reverses when we reciprocate: $-\frac{3}{8} > x$. The two solutions are $\boxed{0 < x \text{ or } -\frac{3}{8} > x}$, which is equivalent to $\boxed{x > 0 \text{ or } x < -\frac{3}{8}}$.

Check: -0.4 is a little less than $-\frac{3}{8}$ (since $\frac{3}{8} = 0.375$ and since -0.4 is more negative

than -0.375), consistent with $-\frac{3}{8} > x$, and $\frac{1}{x} = \frac{1}{-0.4} = -2.5$ is a little more than $-\frac{8}{3}$

(since $\frac{8}{3} \approx 2.667$ and since -2.5 is less negative than -2.667), consistent with $-\frac{8}{3} < \frac{1}{x}$;

also, any positive value of x satisfies $-\frac{8}{3} < \frac{1}{x}$ since a negative number is less than a

positive number

6) Any negative value of x will satisfy this inequality, so one solution is $x < 0$. If x is

positive, both sides are positive, so the inequality reverses when we reciprocate: $\frac{1}{2} \leq x$.

The two solutions are $\boxed{0 > x \text{ or } \frac{1}{2} \leq x}$, which is equivalent to $\boxed{x < 0 \text{ or } x \geq \frac{1}{2}}$.

Check: 0.6 is a little more than $\frac{1}{2}$ (since $\frac{1}{2} = 0.5$), consistent with $\frac{1}{2} \leq x$, and $\frac{1}{x} = \frac{1}{0.6} \approx$

1.667 is a little less than 2, consistent with $2 \geq \frac{1}{x}$; also, any negative value of x satisfies

$2 \geq \frac{1}{x}$ since a positive number is greater than a negative number

7) Any positive value of x will satisfy this inequality, so one solution is $x > 0$. If x is

negative, both sides are negative, so the inequality reverses when we reciprocate: $x <$

$-\frac{6}{5}$. The two solutions are $\boxed{x > 0 \text{ or } x < -\frac{6}{5}}$, which is equivalent to $\boxed{0 < x \text{ or } -\frac{6}{5} > x}$.

Check: -1.3 is a little less than $-\frac{6}{5}$ (since $\frac{6}{5} = 1.2$ and since -1.3 is more negative than -1.2),

consistent with $x < -\frac{6}{5}$, and $\frac{1}{x} = \frac{1}{-1.3} \approx -0.769$ is a little more than $-\frac{5}{6}$ (since $\frac{5}{6} \approx 0.833$ and

since -0.769 is less negative than -0.833), consistent with $\frac{1}{x} > -\frac{5}{6}$

8) Any negative value of x will make the left side negative and therefore less than zero:

$\boxed{x < 0}$.

Check: -0.1 is a little less than zero, consistent with $x < 0$, and $\frac{1}{x} = \frac{1}{-0.1} = -10$ is less

than zero, consistent with $\frac{1}{x} < 0$; clearly, any negative value of x satisfies $\frac{1}{x} < 0$ since

a negative number is less than zero (and since $\frac{1}{x}$ is negative if x is negative)

9) Both sides are negative, so the inequality reverses: $-\frac{2}{7} \leq x$ Since x can't be positive,

we can improve our solution to $\boxed{-\frac{2}{7} \leq x < 0}$, which is equivalent to $\boxed{0 > x \geq -\frac{2}{7}}$.

Check: -0.2 is a little more than $-\frac{2}{7}$ (since $\frac{2}{7} \approx 0.286$ and since -0.2 is less negative than -0.286), consistent with $-\frac{2}{7} \leq x$, and $\frac{1}{x} = \frac{1}{-0.2} = -5$ is less than $-\frac{7}{2}$ (since $\frac{7}{2} = 3.5$ and since -5 is more negative than -3.5), consistent with $-\frac{7}{2} \geq \frac{1}{x}$

10) Both sides are positive, so the inequality reverses: $\frac{4}{9} > x$ Since x can't be negative, we can improve our solution to $\boxed{\frac{4}{9} > x > 0}$, which is equivalent to $\boxed{0 < x < \frac{4}{9}}$.

Check: 0.4 is less than $\frac{4}{9}$ (since $\frac{4}{9} \approx 0.444$), consistent with $\frac{4}{9} > x$, and $\frac{1}{x} = \frac{1}{0.4} = 2.5$ is a little more than $\frac{9}{4}$ (since $\frac{9}{4} = 2.25$), consistent with $\frac{9}{4} < \frac{1}{x}$

11) First multiply both sides by -1 to get $\frac{1}{x} > -\frac{1}{8}$ (recall from Sec. 9.3 that multiplying both sides by a negative number causes the direction of the inequality to reverse) Any positive value of x will satisfy this inequality, so one solution is $x > 0$. If x is negative, both sides are negative, so the inequality reverses when we reciprocate: $x < -8$. The two solutions are $\boxed{x > 0 \text{ or } x < -8}$, which is equivalent to $\boxed{0 < x \text{ or } -8 > x}$.

Check: -8.1 is a little less than -8 (since -8.1 is more negative than -8), consistent with $x < -8$, and $-\frac{1}{x} = -\frac{1}{-8.1} \approx 0.123$ is a little less than $\frac{1}{8}$ (since $\frac{1}{8} \approx 0.125$), consistent with $-\frac{1}{x} < \frac{1}{8}$

12) First multiply both sides by -1 to get $\frac{1}{x} \geq \frac{3}{10}$ (recall from Sec. 9.3 that multiplying both sides by a negative number causes the direction of the inequality to reverse) Both sides of $\frac{1}{x} \geq \frac{3}{10}$ are positive, so the inequality reverses: $x \leq \frac{10}{3}$ Since x can't be negative, we can improve our solution to $\boxed{0 < x \leq \frac{10}{3}}$, which is equivalent to $\boxed{\frac{10}{3} \geq x > 0}$.

Check: 3 is a little less than $\frac{10}{3}$ (since $\frac{10}{3} \approx 3.333$), consistent with $x \leq \frac{10}{3}$, and $-\frac{1}{x} = -\frac{1}{3} \approx -0.333$ is a little less than $-\frac{3}{10}$ (since $\frac{3}{10} = 0.3$ and since -0.333 is more negative than -0.3), consistent with $-\frac{1}{x} \leq -\frac{3}{10}$

Exercise Set 9.5

1) x must be negative (x can't be positive). When we cross multiply, one denominator (x) is negative while the other (3) is positive, which causes the inequality to reverse: $12 > -2x$. When we divide both sides by -2, the inequality reverses again: $-6 < x$. Since x can't be positive, we can improve our solution to $\boxed{-6 < x < 0}$, which is equivalent to $\boxed{0 > x > -6}$.

Check: -5.9 is a little more than -6 (since -5.9 is less negative), consistent with $-6 < x$, and $\frac{4}{x} = \frac{4}{-5.9} \approx -0.678$ is less than $-\frac{2}{3}$ (since $-\frac{2}{3} \approx 0.667$ and since -0.678 is more negative than -0.667), consistent with $\frac{4}{x} < -\frac{2}{3}$

2) x must be positive (x can't be negative). When we cross multiply, the inequality doesn't change: $3 > 2x \rightarrow \frac{3}{2} > x$. Since x can't be negative, we can improve our solution to $\boxed{\frac{3}{2} > x > 0}$, which is equivalent to $\boxed{0 < x < \frac{3}{2}}$.

Check: 1.4 is a little less than $\frac{3}{2}$ (since $\frac{3}{2} = 1.5$), consistent with $\frac{3}{2} > x$, and $\frac{3}{x} = \frac{3}{1.4} \approx 2.14$ is a little more than 2, consistent with $\frac{3}{x} > 2$

3) Any positive value of x will satisfy this inequality, so one solution is $x > 0$. If x is negative, when we cross multiply, one denominator (x) is negative while the other (3) is positive, which causes the inequality to reverse: $18 \leq -4x$. When we divide both sides by -4, the inequality reverses again: $-\frac{18}{4} \geq x \rightarrow -\frac{9}{2} \geq x$. The two solutions are $\boxed{0 < x \text{ or } -\frac{9}{2} \geq x}$, which is equivalent to $\boxed{x > 0 \text{ or } x \leq -\frac{9}{2}}$.

Check: -4.6 is a little less than $-\frac{9}{2}$ (since $\frac{9}{2} = 4.5$ and since -4.6 is more negative than -4.5), consistent with $-\frac{9}{2} \geq x$, and $\frac{6}{x} = \frac{6}{-4.6} \approx -1.304$, is a little more than $-\frac{4}{3}$ (since $\frac{4}{3} \approx 1.333$ and since -1.304 is less negative than -1.333), consistent with $\frac{6}{x} \geq -\frac{4}{3}$

4) Any negative value of x will satisfy this inequality, so one solution is $x < 0$. If x is positive, when we cross multiply, both denominators are positive, so the inequality doesn't change: $25 < x$. The two solutions are $\boxed{0 > x \text{ or } 25 < x}$, which is equivalent to $\boxed{x < 0 \text{ or } x > 25}$.

Check: 25.1 is a little more than 25, which is consistent with $x > 25$, and $\frac{5}{x} = \frac{5}{25.1} \approx$ 0.199 is a little less than $\frac{1}{5}$ (since $\frac{1}{5} = 0.2$), consistent with $\frac{5}{x} < \frac{1}{5}$

5) First multiply both sides by -1 to get $\frac{4}{5} \geq \frac{2}{x}$ (recall from Sec. 9.3 that multiplying both sides by a negative number causes the direction of the inequality to reverse) Any negative value of x will satisfy this inequality, so one solution is $x < 0$. If x is positive, when we cross multiply, both denominators are positive, so the inequality doesn't change: $4x \geq 10 \rightarrow x \geq \frac{10}{4} \rightarrow x \geq \frac{5}{2}$. The two solutions are $\boxed{x < 0 \text{ or } x \geq \frac{5}{2}}$, which is equivalent to $\boxed{0 > x \text{ or } \frac{5}{2} \leq x}$.

Check: 2.6 is a little more than $\frac{5}{2}$ (since $\frac{5}{2} = 2.5$), consistent with $x \geq \frac{5}{2}$, and $-\frac{2}{x} = -\frac{2}{2.6}$ ≈ -0.769 is a little more than $-\frac{4}{5}$ (since $\frac{4}{5} = 0.8$ and since -0.769 is less negative than -0.8), consistent with $-\frac{4}{5} \leq -\frac{2}{x}$

6) We'll see that x may be positive or negative. We'll treat these cases separately, and combine the results together when we finish.

If x is positive, both denominators are positive, so the inequality doesn't change when we cross multiply: $36 > x^2$. Square root both sides. Like we did in Sec. 6.2, we need to consider both $\pm$ roots because $(-6)^2 = 36$ and $6^2 = 36$. We get $\pm 6 > x$. However, when x is positive (as is the case in this paragraph), only $6 > x$ applies. Since x is a positive number in this paragraph, we can narrow this case to $6 > x > 0$.

If x is negative, the left denominator (x) is negative while the other (9) is positive, so the direction of the inequality reverses when we cross multiply: $36 < x^2$. Like we did in Sec. 6.2, we need to consider both $\pm$ roots because $(-6)^2 = 36$ and $6^2 = 36$. We get $\pm 6 < x$. However, when x is negative (as is the case in this paragraph), only the root -6 applies. This negative root is a little tricky, as it causes the inequality to reverse yet again (the proof will be when we check our answers with decimals later): $-6 > x$.

Now we'll combine our solution for positive x, which is $6 > x > 0$, with our solution for negative x, which is $-6 > x$. Our final answers are $\boxed{6 > x > 0}$ and $\boxed{-6 > x}$, which are equivalent to $0 < x < 6$ and $x < -6$.

Check: 5.9 is a little less than 6, consistent with $6 > x > 0$, and $\frac{4}{5.9} \approx 0.678$ is a little more than $\frac{5.9}{9} \approx 0.656$, consistent with $\frac{4}{x} > \frac{x}{9}$. Also, -6.1 is a little less than -6 (since -6.1 is more negative than -6), consistent with $-6 > x$, and $\frac{4}{-6.1} \approx -0.656$ is a little more than $\frac{-6.1}{9} \approx -0.678$ (since -0.656 is less negative than -0.678), also consistent with $\frac{4}{x} > \frac{x}{9}$.

Note that values outside of $0 < x < 6$ and $-6 > x$ don't satisfy $\frac{4}{x} > \frac{x}{9}$. For example, $\frac{4}{6.1} \approx 0.656$ isn't greater than $\frac{6.1}{9} \approx 0.678$. As another example, $\frac{4}{-5.9} \approx -0.678$ isn't greater than $\frac{-5.9}{9} \approx -0.656$ (since -0.678 is more negative than -0.656).

7) Since x^2 can't be negative (for real values of x), the right-hand side is negative. The left-hand side must also be negative (in order to be less than the right-hand side). This requires x to be negative. Therefore, the left denominator (x) is negative while the right denominator (x^2) is positive. (Although we said the "right side" is negative, the "right denominator" is positive. It's the overall minus sign that makes the right side negative.) Since one denominator is negative while the other is positive, when we cross multiply, the direction of the inequality reverses: $3x^2 > -6x$. Add $6x$ to both sides: $3x^2 + 6x > 0$. This inequality is similar to the problems from Sec. 5.11, where it is necessary to factor in order to obtain all of the answers: $3x(x + 2) > 0$. Since we already reasoned (at the beginning of this paragraph) that x must be negative, the only way that $3x(x + 2)$ can be positive is if $3x < 0$ and $x + 2 < 0$. (The product of two negative numbers is positive.) These lead to $x < 0$ and $x < -2$. Both of these inequalities will only be true when $\boxed{x < -2}$.

Check: -2.1 is a little less than -2 (since -2.1 is more negative than -2), consistent with $x < -2$, and $\frac{3}{-2.1} \approx -1.429$ is a little less than $-\frac{6}{(-2.1)^2} \approx -1.361$ (since -1.429 is more negative than -1.361), consistent with $\frac{3}{x} < -\frac{6}{x^2}$.

Note that values outside of $x < -2$ don't satisfy $\frac{3}{x} < -\frac{6}{x^2}$. For example, $\frac{3}{-1.9} \approx -1.579$ isn't less than $-\frac{6}{(-1.9)^2} \approx -1.662$ (since -1.579 is less negative than -1.662).

8) We need to consider whether either (or both) $x - 2$ and $x - 3$ may be negative. (In this problem, we're not worried whether x itself may be negative. We need to know whether the denominators, $x - 2$ and $x - 3$, may be negative.)

Note that it isn't possible in this case for only one of the denominators to be negative. If $x - 3$ is negative while $x - 2$ is positive, the left side will be positive while the right side is negative, which can't satisfy the inequality. Also note that if $x - 2$ is negative, $x - 3$ will also be negative. Therefore, we only need to consider the cases where both denominators are positive or where both denominators are negative. Either way, the inequality won't change when we cross multiply: $4(x - 3) < 5(x - 2)$. Distribute: $4x - 12 < 5x - 10$ Add 10 to both sides: $4x - 2 < 5x$ Subtract $4x$ from both sides: $-2 < x$ We're not finished yet because this inequality, $-2 < x$, includes values of x that would make exactly one of the denominators negative (which we already ruled out). Exactly one denominator would be negative when x lies between 2 and 3, so we need to exclude this range from our solution. Also note that x can't equal 2 or 3 because that would make one denominator undefined. Therefore, the full solution to this problem is $\boxed{-2 < x < 2 \text{ or } 3 < x}$. (The latter inequality is equivalent to $x > 3$.)

Check: -1.9 is a little more than -2 (since -1.9 is less negative than -2), consistent with $-2 < x$, and $\frac{4}{x-2} = \frac{4}{-1.9-2} \approx -1.026$ is a little less than $\frac{5}{x-3} = \frac{5}{-1.9-3} \approx -1.020$ (since -1.026 is more negative than -1.020), consistent with $\frac{4}{x-2} < \frac{5}{x-3}$. Also, 1.9 is a little less than 2, consistent with $x < 2$, and $\frac{4}{x-2} = \frac{4}{1.9-2} \approx -40$ is less than $\frac{5}{x-3} = \frac{5}{1.9-3} \approx -4.545$ (since -40 is more negative than -4.545). Finally, 3.1 is a little more than 3, consistent with $3 < x$, and $\frac{4}{x-2} = \frac{4}{3.1-2} \approx 3.636$ is less than $\frac{5}{x-3} = \frac{5}{3.1-3} \approx 50$, consistent with $\frac{4}{x-2} < \frac{5}{x-3}$.

Note that values outside of $-2 < x < 2$ or $3 < x$ don't satisfy $\frac{4}{x-2} < \frac{5}{x-3}$. For example, $\frac{4}{2.1-2} \approx 40$ isn't less than $\frac{5}{2.1-3} \approx -5.556$. Also, $\frac{4}{2.9-2} \approx 4.444$ isn't less than $\frac{5}{2.9-3} = -50$. Also, $\frac{4}{-2.1-2} \approx -0.976$ isn't less than $\frac{5}{-2.1-3} \approx -0.980$ (since -0.976 is less negative than -0.980).

Exercise Set 9.6

1) x = one number and $4x$ = the other number

$x + 4x \geq 20$ Combine like terms: $5x \geq 20 \rightarrow \boxed{x \geq 4}$ and $\boxed{4x \geq 16}$

2) A = the area of the rectangle, L = the length of the rectangle, and W = the width of the rectangle: $A = 12$ square yards, $L > 4$ yards, a rectangle has an area equal to $A = LW$, divide by W on both sides: $\frac{A}{W} = L$ Replace L with $\frac{A}{L}$ in $L > 4$ to get $\frac{A}{W} > 4$

Plug $A = 12$ into the previous inequality: $\frac{12}{W} > 4$ Multiply by W on both sides: $12 > 4W$

Divide by 4 on both sides: $\boxed{3 > W}$ The width must be less than 3 yards.

3) d = the distance traveled, r = the speed (which is a rate), and t = the time

$t = 60$ seconds (convert one minute to seconds to match the units of the rates)

$4 < r < 5$ (the rate in m/s lies in this range) Recall the rate formula from Sec. 7.9:

$d = rt = 60r$ (since $t = 60$ seconds) Multiply r by 60 to find d: $\boxed{240 < d < 300}$ meters

4) J = Jenny's age, M = Jenny's mother's age, and F = Jenny's father's age

$M \geq 30$ and $F \leq 40$, $M < F$, $\frac{M}{2} \leq J < \frac{F}{2}$, $\frac{30}{2} \leq J < \frac{40}{2}$, $\boxed{15 \leq J < 20}$

Note: Jenny might be half as old as her mother (who could be as young as 30), so there is a less than or equal to sign for the lower limit. Jenny isn't half as older as her father (who could be as old as 40), so there is a strictly less than sign at the upper limit (not less than or equal to).

GLOSSARY

absolute value: the value that a number has without its sign, indicated by vertical lines around the number. For example, $|-4| = 4$.

additive inverse: the number that is the opposite to adding a number. For example, -3 is the additive inverse of 3 since $3 + (-3) = 0$. In general, $x + (-x) = x - x = 0$.

algebra: a branch of mathematics that uses letters (like x or y) to represent unknown quantities, which provides a system of rules for determining the unknowns.

array: numbers arranged systematically in rows and columns.

associative property of addition: $(x + y) + z = x + (y + z)$.

associative property of multiplication: $(xy)z = x(yz)$.

base: a number that has an exponent (or power). For example, in 8^5 the base is 8.

binomial: an expression containing two terms separated by a plus or minus sign, like $3x - 4$ or like $x + y$. It may be raised to a power, like $(x + 7)^2$.

binomial expansion: the expression that results from multiplying out a binomial that is raised to a power. For example, $(x + 2)^3 = x^3 + 6x^2 + 12x + 8$.

coefficient: a number multiplying a variable. For example, the coefficient of $5x$ is 5.

colon: the symbol appears between numbers in a ratio. For example, 2:5 represents the ratio of two to five.

combine like terms: add the coefficients of terms with the same variable raised to the same power, like $3x^2 + 2x^2 = 5x^2$. (This concept applies the principal of factoring.)

common denominator: fractions expressed with the same denominator, like $\frac{2}{3}$ and $\frac{5}{3}$.

commutative property of addition: $x + y = y + x$.

commutative property of multiplication: $xy = yx$.

complex number: a number with real and imaginary parts, like $1 + i$.

constant: a fixed value. For example, in $x + 7$ the number 7 is a constant.

conversion: the process of changing the form of a quantity, such as converting units (like feet to inches) or converting a percent to a decimal.

Cramer's rule: a method that uses determinants to solve a system of equations.

cross multiply: multiply along the diagonals for equated fractions. For example, $\frac{2}{x} = \frac{5}{8}$ becomes $2(8) = 5x$.

cube: raise a number to the power (or exponent) of three. For example, 5^3 is 5 cubed.

cube root: determine which number cubed equals a specified value. For example, $\sqrt[3]{64}$ is the cube root of 64. Note that $\sqrt[3]{64} = 4$ because $4^3 = 64$.

decimal: a fraction where the denominator is a power of ten. For example, $0.87 = \frac{87}{100}$.

decimal point: a low dot (.) that appears between the whole part and the fractional part. For example, 2.7 is the same as $2 + .7$ (or $2 + 0.7$) and as $2 + \frac{7}{10}$.

decimal position: the location of the decimal point. For example, 2.34 and 23.4 differ by a factor of 10 because of the difference in their decimal positions.

degree: the largest exponent of a polynomial. For example, $5x^4 + 8x^3$ is degree four.

denominator: the number at the bottom of a fraction. For example, the denominator of $\frac{3}{8}$ is 8.

determinant: a single number obtained by multiplying the diagonals of a matrix.

difference of squares: $x^2 - y^2 = (x + y)(x - y)$.

discriminant: the expression $b^2 - 4ac$ appearing inside the radical of the quadratic formula. The discriminant helps to determine the nature of the answers to a quadratic equation, such as whether the solutions are real or complex.

distance: the length between two points, or how far an object travels.

distinct: quantities that are neither the same nor equivalent.

distributive property: $x(y + z) = xy + xz$.

division symbol: the $\div$ or $/$ symbol that appears between numbers, like $18 \div 6$ or $18/3$. If a variable is involved, it is usually expressed as a fraction, like $\frac{x}{2}$.

elapsed time: the amount of time that has passed, such as the duration of time for which an object travels.

equal sign: the $=$ symbol appearing between expressions or numbers.

equation: a mathematical statement that sets two expressions or numbers equal. An equation always contains an equal sign, such as $3x + 2 = 8$.

evaluate: plug numbers into an expression. For example, when $x = 3$, the expression $4x - 5$ is equal to $4(3) - 5 = 12 - 5 = 7$.

even number: a number that is evenly divisible by 2, such as 2, 4, 6, 8, 10, 12, 14, etc.

even root: a root $\sqrt[n]{x}$ where n is even, like a square root ($\sqrt{x}$) or fourth root ($\sqrt[4]{x}$).

expansion of powers: the expression that results from multiplying an expression that is raised to a power, like $(x + 3)^3 = x^3 + 9x^2 + 27x + 27$.

exponent: a number appearing to the top right of a base, indicating the number of times that the base is multiplied together. For example, in 2^4 the exponent is 4 and means $2 \times 2 \times 2 \times 2$. Another word for exponent is power.

expression: a mathematical idea involving constants, variables, and operators, which does not have an equal sign or inequal sign (like $<$ or $>$). An example is $4x^2 - 9$.

factor: a number being multiplied, like the 3 and 8 in $(3)(8) = 24$.

factor an expression: distribute in reverse, like $4x^3 + 6x^2 = 2x^2(2x + 3)$.

factor a perfect square: pull a perfect square out of a radical, like $\sqrt{18} = \sqrt{9}\sqrt{2} = 3\sqrt{2}$.

factorial: Multiply by successively smaller integers until reaching one. A factorial is represented by an exclamation mark (!). For example, $3! = 3(2)(1) = 6$.

factorization: numbers that multiply together to make another number. For example, the prime factorization of 12 is written as $2(2)(3)$.

FOIL: an abbreviation which stands for "first, outside, inside, last" to help remember that $(w + x)(y + z) = wy + wz + xy + xz$.

formula: an equation written in symbols used to calculate a quantity. For example, the formula for the perimeter of a rectangle is $P = 2L + 2W$.

fraction: a number of the form $\frac{3}{5}$ or $3/5$.

greater than sign: the symbol $>$ indicating that the left value is larger than the right value, as in $9 > 5$.

greatest common expression: the largest power of the variable and the greatest factor of the coefficients that is common to each term. For example, the greatest common expression for $6x^3 - 9x^2$ is $3x^2$.

greatest common factor: the largest factor that is common to two different integers. For example, 20 is the greatest common factor of 60 and 80.

identity property of addition: $x + 0 = x$.

identity property of multiplication: $1x = x$.

imaginary number: the square root of minus one $\left(\sqrt{-1}\right)$, indicated by the symbol i.

improper fraction: a fraction where the numerator exceeds the denominator, like $\frac{4}{3}$.

independent: when no equation in a system is a linear combination of other equations in the system. For example, $x + y = 4$ and $5x + 5y = 20$ aren't independent because the second equation can be formed by multiplying both sides of the first equation by 5.

indeterminate: when enough information isn't available to determine the answer.

inequality: a mathematical statement where the two sides aren't equal (like $x \neq 1$) or where one side is less than or greater than the other (like $x < 6$).

inequal sign: the $\neq$ sign (or a less than or greater than sign).

integer: numbers like 0, 1, 2, 3, 4, etc. and their negatives $(-1, -2, -3, -4, \text{etc.})$.

inverse property: the opposite of an operation. For example, $x + (-x) = 0$ represents the inverse property of addition and $x\left(\frac{1}{x}\right) = \frac{x}{x} = 1$ represents the inverse property of multiplication.

irrational number: a number that can't be expressed in the form of an integer or as the ratio of two integers, such as $\sqrt{5}$ (but not like $\sqrt{9}$ since $\sqrt{9} = 3$).

isolate the unknown: apply operations to both sides of an equation and combine like terms in order to get the variable all by itself on one side of the equation.

keyword: a word or phrase in a word problem that helps to relate the language to the mathematics. For example, the word "sum" is a keyword for addition.

leading zero: a zero that comes after a decimal point and before nonzero digits. For example, 0.0063 has 2 leading zeroes. (The 0 *before* the decimal point doesn't count.)

least common denominator: the smallest common denominator that two fractions can make. For example, the least common denominator for $\frac{5}{8}$ and $\frac{7}{12}$ is 24.

less than sign: the symbol $<$ indicating that the left value is smaller than the right value, as in $4 < 7$.

like terms: terms in an expression, equation, or inequality where the same variable is raised to the same power, like the two terms in $3x^2 + 8x^2$ (since both involve x^2).

matrix: an array of numbers appearing between parentheses.

middle dot: the symbol $\cdot$ used when numbers are multiplied, like $2 \cdot 3 = 6$.

minus sign: the $-$ symbol used to indicate subtraction or a negative value.

mixed number: a number that includes an integer plus a fraction.

moving the decimal point: multiplying by a power of 10 to move the decimal point to the right, like $4.2 \times 10^3 = 4200$, or dividing by a power of 10 to move the decimal point to the left, like $3.5 \times 10^{-2} = 0.035$.

multiple solutions: when two distinct solutions solve the same equation or system.

multiplication symbol: the times symbol $(\times)$ is seldom used in the context of algebra because it could easily be confused with the variable x. For numbers, multiplication is represented with a middle dot, like $4 \cdot 5 = 20$, or parentheses, like $4(5) = 20$ or $(4)(5) = 20$. When multiplying variables, no symbol is used, such as $5xy$.

multiplicative inverse: the number that is the opposite to multiplying by a number. For example, $\frac{1}{6}$ is the reciprocal of 6 since $6\left(\frac{1}{6}\right) = 1$. In general, $x\left(\frac{1}{x}\right) = \frac{x}{x} = x \div x = 1$.

negative exponent: an exponent with a minus sign, equivalent to raising the reciprocal of the base to the absolute value of the exponent. For example, $3^{-2} = \left(\frac{1}{3}\right)^2 = \frac{1}{3^2}$.

negative number: a number with a minus sign before it, like -5, which is opposite to a corresponding positive number on the number line.

negative root: a negative answer to a root (or the negative solution to an equation). For example, -3 is the negative root to $\sqrt{9}$. Note that $(-3)^2 = (-3)(-3) = 9$.

nonnegative: not negative. A nonnegative number isn't necessarily positive because it could be zero.

nonpositive: not positive. A nonpositive number could be negative or zero.

nonzero: not zero. A nonzero number could be positive or negative.

no solution: when the solution to an equation or system doesn't exist. The phrase "does not exist" is sometimes used in place of the phrase "no solution."

numerator: the number at the top of a fraction. For example, in $\frac{5}{6}$ the numerator is 5.

odd number: a number that isn't evenly divisible by 2, such as 1, 3, 5, 7, 9, 11, 13, etc.

odd root: a root $\sqrt[n]{x}$ where n is odd, like a cube root ($\sqrt[3]{x}$) or fifth root ($\sqrt[5]{x}$).

operation: a mathematical process such as addition, subtraction, multiplication, or division.

operator: a mathematical symbol (or expression) indicating a process to be carried out. For example, the slash (/) symbol in 24/6 represents the division operator.

order of operations: the order for carrying out arithmetic operations: parentheses first, then exponents, then multiplication and division from left to right, and then addition and subtraction from left to right.

parentheses: the symbols (and) placed around an expression, like $4(x + 5)$. The singular form of the word is parenthesis, whereas the plural form is parentheses.

part to part: a ratio between two parts, such as the ratio of girls to boys (since girls and boys are both parts of the total population).

part to whole: a ratio between one part and the whole, such as the ratio of girls to the total number of students.

Pascal's triangle: a triangle formed by working out the binomial expansion of $(x + y)^n$ for successive integer values of n.

PEMDAS: an abbreviation for "parentheses, exponents, multiply/divide from left to right, and add/subtract from left to right" used to help remember the order of operations.

percent: a specified fraction of 100. For example, 92% means 92 out of 100.

percentage: an unspecified amount, like "a percentage of the students."

perfect square: an integer that equals the square of another integer. For example, 36 is a perfect square because $6^2 = 6 \times 6 = 36$.

place value: the position of a digit in a number. For example, in 12.345, the 1 is in the tens place, the 2 is in the units place, the 3 is in the tenths place, the 4 is in the hundredths place, and the 5 is in the thousandths place.

plus or minus sign: the $\pm$ symbol, indicating two possible values, one for each sign. For example, $x = \pm 8$ means $x = 8$ or $x = -8$.

plus sign: the $+$ symbol used to indicate addition between expressions or numbers.

polynomial: terms of the form ax^n that are added together (or subtracted), where the exponents are nonnegative integers. An example of a polynomial is $5x^2 + 8x - 7$. The last term corresponds to $n = 0$ since $x^0 = 1$.

positive number: a number that is neither zero nor negative.

positive root: a positive answer to a root (or the positive solution to an equation). For example, $x^2 = 9$ has two roots: $x = \pm 3$ because $(-3)^2 = 9$ and $3^2 = 9$. In this case, the positive root is $x = 3$.

power: a number appearing to the top right of a base, indicating the number of times that the base is multiplied together. For example, in 2^4 the power is 4 and means $2 \times 2 \times 2 \times 2$. Another word for power is exponent.

power of ten: an exponent of ten. For example, 1,000,000 is a power of ten since $10^6 = 1{,}000{,}000$.

product: the result of multiplying numbers together. For example, the product of 5 and 3 equals 15.

product rule: $x^m x^n = x^{m+n}$.

proper fraction: a fraction where the numerator is smaller than the denominator, like $\frac{3}{4}$.

proportion: an equality between two ratios or rates.

quadratic equation: an equation with a quadratic term (with the variable squared), a linear term (with the variable not raised to a power, meaning that it is raised to the first power since $x^1 = x$), and a constant term (with no variable). An example of a quadratic equation is $3x^2 + 6x - 9 = 0$.

quadratic expression: an expression with a quadratic term (with the variable squared), a linear term (with the variable not raised to a power, meaning that it is raised to the first power since $x^1 = x$), and a constant term (with no variable). An example of a quadratic expression is $3x^2 + 6x - 9$. (Unlike an expression, an equation has an equal sign.)

quadratic formula: $x = \frac{-b \pm \sqrt{b^2 - 4ac}}{2a}$, which is the solution to $ax^2 + bx + c = 0$.

quotient: the result of dividing two numbers. For example, the quotient of 18 divided by 3 equals 6. The fraction indicating the division is also referred to as the quotient, which in this case is $\frac{18}{3}$.

quotient rule: $\frac{x^m}{x^n} = x^{m-n}$.

radical: the $\sqrt{}$ symbol used to indicate a square root or other root (such as $\sqrt[3]{}$).

rate: a fraction made by dividing two quantities that have different units, like $\frac{200 \text{ miles}}{7 \text{ hours}}$.

ratio: a fixed relationship expressed in the form $x{:}y$. For example, the ratio of fingers to hands is 10:2 (which reduces to 5:1) for a typical human being.

rational number: a number that can be expressed in the form of an integer or as the ratio of two integers, such as 3 or $\frac{2}{5}$ (but not $\sqrt{7}$).

real number: a number which does not have an imaginary part. For example, 5, $\sqrt{2}$, and $1 + \sqrt{3}$ are real numbers (whereas $\sqrt{-1}$ is imaginary because no real number multiplied by itself can be negative).

reciprocal: one divided by a number. For a fraction, this means to swap the numerator and denominator. For example, $\frac{7}{4}$ is the reciprocal of $\frac{4}{7}$, and $\frac{1}{6}$ is the reciprocal of 6.

reciprocate: to find take the reciprocal of a number or expression.

reduced fraction: the simplest form of a fraction, where the numerator and denominator do not share a common factor. For example, $\frac{2}{3}$ is the reduced form of $\frac{8}{12}$.

reflexive property: $x = x$. If $x = y$, it follows that $y = x$.

root: the opposite of a power. A general root asks, "Which number raised to the given power equals the value under the radical?" For example, $\sqrt[3]{64} = 4$ because $4^3 = 64$.

scientific notation: a power of ten used to position a decimal point immediately after the first digit. For example, in scientific notation 2500 is expressed as 2.5×10^3.

simplify: make an expression simpler. For example, $4x - 5 + 3x$ simplifies to $7x - 5$.

simultaneous equations: a method of solving a system of equations where each equation is multiplied by the factor needed to make equal and opposite coefficients.

slash: the / symbol sometimes used to indicate division or a fraction like 4/7.

solve: determine the values of the variables by following a procedure (like isolating the unknown or like applying Cramer's rule).

special solution: a solution to an equation or system that isn't unique, such as when the solution is indeterminate or such as when the solution doesn't exist.

speed: a measure of how fast an object moves, like 35 mph (miles per hour).

square: raise a number to the power of two, like 5^2 (which equals 25).

square root: a number that when multiplied by itself makes the value under the radical. For example, $\sqrt{9} = \pm 3$ because $3^2 = 9$ and $(-3)^2 = 9$.

standard form: when a quadratic equation is expressed in the form $ax^2 + bx + c = 0$.

substitution: a method for solving a system of equations where one variable is isolated in one equation and then substituted into the other equations.

sum: the total of adding numbers together. For example, the sum of 5 and 3 equals 8.

system of equations: two or more equations with two or more variables, where the answers for the variables must satisfy all of the equations.

term: an expression separated from other expressions by plus ($+$), minus ($-$), equal ($=$), or inequal ($<$ or $>$) signs in an algebraic statement. For example, $2x + 8 = 6x$ consists of three terms ($2x$, 8, and $6x$).

times symbol: a symbol used to indicate multiplication. The standard times symbol ($\times$) ordinarily isn't used in algebra because it could cause confusion with x. Instead, a middle dot, like $6{\cdot}5$, or parentheses, like $(6)(5)$ or $6(5)$, is used when numbers are multiplied. No symbol is used to multiply variables, like $7xy$.

trailing zero: a zero that comes at the end of a number (and after a decimal point). For example, 0.0046000 has 3 trailing zeroes (which come after the 6).

transitive property: if $x = z$ and $y = z$, it follows that $x = y$.

undefined: a problem where a finite answer doesn't make sense, like one divided by zero (since no number times zero will equal one).

unique: when the answer (or set of answers) is the only one that satisfies the given equation (or system of equations).

unit: a standard value for measurement, such as a meter, kilometer, or a second.

unity: the number one.

unknown: a quantity like x (which is a variable) that needs to be solved for.

unlike terms: terms with different powers of the variable, such as $5x^2$ and $4x$.

variable: an unknown quantity represented by a symbol, like x or y.

whole number: a number that is whole like 1, 2, 3, etc.

whole to part: a ratio between the whole and one part, such as the ratio of the total number of students to the number of girls.

INDEX

WAS THIS BOOK HELPFUL?

A great deal of effort and thought was put into this book, such as:

- Breaking down the solutions to help make the math easier to understand.
- Careful selection of examples and problems for their instructional value.
- Explanations of the ideas behind the math.
- An introductory chapter about what algebra is and a glossary in the back.
- Coverage of a variety of essential algebra topics and skills.
- Full solutions to the exercises included in the answer key.

If you appreciate the effort that went into making this book possible, there is a simple way that you could show it:

Please take a moment to post an honest review.

For example, you can review this book at Amazon.com or Goodreads.com.

Even a short review can be helpful and will be much appreciated. If you're not sure what to write, following are a few ideas, though it's best to describe what's important to you.

- How much did you learn from reading and using this workbook?
- Were the solutions at the back of the book helpful?
- Were you able to understand the solutions?
- Was it helpful to follow the examples while solving the problems?
- Would you recommend this book to others? If so, why?

Do you believe that you found a mistake? Please email the author, Chris McMullen, at greekphysics@yahoo.com to ask about it. One of two things will happen:

- You might discover that it wasn't a mistake after all and learn why.
- You might be right, in which case the author will be grateful and future readers will benefit from the correction. Everyone is human.

ABOUT THE AUTHOR

Dr. Chris McMullen has over 20 years of experience teaching university physics in California, Oklahoma, Pennsylvania, and Louisiana. Dr. McMullen is also an author of math and science workbooks. Whether in the classroom or as a writer, Dr. McMullen loves sharing knowledge and the art of motivating and engaging students.

The author earned his Ph.D. in phenomenological high-energy physics (particle physics) from Oklahoma State University in 2002. Originally from California, Chris McMullen earned his Master's degree from California State University, Northridge, where his thesis was in the field of electron spin resonance.

As a physics teacher, Dr. McMullen observed that many students lack fluency in fundamental math skills. In an effort to help students of all ages and levels master basic math skills, he published a series of math workbooks on arithmetic, fractions, long division, algebra, geometry, trigonometry, and calculus entitled *Improve Your Math Fluency*. Dr. McMullen has also published a variety of science books, including astronomy, chemistry, and physics workbooks.

Author, Chris McMullen, Ph.D.

MATH

This series of math workbooks is geared toward practicing essential math skills:

- Prealgebra
- Algebra
- Geometry
- Trigonometry
- Logarithms and exponentials (precalculus)
- Calculus
- Fractions, decimals, and percents
- Long division (with remainders)
- Multiplication, division, addition, and subtraction
- Test your math knowledge
- Roman numerals

www.improveyourmathfluency.com

PUZZLES

The author of this book, Chris McMullen, enjoys solving puzzles. His favorite puzzle is Kakuro (kind of like a cross between crossword puzzles and Sudoku). He once taught a three-week summer course on puzzles. If you enjoy mathematical pattern puzzles, you might appreciate:

300+ Mathematical Pattern Puzzles

Number Pattern Recognition & Reasoning
- Pattern recognition
- Visual discrimination
- Analytical skills
- Logic and reasoning
- Analogies
- Mathematics

THE FOURTH DIMENSION

Are you curious about a possible fourth dimension of space?

- Explore the world of hypercubes and hyperspheres.
- Imagine living in a two-dimensional world.
- Try to understand the fourth dimension by analogy.
- Several illustrations help to try to visualize a fourth dimension of space.
- Investigate hypercube patterns.
- What would it be like to be a 4D being living in a 4D world?
- Learn about the physics of a possible four-dimensional universe.

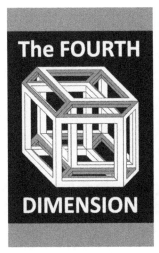
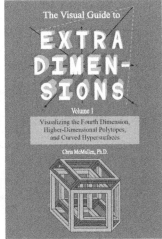
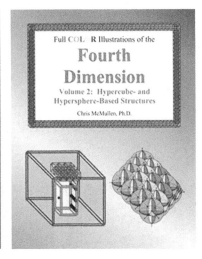

SCIENCE

Dr. McMullen has published a variety of **science** books, including:

- Basic astronomy concepts
- Basic chemistry concepts
- Balancing chemical reactions
- Calculus-based physics textbooks
- Calculus-based physics workbooks
- Calculus-based physics examples
- Trig-based physics workbooks
- Trig-based physics examples
- Creative physics problems
- Modern physics
- Test your science knowledge

www.monkeyphysicsblog.wordpress.com

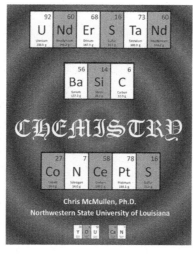

Made in the USA
Las Vegas, NV
31 January 2023

66529115R10214